U.S. Climate Action Report – 2002

Third National Communication of the United States of America
Under the United Nations Framework Convention on Climate Change

You may electronically download this document from the following U.S. Environmental Protection Agency Web site:
http://www.epa.gov/globalwarming/publications/car/index.html.

To purchase copies of this report, visit the U.S. Government Printing Office Web site at http://bookstore.gpo.gov.
Phone orders may be submitted at 1-866-512-1800 (toll-free) or 1-202-512-1800 between 7:30 a.m. and 4:30 p.m., Eastern Time.

This document may be cited as follows: U.S. Department of State, U.S. Climate Action Report 2002, Washington, D.C., May 2002.

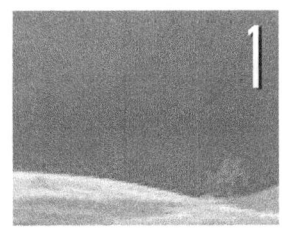

Policies and Measures   **50**

Projected Greenhouse Gas Emissions   **70**

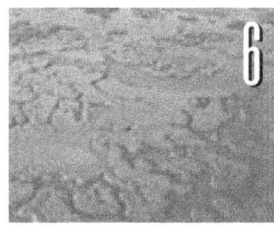

Impacts and Adaptation   81

Financial Resources and Transfer of Technology   113

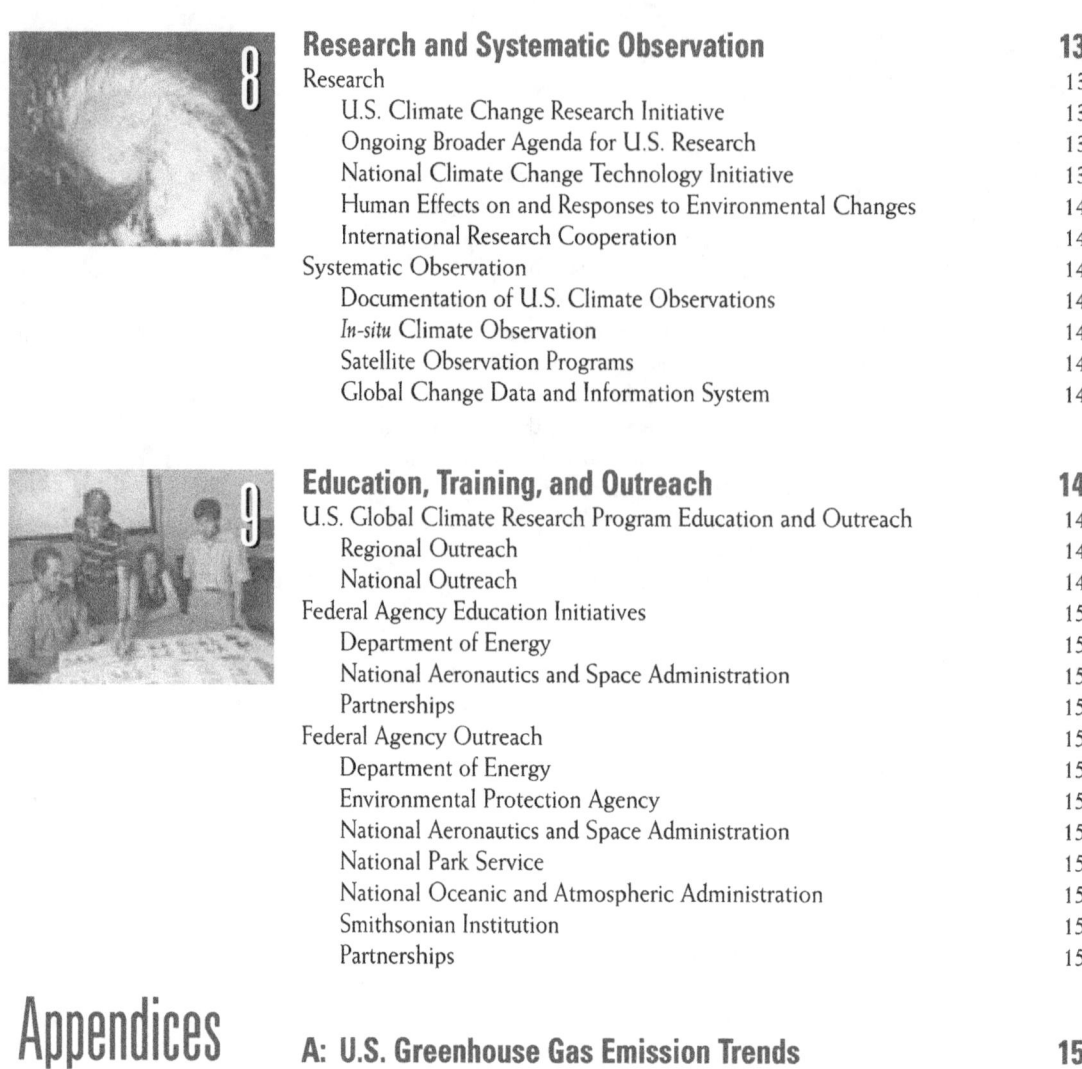

"The Earth's well-being is ... an issue important to America—and it's an issue that should be important to every nation and in every part of the world. My Administration is committed to a leadership role on the issue of climate change. We recognize our responsibility, and we will meet it—at home, in our hemisphere, and in the world."—George W. Bush, June 2001

Chapter 1
Introduction and Overview

With this pledge, President Bush reiterated the seriousness of climate change and ordered a Cabinet-level review of U.S. climate change policy. He requested working groups to develop innovative approaches that would: (1) be consistent with the goal of stabilizing greenhouse gas concentrations in the atmosphere; (2) be sufficiently flexible to allow for new findings; (3) support continued economic growth and prosperity; (4) provide market-based incentives; (5) incorporate technological advances; and (6) promote global participation.

The President's decision to take a deeper look at climate change policy arose from the recognition that the international dialogue begun to date lacked the requisite participatory breadth for a global response to climate change. At the 1992 Earth Summit in Rio de Janeiro, the United Nations Framework Convention on Climate Change (UNFCCC) was adopted, with the ultimate objective of providing a higher quality of life for future generations. Signatories pledged to:

> achieve...stabilization of greenhouse gas concentrations in the atmosphere at a level that would prevent dangerous anthropogenic interference with the climate system. Such a level should be achieved within a timeframe sufficient to allow ecosystems to adapt naturally to climate change, to ensure that food production is not threatened, and to enable economic development to proceed in a sustainable manner.

In Rio, ambitious plans were set in motion to address climate change. However, participation in constructing measures for adapting to and mitigating the effects of climate change fell short of the breadth necessary to confront a problem that President Bush recently said has "the potential to impact every corner of the world." A global problem demands a truly participatory global response, while at the same time taking near-term action that would reduce projected growth in emissions cost-effectively and enhance our ability to cope with climate change impacts.

Based on his Cabinet's review and recommendation, President Bush recently announced a commitment to reduce greenhouse gas intensity in the United States by 18 percent over the next decade through a combination of voluntary, incentive-based, and existing mandatory measures. This represents a 4.5 percent reduction from forecast emissions in 2012, a serious, sensible, and science-based response to this global problem—despite the remaining uncertainties concerning the precise magnitude, timing, and regional patterns of climate change. The President's commitment also emphasized the need for partners in this endeavor. All countries must actively work together to achieve the long-term goal of stabilizing greenhouse gas concentrations at a level that will prevent dangerous interference with the climate system.

For our part, the United States intends to continue to be a constructive and active Party to the Framework Convention. We are leading global research efforts to enhance the understanding of the science of climate change, as called for under the Framework Convention. We lead the world in investment in climate science and in recent years have spent $1.7 billion on federal research annually. Since 1990, the United States has provided over $18 billion for climate system research—more resources than any other country. In June 2001, President Bush announced a new Climate Change Research Initiative to focus on key remaining gaps in our understanding of anthropogenic climate change and its potential impacts.

As envisioned by the Framework Convention, we are helping to develop technologies to address climate change. The President has pledged to reprioritize research budgets under the National Climate Change Technology Initiative so that funds will be available to develop advanced energy and sequestration technologies. Energy policies improve efficiency and substitute cleaner fuels, while sequestration technologies will promote economic and environmentally sound methods for the capture and storage of greenhouse gases.

We plan to increase bilateral support for climate observation systems and to finance even more demonstration projects of advanced energy technologies in developing countries. President Bush's Western Hemisphere Initiative—created to enhance climate change cooperation with developing countries in the Americas and elsewhere—will also strengthen implementation of our Framework Convention commitments. In line with those commitments, we have provided over $1 billion in climate change-related assistance to developing countries over the last five years. All of this is just the beginning: we intend to strengthen our cooperation on climate science and advanced technologies around the world whenever and wherever possible.

We continue to make progress in limiting U.S. emissions of greenhouse gases by becoming more energy efficient. In the last decade, we have seen tremendous U.S. economic growth, and our level of emissions per unit of economic output has declined significantly. The President has committed the United States to continue this improvement and reduce intensity beyond forecast levels through enhanced voluntary measures. The United States is a world leader in addressing and adapting to a variety of national and global scientific problems that could be exacerbated by climate change, including malaria, hunger, malnourishment, property losses due to extreme weather events, and habitat loss and other threats to biological diversity.

Climate change is a long-term problem, decades in the making, that cannot be solved overnight. A real solution must be durable, science-based, and economically sustainable. In particular, we seek an environmentally sound approach that will not harm the U.S. economy, which remains a critically important engine of global prosperity. We believe that economic development is key to protecting the global environment. In the real world, no one will forego meeting basic family needs to protect the global commons. Environmental protection is neither achievable nor sustainable without opportunities for continued development and greater prosperity. Our objective is to ensure a long-term solution that is environmentally effective, economically efficient and sustainable, and appropriate in terms of addressing the urgent problems of today while enhancing our ability to deal with future problems. Protecting the global environment is too important a responsibility for anything less.

In this *U.S. Climate Action Report*, we provide our third formal national communication under the Framework Convention, as envisioned under Articles 4 and 12 of the Convention. We describe our national circumstances, identify existing and planned policies and measures, indicate future trends in greenhouse gas emissions, outline expected impacts and adaptation measures, and provide information on financial resources, technology transfer, research, and systematic observations.[1]

---

[1] Some sections of this report (e.g., the projections in Chapter 5) are included, despite the absence of a binding requirement to do so under the Convention. Note that these projections do not include the impact of the President's climate change initiative announced in February 2002, nor do they include the effects of measures in the *National Energy Policy* that have not yet been implemented.

The Science

Greenhouse gases are accumulating in Earth's atmosphere as a result of human activities, causing global mean surface air temperature and subsurface ocean temperature to rise. While the changes observed over the last several decades are likely due mostly to human activities, we cannot rule out that some significant part is also a reflection of natural variability.

Reducing the wide range of uncertainty inherent in current model predictions will require major advances in understanding and modeling of the factors that determine atmospheric concentrations of greenhouse gases and aerosols, and the feedback processes that determine the sensitivity of the climate system. Specifically, this will involve reducing uncertainty regarding:

- the future use of fossil fuels and future emissions of methane,

- the fraction of the future fossil fuel carbon that will remain in the atmosphere and provide radiative forcing versus exchange with the oceans or net exchange with the land biosphere,

- the feedbacks in the climate system that determine both the magnitude of the change and the rate of energy uptake by the oceans,

- the impacts of climate change on regional and local levels,

- the nature and causes of the natural variability of climate and its interactions with forced changes, and

- the direct and indirect effects of the changing distributions of aerosols.

Knowledge of the climate system and of projections about the future climate is derived from fundamental physics, chemistry, and observations. Data are then incorporated in global circulation models. However, model projections are limited by the paucity of data available to evaluate the ability of coupled models to simulate important aspects of climate. To overcome these limitations, it is essential to ensure the existence of a long-term observing system and to make more comprehensive regional measurements of greenhouse gases.

Evidence is also emerging that black carbon aerosols (soot), which are formed by incomplete combustion, may be a significant contributor to global warming, although their relative importance is difficult to quantify at this point. These aerosols have significant negative health impacts, particularly in developing countries.

While current analyses are unable to predict with confidence the timing, magnitude, or regional distribution of climate change, the best scientific information indicates that if greenhouse gas concentrations continue to increase, changes are likely to occur. The U.S. National Research Council has cautioned, however, that "because there is considerable uncertainty in current understanding of how the climate system varies naturally and reacts to emissions of greenhouse gases and aerosols, current estimates of the magnitude of future warmings should be regarded as tentative and subject to future adjustments (either upward or downward)." Moreover, there is perhaps even greater uncertainty regarding the social, environmental, and economic consequences of changes in climate.

Source: NRC 2001a.

The remainder of this chapter provides a brief description of the climate system science that sets the context for U.S. action, as well as an overview of the U.S. program that is the focus of this report.

NATIONAL CIRCUMSTANCES: THE U.S. CONTEXT

The perspective of the United States on climate change is informed by our economic prosperity, the rich diversity of our climate conditions and natural resources, and the demographic trends of over 280 million residents. Because of our diverse climatic zones, climate change will not affect the country uniformly. This diversity will also enhance our economy's resilience to future climate change.

Higher anthropogenic greenhouse gas emissions are a consequence of robust economic growth: higher incomes traditionally promote increased expenditures of energy. During the 1990s, investments in technology led to increases in energy efficiency, which partly offset the increases in greenhouse gas emissions that would normally attend strong economic growth. In addition, much of the economic growth in the United States has occurred in less energy-intensive sectors (e.g., computer technologies). Consequently, in the 1990s the direct and proportionate correlation between economic growth and greenhouse gas emissions was altered.

While the United States is the world's largest consumer of energy, it is also the world's largest producer of energy, with vast reserves of coal, natural gas, and crude oil. Nevertheless, our energy use per unit of output—i.e., the energy intensity of our economy—compares relatively well with the rest of the world. The President's new *National Energy Policy* (NEP) includes recommendations that would reduce greenhouse gas emissions by expanded use of both existing and developing technologies (NEPD Group 2001). The NEP's recommendations address expanded nuclear power generation; improved energy efficiency for vehicles, buildings, appliances, and industry; development of hydrogen fuels and renewable technologies; increased access to federal lands and expedited licensing practices; and expanded use of cleaner fuels, including initiatives for coal and natural gas. Tax incentives recommended in the NEP and the President's FY 2003 Budget will promote use of renewable energy forms and combined heat-and-power systems and will encourage technology development.

The nation's response to climate change—our vulnerability and our ability to adapt—is also influenced by U.S. governmental, economic, and social structures, as well as by the concerns of U.S. citizens. The political and institutional systems participating in the development and protection of environmental and natural resources in the United States are as diverse as the resources themselves.

President Bush said last year that technology offers great promise to significantly and cost-effectively reduce emissions in the long term. Our national circumstances—our prosperity and our diversity—may shape our response to climate change, but our commitment to invest in innovative technologies and research will ensure the success of our response.

GREENHOUSE GAS INVENTORY

This report presents U.S. anthropogenic greenhouse gas emission trends from 1990 through 1999 and fulfills the U.S. commitment for 2001 for an annual inventory report to the UNFCCC. To ensure that the U.S. emissions inventory is comparable to those of other UNFCCC signatory countries, the emission estimates were calculated using methodologies consistent with those recommended in the *Revised 1996 IPCC Guidelines for National Greenhouse Gas Inventories* (IPCC/UNEP/OECD/IEA 1997).

Naturally occurring greenhouse gases—that is, gases that trap heat—include water vapor, carbon dioxide (CO_2), methane (CH_4), nitrous oxide (N_2O), and ozone (O_3). Several classes of halogenated substances that contain fluorine, chlorine, or bromine are also greenhouse gases, but for the most part, they are solely a product of industrial activities. Chlorofluorocarbons (CFCs), hydrochlorofluorocarbons (HCFCs), and bromofluorocarbons (halons) are stratospheric ozone-depleting substances covered under the *Montreal Protocol on Substances That Deplete the Ozone Layer* and, hence, are not included in national greenhouse gas inventories. Some other halogenated substances—hydrofluorocarbons (HFCs), perfluorocarbons (PFCs), and sulfur hexafluoride

(SF_6)—do not deplete stratospheric ozone but are potent greenhouse gases and are accounted for in national greenhouse gas inventories.

Although CO_2, CH_4, and N_2O occur naturally in the atmosphere, their atmospheric concentrations have been affected by human activities. Since pre-industrial time (i.e., since about 1750), concentrations of these greenhouse gases have increased by 31, 151, and 17 percent, respectively (IPCC 2001d). This increase has altered the chemical composition of the Earth's atmosphere and has likely affected the global climate system.

In 1999, total U.S. greenhouse gas emissions were about 12 percent above emissions in 1990. A somewhat lower (0.9 percent) than average (1.2 percent) annual increase in emissions, especially given the robust economic growth during this period, was primarily attributable to the following factors: warmer than average summer and winter conditions, increased output from nuclear power plants, reduced CH_4 emissions from coal mines, and reduced HFC-23 by-product emissions from the chemical manufacture of HCFC-22.

As the largest source of U.S. greenhouse gas emissions, CO_2 accounted for 82 percent of total U.S. greenhouse gas emissions in 1999. Carbon dioxide from fossil fuel combustion was the dominant contributor. Emissions from this source category grew by 13 percent between 1990 and 1999.

Methane accounted for 9 percent of total U.S. greenhouse gas emissions in 1999. Landfills, livestock operations, and natural gas systems were the source of 75 percent of total CH_4 emissions. Nitrous oxide accounted for 6 percent of total U.S. greenhouse gas emissions in 1999, and agricultural soil management represented 69 percent of total N_2O emissions. The main anthropogenic activities producing N_2O in the United States were agricultural soil management, fuel combustion in motor vehicles, and adipic and nitric acid production processes. HFCs, PFCs, and SF_6 accounted for 2 percent of total U.S. greenhouse gas emissions in 1999, and

substitutes for ozone-depleting substances comprised 42 percent of all HFC, PFC, and SF_6 emissions.

Evidence is also emerging that black carbon aerosols (soot), which are formed by incomplete combustion, may be a significant anthropogenic agent. Although the U.S. greenhouse gas inventory does not cover emissions of these particles, we anticipate that U.S. research will focus more on them in coming years.

POLICIES AND MEASURES

U.S. climate change programs reduced the growth of greenhouse gas emissions by an estimated 240 teragrams (million metric tons) of CO_2 equivalent in 2000 alone. This reduction helped to significantly lower (17 percent since 1990) greenhouse gases emitted per unit of gross domestic product (GDP), and thus ranks as a step forward in addressing climate change.

However, the U.S. effort was given a potentially greater boost in June 2001, when President Bush announced major new initiatives to advance climate change science and technology. These initiatives came about after government consultation with industry leaders, the scientific community, and environmental advocacy groups indicated that more could and should be done to address scientific uncertainties and encourage technological innovation.

In February 2002, the President announced a new U.S. approach to the challenge of global climate change. This approach contains policies that will harness the power of markets and technology to reduce greenhouse gas emissions. It will also create new partnerships with the developing world to reduce the greenhouse gas intensity of both the U.S. economy and economies worldwide through policies that support the economic growth that makes technological progress possible.

The U.S. plan will reduce the greenhouse gas intensity of the U.S. economy by 18 percent in ten years. This reduction exceeds the 14 percent projected reduction in greenhouse gas intensity in the absence of the additional proposed policies and measures.

The new measures include an enhanced emission reduction registry; creation of transferable credits for emission reduction; tax incentives for investment in low-emission energy equipment; support for research for energy efficiency and sequestration technology; emission reduction agreements with specific industrial sectors, with particular attention to reducing transportation emissions; international outreach, in tandem with funding, to promote climate research globally; carbon sequestration on farms and forests; and, most important, review of progress in 2012 to determine if additional steps may be needed—as the science justifies—to achieve further reductions in our national greenhouse gas emission intensity.

The above strategies are expected to achieve emission reductions comparable to the average reductions prescribed by the Kyoto agreement, but without the threats to economic growth that rigid national emission limits would bring. The registry structure for voluntary participation of U.S. industry in reducing emissions will seek compatibility with emerging domestic and international approaches and practices, and will include provisions to ensure that early responders are not penalized in future climate actions. Furthermore, the President's approach provides a model for developing nations, setting targets that reduce greenhouse gas emissions without compromising economic growth.

PROJECTED GREENHOUSE GAS EMISSIONS

Forecasts of economic growth, energy prices, program funding, and regulatory developments were integrated to project greenhouse gas emissions levels in 2005, 2010, 2015, and 2020. When sequestration is accounted for, total U.S. greenhouse gas emissions are projected to increase by 43 percent between 2000 and 2020. This increased growth in absolute emissions will be accompanied by a decline in emissions per unit of GDP. Note that these forecasts exclude the impact of the

President's climate change initiative announced in February 2002.

Despite best efforts, the uncertainties associated with the projected levels of greenhouse gas emissions are primarily associated with forecast methodology, meteorological variations, and rates of economic growth and technological development. In addition, since the model used to generate these projections does not completely incorporate all current and future policies and measures to address greenhouse gas emissions, these measures, as well as legislative or regulatory actions not yet in force, add another layer of uncertainty to these projections.

IMPACTS AND ADAPTATION

One of the weakest links in our knowledge is the connection between global and regional projections of climate change. The National Research Council's response to the President's request for a review of climate change policy specifically noted that fundamental scientific questions remain regarding the specifics of regional and local projections (NRC 2001a). Predicting the potential impacts of climate change is compounded by a lack of understanding of the sensitivity of many environmental systems and resources—both managed and unmanaged—to climate change.

Chapter 6 provides an overview of potential negative and positive impacts and possible response options, based primarily on *Climate Change Impacts on the United States: The Potential Consequences of Climate Variability and Change* (NAST 2000). This assessment used historical records, model simulations, and sensitivity analyses to explore our potential vulnerability to climate change and highlighted gaps in our knowledge.

The United States is engaged in many efforts that will help our nation and the rest of the world—particularly the developing world—reduce vulnerability and adapt to climate change. By and large these efforts address public health and environmental problems that are of urgent concern today and that may be exacerbated by climate

change. Examples include reducing the spread of malaria, increasing agricultural and forest productivity, reducing the damages from extreme weather events, and improving methods to forecast their timings and locations. Besides benefiting society in the short term, these efforts will enhance our ability to adapt to climate change in the longer term.

Challenges associated with climate change will most likely increase during the 21st century. Although changes in the environment will surely occur, our nation's economy should continue to provide the means for successful adaptation to climate changes.

FINANCIAL RESOURCES AND TRANSFER OF TECHNOLOGY

To address climate change effectively, developed and developing countries must meet environmental challenges together. The United States is committed to helping developing countries and countries with economies in transition meet these challenges in ways that promote economic well-being and protect natural resources. This commitment has involved many players, ranging from government to the private sector, who have contributed significant resources to developing countries. As recognized in the UNFCCC guidelines, this assistance can take the form of hard and/or soft technology transfer.

Projects targeting hard technology transfer, such as equipment to control emissions and increase energy efficiency, can be particularly effective in reducing emissions. And projects that target the transfer of soft technologies, such as capacity building and institution strengthening through the sharing of technical expertise, can help countries reduce their vulnerability to the impacts of climate change. But whether hard or soft, technology transfer programs are most effective when they are approached collaboratively and are congruent with the development objectives and established legal framework of the target country. To this end, the United States works closely with beneficiary countries to ensure a good fit between the resources it provides and the country's needs.

RESEARCH AND SYSTEMATIC OBSERVATION

The United States leads the world in research on climate and other global environmental changes, funding approximately half of the world's climate change research expenditures. We intend to continue funding research in order to ensure vigorous, ongoing programs aimed at narrowing the uncertainties in our knowledge of climate change. These research programs will help advance the understanding of climate change.

The President's major new initiatives directed at addressing climate change are informed by a wealth of input and are intended to result in significant improvements in climate modeling, observation, and research efforts. The long-term vision embraced by the new initiatives is to help government, the private sector, and communities make informed management decisions regarding climate change in light of persistent uncertainties.

EDUCATION, TRAINING, AND OUTREACH

The United States undertakes and supports a broad range of activities aimed at enhancing public understanding and awareness of climate change. These activities range from educational initiatives sponsored by federal agencies to cooperation with independent research and academic organizations. Nongovernmental organizations, industry, and the press also play active roles in increasing public awareness and interest in climate change.

The goal of all of these endeavors—education, training, and public awareness—is to create an informed populace. The United States is committed to providing citizens with access to the information necessary to critically evaluate the consequences of policy options to address climate change in a cost-effective manner that is sustainable and effective in achieving the Framework Convention's long-term goal.

Chapter 2
National Circumstances

During the 1990s, greenhouse gas emissions per unit of gross domestic product (GDP) declined steadily due to continued investments in new energy-efficient technologies and an increase in the portion of GDP attributable to the nonmanufacturing and less energy-intensive manufacturing sectors. However, aggregate U.S. greenhouse gas emissions have continued to increase over the past few years, primarily as a result of economic growth and the accompanying rise in demand for energy.

U.S. energy needs and, hence, emissions of greenhouse gases are also heavily influenced by a number of other factors, including climate, geography, land use, resource base, and population growth. How the nation responds to the issue of climate change is affected by U.S. governmental, economic, and social structures, as well as by the availability of technologies and wealth, which allows such technologies to be employed. All of these factors also affect the nation's vulnerability to climate change and its ability to adapt to a changing natural environment.

Global climate change presents unique challenges and opportunities for the United States. This chapter describes U.S. national circumstances as they relate to climate change: historical developments, current conditions, and trends in those conditions.

CLIMATE PROFILE

The diverse U.S. climate zones, topography, and soils support many ecological communities and supply renewable resources for many human uses. The nature and distribution of these resources have played a critical role in the development of the U.S. economy, thus influencing the pattern of U.S. greenhouse gas emissions.

U.S. climate conditions are representative of all the major regions of the world, except the ice cap. Average annual temperatures range from –1 to +4°C (30–40°F) in the North to 21–27°C (70–80°F) in the South, and

have significant implications for energy demand across the country. In the North, heating needs dominate cooling needs, while the reverse is true in the South. The number of heating and cooling degree-days across U.S. regions illustrates this climatic diversity (Figure 2-1). Because of this diversity of climate and ecological zones, describing the effects of climate change on the nation as either positive or negative overall is an oversimplification.

U.S. baseline rainfall levels also vary significantly by region, with most of the western states being arid. Although the eastern states only rarely experience severe drought, they are increasingly vulnerable to flooding and storm surges as sea level rises, particularly in increasingly densely populated coastal areas. In recent years, although deaths due to tornadoes, floods, and tropical storms have declined substantially, insurance losses have increased. If extreme weather events of this kind were to occur with greater frequency or intensity (which may or may not happen), damages could be extensive.

GEOGRAPHIC PROFILE

The federal government owns slightly more than 20 percent of the total U.S. land area of nearly 920 million hectares (over 2 billion acres). By contrast, the federal government owns over 65 percent of Alaska's nearly 150 million hectares (370 million acres), the state government owns nearly 25 percent, private ownership accounts for about 10 percent, and lands held in trust by the Bureau of Indian Affairs account only for about one-third of 1 percent.

The private sector plays a primary role in developing and managing U.S. natural resources. However, federal, state, and local governments also manage and protect these resources through regulation, economic incentives, and education. Governments and private interests also manage lands set aside for forests, parks, wildlife reserves, special research areas, recreational areas, and suburban and urban open spaces. Table 2-1 and Figure 2-2 illustrate the composition and share of the individual components of U.S. land resources in 1997. This snapshot is discussed in greater detail later in this chapter.

FIGURE 2-1 Climatic Diversity in the Contiguous U.S.

Regions of the country with cooler climates may benefit from climate change through reduced demand for heating, while energy consumption for cooling may increase in warmer regions, which could result in higher emissions of greenhouse gases.

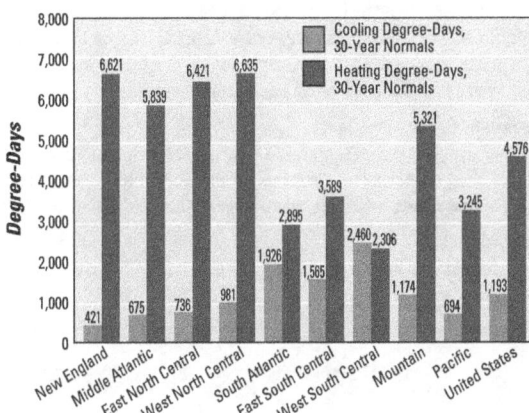

Notes:

• Cooling and heating degree-days represent the number of degrees that the daily average temperature is above (cooling) or below (heating) 65°F. The daily average temperature is the mean of the maximum and minimum temperatures for a 24-hour period. For example, a weather station recording a mean daily temperature of 40°F would report 25 heating degree-days.

• Degree-day normals are simple arithmetic averages of annual degree-days from 1961 to 1990.

• Data for the Pacific region exclude Alaska and Hawaii.

Source: U.S. DOE/EIA 2000a.

POPULATION PROFILE

Population levels and growth rates drive a nation's consumption of energy and other resources, as more people require more energy services. The population dispersion in the United States increases the need for transportation services, and population density and household size influence housing sizes. Settlement patterns and population density also affect the availability of land for various uses.

With a population of just over 280 million in 2000, the United States is the third most populous country in the world, after China and India. U.S. population density, however, is relatively low (Figure 2-3). Population density also varies widely within the United States, and those patterns are changing as people move not only from rural to metropolitan areas, but also from denser city cores to surrounding suburbs. In addition, populations in the warmer parts of the country—the Sunbelt in the South and Southwest—are growing more rapidly than in other parts, showing a preference for warmer climates.

Overall, the annual rate of U.S. population growth has fallen from slightly over 1 percent in 1990 to about 1 percent in 2000. But this is still high by the standards of the Organization of Economic Cooperation and Development (OECD)—about five times the rate in Japan, and more than three times the

TABLE 2-1 AND FIGURE 2-2 U.S. Land Use: 1997

Land is used in many different ways in the United States. Much of the land is forested or used for agricultural purposes.

Land Use	Hectares	Acres
		(in millions)
Urban Land Residential, industrial, commercial, and institutional land. Also includes land for construction sites; sanitary landfills; sewage treatment plants; water control structures and spillways; and airports, highways, railroads, and other transportation facilities.	25	65
Forest-Use Land At least 10 percent stocked by single-stemmed forest trees of any size, which will be at least 4 meters (13 feet) tall at maturity. When viewed vertically, canopy cover is 25 percent or greater.	260	640
Cropland Used for Crops Areas used for the production of adapted crops for harvest.	140	350
Cropland Idled, including Conservation Reserve Program Includes land in cover and soil improvement crops, and completely idle cropland. Some cropland is idle each year for various physical and economic reasons. Acreage diverted from crops to soil-conserving uses under federal farm programs is included in this component. For example, cropland enrolled in the Federal Conservation Reserve Program is included.	15	40
Cropland Used for Pasture Generally considered as being tilled, planted in field crops, and then reseeded to pasture at varying intervals. However, some cropland pasture is marginal for crop uses and may remain in pasture indefinitely. Also includes some land that was used for pasture before crops reached maturity and some land that could have been cropped without additional improvement.	30	70
Grassland Pasture and Range Principally native grasses, grasslike plants, forbs or shrubs suitable for grazing and browsing, and introduced forage species that are managed with little or no chemicals or fertilizer being applied. Examples include grasslands, savannas, many wetlands, some deserts, and tundra.	235	580
Special Uses Includes national and state parks and wildlife areas, defense installations, and rural transportation.	115	285
Miscellaneous Other Land Includes rural residential, marshes, open swamps, deserts, tundra, and other areas not inventoried.	95	235
TOTAL LAND, 50 STATES	**915**	**2,265**

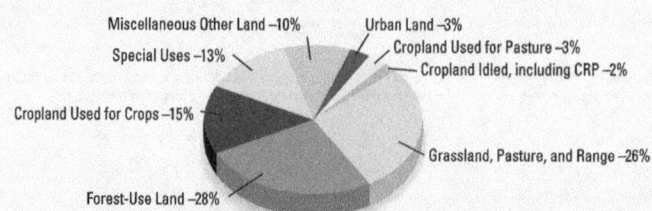

Miscellaneous Other Land –10%
Special Uses –13%
Cropland Used for Crops –15%
Forest-Use Land –28%
Urban Land –3%
Cropland Used for Pasture –3%
Cropland Idled, including CRP –2%
Grassland, Pasture, and Range –26%

Note: Individual land uses may not sum to total land due to rounding.
Source: USDA/NRCS 2001.

rate in the European Union. Among the OECD countries, the United States has been and continues to be one of the largest recipients of immigrants (in absolute terms). Net immigration contributes about one-third of the total annual population growth, and natural increase (births minus deaths) contributes the remaining two-thirds.

The U.S. population is aging. The current median age is about 35 years, compared to about 33 in 1990 and 28 in 1970. This change in median age has been a result of both an increase in life expectancy, which now stands at 77 years, and reduced fertility rates. Along with an aging population, trends also indicate a steady reduction in average household size, as people marry later, have fewer children, are more likely to divorce, and are more likely to live alone as they age. Thus, between 1970 and 2000, while the population has grown by nearly 40 percent, the number of households has grown by over 65 percent.

Although the average household size has declined, the average size of housing units has been increasing. Between 1978 and 1997, the proportion of smaller housing units (with four or fewer rooms) has decreased from about 35 to 30 percent, and the proportion of large housing units (with seven or more rooms) has increased from about 20 to nearly 30 percent. In general, larger housing units result in increased demands for heating, air conditioning, lighting, and other energy-related needs.

The share of the total U.S. population living in metropolitan areas of at least one million people has increased to nearly 60 percent in 2000, up from nearly 30 percent in 1950. This growth has been concentrated in suburbs, rather than in city centers. In fact, most major cities have experienced declines in population, as crime, congestion, high taxes, and the desire for better schools have led people to move to the suburbs. As a result, population densities in the U.S. metropolitan areas are far lower than in metropolitan areas around the world, and they continue to decline. For example, the ten largest European cities, on average, have population densities four times greater than the ten largest U.S. cities. The increased concentration of the U.S population in the suburbs has resulted in both greater reliance on decentral-ized travel modes, such as the automobile, and relatively high per capita energy use.

Another factor leading to higher emissions is the increasing mobility of the U.S. population. The average U.S. citizen tends to move more than ten times in his or her lifetime. According to the 1990 census, nearly 40 percent of U.S. residents do not live in the state where they were born, as compared to about 30 percent in 1980 and about 25 percent in 1970. Families are often dispersed across the country for education, career, or personal reasons. All of these factors have led to an ever-growing need for transportation services.

GOVERNMENT STRUCTURE

The U.S. political and institutional systems participating in the development and protection of environmental and natural resources are as varied as the resources themselves. These systems span federal, state, and local government jurisdictions, and include legislative, regulatory, judicial, and executive institutions.

The U.S. government is divided into three separate branches: the executive branch, which includes the Executive

---

FIGURE 2-3 U.S. Population Density: 2000

Though the United States is the third most populous country in the world, U.S. population density is relatively low. This combination tends to have negative implications for energy and automobile use and, hence, emissions of greenhouse gases.

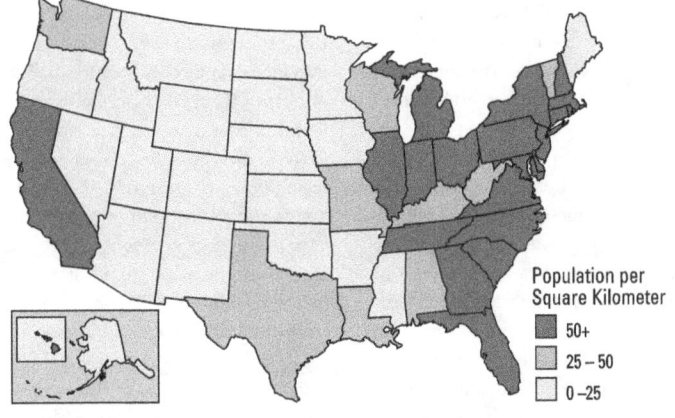

Population per Square Kilometer
- 50+
- 25–50
- 0–25

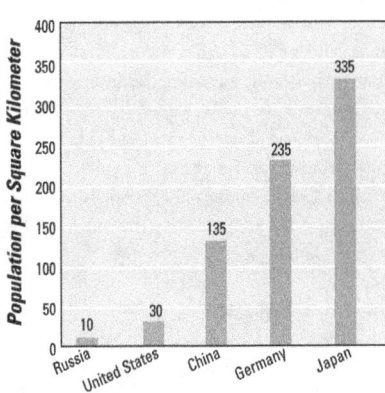

Note: International population density comparisons have been rounded.

Sources: U.S. DOC/Census 2000 and World Bank 2000.

Office of the President, executive departments, and independent agencies; the legislative branch (the U.S. Congress); and the judicial branch (the U.S. court system). The distinct separation of powers in this tripartite system is quite different from parliamentary governments.

Federal Departments and Agencies

The executive branch is comprised of 14 executive departments, 7 agencies, and a host of commissions, boards, other independent establishments, and government corporations. The traditional functions of a department or an agency are to help the President propose legislation; to enact, administer, and enforce regulations and rules implementing legislation; to implement Executive Orders; and to perform other activities in support of the institution's mission, such as encouraging and funding the research, development, and demonstration of new technologies.

No single department, agency, or level of government in the United States has sole responsibility for the panoply of issues associated with climate change. In many cases, the responsibilities of federal agencies are established by law, with limited administrative discretion. At the federal level, U.S. climate change policy is determined by an interagency coordinating committee, chaired from within the Executive Office of the President, and staffed with members of the executive offices and officials from the relevant departments and agencies, including the Departments of Agriculture, Commerce, Defense, Energy, Justice, State, Transportation, and Treasury, as well as the U.S. Environmental Protection Agency and the U.S. Agency for International Development.

The U.S. Congress

As the legislative branch of the U.S. government, Congress also exercises responsibility for climate change and other environmental and natural resource issues at the national level. It influences environmental policy through two principal vehicles: creation of laws and oversight of the federal executive branch. Thus, Congress can enact laws establishing regulatory regimes for environmental purposes, and can pass bills to appropriate funds for environmental purposes. Under its constitutional authority, Congress ratifies international treaties, such as the United Nations Framework Convention on Climate Change.

The U.S. Congress comprises two elected chambers—the Senate and the House of Representatives—having generally equal functions in lawmaking. The Senate has 100 members, elected to six-year terms, with two representatives for each of the 50 states. The House has 435 members, elected to two-year terms, each of whom represents an electoral district of roughly equal population. The less populated but often resource-rich regions of the country, therefore, have proportionately greater representation in the Senate than in the House.

Environmental proposals, like most other laws, may be initiated in either chamber of the U.S. Congress. After their introduction, proposals or "bills" are referred to specialized committees and subcommittees, which hold public hearings on the bills to receive testimony from interested and expert parties. After reviewing the testimony, the committees and subcommittees deliberate and revise the bills, and then submit them for debate by the full membership of that chamber. Differences between bills originating in either the House or the Senate are resolved in a formal conference between the two chambers. To become a law, a bill must be approved by the majorities of both chambers, and then must be signed by the President. The President may oppose and veto a bill, but Congress may override a veto with a two-thirds majority from each chamber.

As a rule, spending bills must go through this process twice. First, the committee responsible for the relevant issue must submit a bill to authorize the expenditure. Then, once both chambers pass the authorization bill, the Appropriations Committee, in a separate process, must submit a bill appropriating funds from the budget. The funds that are actually appropriated often are less than the authorized amount.

States, Tribes, and Local Governments

States, Native American tribal organizations, localities, and even regional associations also exert significant influence over the passage, initiation, and administration of environmental, energy, natural resource, and other climate-related programs. For example, the authority to regulate electricity production and distribution lies with state and local public utility commissions. In addition, the regulation of building codes—strongly tied to the energy efficiency of buildings—is also controlled at the state and local levels.

Although the federal government promulgates and oversees environmental regulations at the national level, the states and tribes often are delegated the authority to implement some federal laws by issuing permits and monitoring compliance with regulatory standards. The states also generally have the discretion to set environmental standards that are more stringent than the national standards. Individual states also enjoy autonomy in their approach to managing their environmental resources that are not subject to federal laws. In addition to regulation, some states and localities have developed voluntary and incentive programs that encourage energy efficiency and conservation, and/or mitigate greenhouse gas emissions.

Local power to regulate land use is derived from a state's power to enact legislation to promote the health, safety, and welfare of its citizens. States vary in the degree to which they delegate these "powers" to local governments, but land use is usually controlled to a considerable extent by local governments (county or city). This control may take the form of authority to adopt comprehensive land-use plans to enact zoning ordinances and subdivision regulations or to restrict shoreline, floodplain, or wetland development.

The U.S. Court System

The U.S. court system is also crucial to the disposition of environmental issues. Many environmental cases are litigated in the federal courts. The role of the courts is to settle disagreements on how to interpret the law. The federal court system is three-tiered: the district court level; the first appellate (or circuit) court level; and the second and final appellate level (the U.S. Supreme Court). There are 94 federal district courts, organized into federal circuits, and 13 federal appeals courts.

Cases usually enter the federal court system at the district court level, though some challenges to agency actions are heard directly in appellate courts, and disputes between states may be brought directly before the U.S. Supreme Court. Generally, any person (regardless of citizenship) may file a complaint alleging a grievance. In civil enforcement cases, complaints are brought on behalf of the government by the U.S. attorney general and, in some instances, may be filed by citizens as well.

Sanctions and relief in civil environmental cases may include monetary penalties, awards of damages, and injunctive and declaratory relief. Courts may direct, for example, that pollution be controlled, that contaminated sites be cleaned up, or that environmental impacts be assessed before a project is initiated. Criminal cases under federal environmental laws may be brought only by the government—i.e., the attorney general or state attorneys general. Criminal sanctions in environmental cases may include fines and imprisonment.

ECONOMIC PROFILE

The U.S. is endowed with a large and dynamic population, bountiful land and other natural resources, and vibrant competition in a market economy. These factors have contributed to making the U.S. economy (in terms of its real GDP) the largest in the world, accounting for over one-fourth of the global economy.

Government and the Market Economy

A number of principles, institutions, and technical factors have played a role in the evolution of the U.S. market economy. The first of these is the respect for property rights, which includes the right to own, use, and transfer private property to one's own advantage. The U.S. economic system is also underpinned by a reliance on market forces, as opposed to tradition or force, as the most efficient means of organizing economic activity. In other words, in a well-functioning market, relative prices are the primary basis on which economic agents within the U.S. economy make decisions about production and consumption. Ideally, the price system, combined with a system of well-defined and well-protected private property rights, allocates the resources of an economy in a way that produces the greatest possible economic welfare.

However, in some cases, due to imperfect information, lack of clearly defined property rights for public goods (such as air and water), and/or other market imperfections, the production of goods and services creates externalities (i.e., costs or benefits) that are not borne directly by the producers and consumers of those goods and services. For example, if the production of a good has environmental costs that are not borne by its producers or consumers, that product may be priced too low, thereby stimulating excess demand and pollution. Alternatively, research and development (R&D) may produce benefits to society beyond those that accrue to the firm doing the research, but if those benefits are not captured in the price, firms will underinvest in R&D. Under such circumstances, the U.S. government intervenes to alter the allocation of resources.

Government intervention may include limiting the physical quantity of pollution that can be produced, or charging polluters a fee for each unit of pollution emitted. As a practical matter, however, accurately establishing the cost of the externality to be internalized by a fee, a tax, or a regulation can be very difficult. There is also a risk that government intervention could have other, unintended consequences. For these reasons, the U.S. government tends to be cautious in its interventions, although it does take actions necessary to protect the economy, the environment, human health, natural resources, and national security.

In addition, many government interventions are intended to correct market imperfections and facilitate smooth functioning of markets. By protecting property rights, producing public goods such as roads and other types of infrastructure, formulating policies that internalize external costs (e.g., environmental policies), and enacting legislation to ensure a minimum standard of living for all of its citizens, the U.S. government fosters an environment in which market forces can function effectively. Finally, the government inevitably influences the economy through regulatory and fiscal processes, which in turn affect the functioning of markets.

Composition and Growth

Robust economic growth typically leads to higher greenhouse gas emissions and degradation of environmental resources in general. Nonetheless, it is often the case that as the health of the economy improves and concerns about unemployment and economic growth lessen, greater emphasis is placed on environmental issues.

From 1960 to 2000, the U.S. economy grew at an average annual rate of over 3 percent, raising real GDP from about $2 trillion to over $9 trillion (in 1996 constant dollars). This implies that, with population growth averaging about 1 percent over the same period, real GDP per capita has increased at an average annual rate of over 2 percent to about $32,800 in 2000 from nearly $13,300 in 1960 (all in 1996 constant dollars).

Between 1960 and 2000, the labor force more than doubled from slightly over 65 million to about 135 million, as the influx of women into the work force raised the overall labor participation rate from nearly 60 percent to over 65 percent. The rapid growth in the size of the labor force has been led by the service sector (which includes communications, utilities, finance, insurance, and real estate), as shown in Figure 2-4. While the size of the service sector labor force more than doubled between 1970 and 2000, its sectoral share in the U.S. labor force increased by more than 40 percent over the same period. Employment in several other industries, such as construction, trade, and finance also increased significantly, along with their sectoral shares in the U.S. labor force. In contrast, employment in agriculture, along with its sectoral share in the U.S. labor force, declined during the same period.

From the latter part of 1991 through 2000, the United States experienced the longest peacetime economic expansion in history. The average annual U.S. economic growth (in terms of real GDP) was about 3 percent per year between 1991 and 1995 and more than 4 percent per year between 1996 and 2000. During the second half of 2000, the economy, nonetheless, showed signs of moderating, with real annual GDP growth registering at a little over 3 percent in 2000, relative to the previous year. Overall, unemployment was reduced to about 4 percent in 2000, while producing healthy increases in real wages and real disposable income. Both personal consumption and industrial production have increased as a result of this economic growth and have, therefore, contributed to greater energy con-

sumption and fossil fuel-related carbon dioxide emissions. Much of this economic growth, however, has occurred in sectors of the economy that are less energy-intensive (e.g., computer technologies), which in turn has lowered the energy intensity of the U.S. economy.

ENERGY PRODUCTION AND CONSUMPTION

The United States continues to be the world's largest energy producer and consumer. The nation's patterns of energy use are determined largely by its economic and population growth, large land area, climate regimes, population dispersion, average size of household, other population characteristics, and availability of indigenous resources. Much of the infrastructure of U.S. cities, highways, and industries was developed

---

FIGURE 2-4 U.S. Employment by Industry: 1970–2000

Between 1970 and 2000, employment rose most rapidly in the construction, trade, financial, utilities, and services sectors. The service sector is by far the largest in the United States, employing more than one-third of the population.

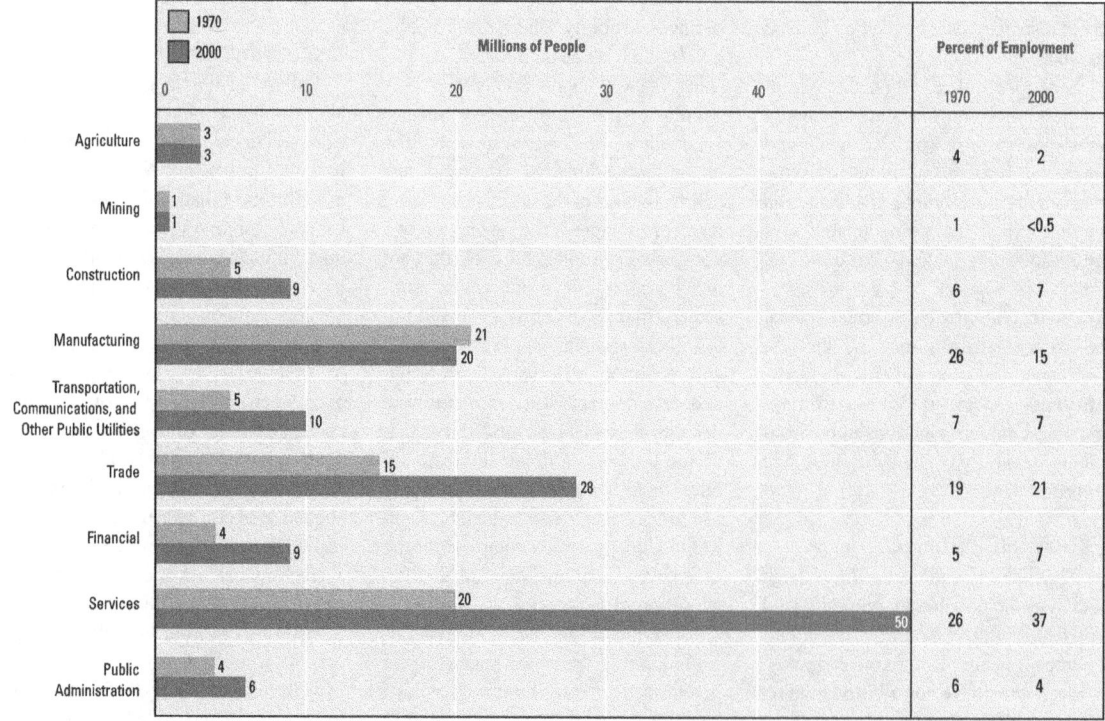

Note: All numbers are rounded to the nearest integer.

Sources: U.S. DOC/Census 1999 and U.S. DOL/BLS 2000.

in response to abundant and relatively inexpensive energy resources. Figure 2-5 provides a comprehensive overview of the energy flows through the U.S. economy in 2000.

Different regions of the country rely on different mixes of energy resources (reflecting their diverse resource endowments) to generate power and meet other energy needs. For example, the Pacific Northwest and Tennessee Valley have abundant hydropower resources, while the Midwest relies heavily on coal for power generation and industrial energy needs.

Resources

The vast fossil fuel resources of the United States have contributed to low prices and specialization in relatively energy-intensive activities. Coal, which has the highest emissions of greenhouse gases per unit of energy, is particularly abundant, with current domestic recoverable reserves estimated at nearly 460 billion metric tons (about 503 billion short tons)—enough to last for over 460 years at current recovery rates. Recent gains in mining productivity, coupled with increased use of less-expensive western coal made possible by railroad deregulation, have led to a continual decline in coal prices over the past two decades. As a result, the low cost of coal on a Btu basis has made it the preferred fuel for power generation, supplying over half of the energy consumed to generate electricity.

Proved domestic reserves of oil (nearly 4 trillion liters or over 20 billion barrels at the start of 2000) have been on a downward trend ever since the addition of reserves under Alaska's North Slope in 1970. Restrictions on exploration in many promising but ecologically sensitive areas have constrained additions to reserves. Reserves of natural gas were nearly 5 trillion cubic meters (nearly 170 trillion cubic feet) at the start of 2000. The estimated natural gas resources of nearly 37 trillion cubic meters (nearly 1,300 trillion cubic feet) are expected to last for more than 65 years at current rates of production. U.S. energy resources also include over 120 million kg (about 270 million pounds) of uranium oxide, recoverable at about $65 per kilogram ($30 per pound) or less (in 2000 current dollars). Hydroelectric resources are abundant in certain areas of the country, where they have already largely been exploited.

FIGURE 2-5 Energy Flow Through the U.S. Economy: 2000 (Quadrillion Btus)

The U.S. energy system is the largest in the world and is composed of multiple energy sources and end users. Most of the energy is produced and consumed domestically, although imports constitute a significant portion, and a small fraction of energy is exported.

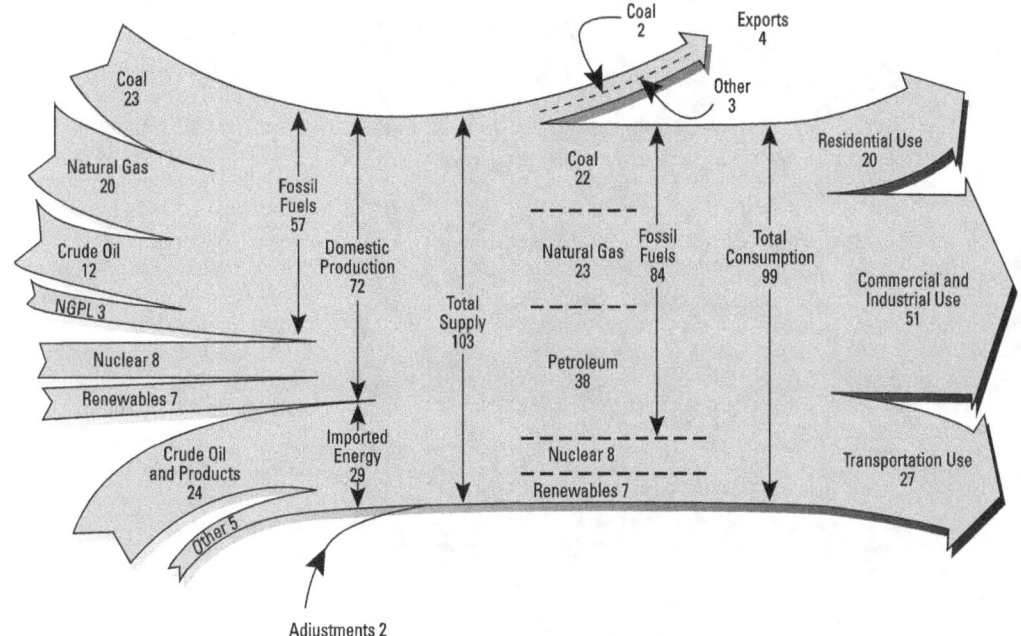

Note: Shares may not sum to totals due to rounding.
Source: U.S. DOE/EIA 2000a.

National Energy Policy Goals

In May 2001, the Bush Administration published the *National Energy Policy* (NEP). This long-term, comprehensive strategy was primarily designed to assist the private sector, states, and local governments in promoting "dependable, affordable, and environmentally sound production and distribution of energy for the future"(NEPD Group 2001). The NEP seeks to promote new, environmentally friendly technologies to increase energy supplies and to encourage cleaner, more efficient energy use. It also seeks to raise the living standards of Americans by fully integrating national energy, environmental, and economic policies. The following goals are the NEP's guiding principles.

Modernize Conservation

This NEP goal seeks to increase energy efficiency by applying new technology, which is expected to raise productivity, reduce waste, and trim costs. Some of the recommendations include: increased funding for renewable energy and energy efficiency research and development programs; income tax credits for the purchase of hybrid and fuel cell vehicles; extension of the ENERGY STAR® efficiency program; and tax incentives and streamlined permitting to promote clean combined heat and power (CHP) technology.

Modernize Energy Infrastructure

This NEP goal seeks to modernize and expand the national energy infrastructure such that energy supplies can be safely, reliably, and affordably transported to homes and businesses. Some of the recommendations include: improving pipeline safety and expediting pipeline permitting; expanding research and development on transmission reliability and superconductivity; and enacting comprehensive electricity legislation that promotes competition, encourages new generation, protects consumers, enhances reliability, and promotes renewable energy.

Increase Energy Supplies

This NEP goal seeks to increase and diversify the nation's traditional and alternative fuel sources so as to provide 'families and businesses with reliable and affordable energy, to enhance national security, and to improve the environment." Some of the recommendations include: environmentally regulated exploration and production of oil using leading-edge technology in the Arctic National Wildlife Refuge (ANWR); regulated increase in oil and natural gas development on other federal lands; fiscal incentives for selected renewable power generation technologies; and streamlining the relicensing of hydropower and nuclear facilities.

Accelerate Protection and Improvement of the Environment

This NEP goal seeks to integrate long-term national energy policy with national environmental goals. Some of the recommendations include multi-pollutant legislation to establish a flexible, market-based program to significantly reduce and cap emissions of sulfur dioxide, nitrogen oxides, and mercury from electric power generators; land conservation efforts; and new guidelines to reduce truck-idling emissions at truck stops.

Increase Energy Security

This NEP goal seeks to lessen the impact of energy price volatility and supply uncertainty on the American people. Some of the recommendations include increasing funding for the Low-Income Home Energy Assistance Program; preparing the Federal Emergency Management Administration for managing energy-related emergencies; and streamlining and expediting permitting procedures to expand and accelerate cross-border energy investment, oil and gas pipelines, and electricity grid connections with Mexico and Canada.

Production

Coal, natural gas, and crude oil constitute the bulk of U.S. domestic energy production. In 1960, these fossil fuels accounted for nearly 95 percent of production. By 2000 their contribution had fallen to about 80 percent, with the nuclear electric power displacing some of the fossil fuel production (Figure 2-6). Further displacement will most likely be limited, however, due to uncertainties related to deregulation of the electric industry, difficulty in siting new nuclear facilities, and management of commercial spent fuel. Renewable resources contribute a small but growing share.

Crude Oil

Before 1970, the United States imported only a small amount of energy, primarily in the form of petroleum. Beginning in the early 1970s, however, lower acquisition costs for imported crude oil and rising costs of domestic production put domestic U.S. oil producers at a comparative disadvantage, leading to a divergence in trends of energy production and consumption. In 2000, the United States produced over 70 quadrillion Btus of energy and exported 4 quadrillion Btus, over 35 percent of which was coal. Consumption totaled nearly 100 quadrillion Btus, requiring imports of nearly 30 quadrillion Btus. Domestic crude oil production is projected to remain relatively stable through 2003 as a result of a favorable price environment and increased success of offshore drilling. A decline in production is projected from 2004 through 2010, followed by another period of projected stable production levels through 2020 as a result of rising prices and continuing improvements in technology. In 2020, the projected domestic production level of slightly over 5 million barrels per day would be almost one million barrels per day less than the 1999 level. In 2000, net imports of petroleum accounted for over 60 percent of domestic petroleum consumption. Continued dependence on petroleum imports is projected, reaching about 65 percent in 2020.

Coal is the largest source of domestic energy, followed by natural gas and oil. Since 1970, the production of coal, nuclear, and renewables has risen to offset the decline in oil and natural gas production.

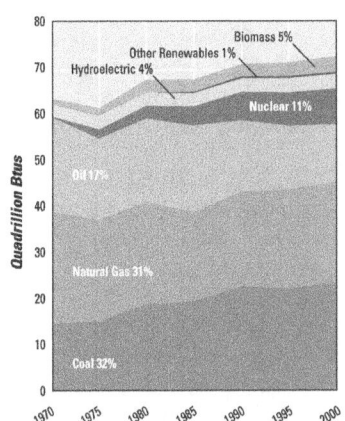

Notes: Fuel share estimates correspond to 2000 data. Shares may not sum to 100 percent due to rounding

Source: U.S. DOE/EIA 2000a.

Coal

Coal is the largest source of domestically produced energy. As the only fossil fuel for which domestic production exceeds consumption, coal assumed a particularly important role in the wake of the oil shocks in the 1970s. Between 1991 and 2000, U.S coal production increased by about 8 percent. However, more recently (between 1998 and 2000), coal production has declined by nearly 4 percent from slightly over one billion metric tons in 1998. This decline was primarily attributed to a large drop in coal exports and a smaller than usual growth in coal consumption for power generation.

From 1996 to 2000, U.S. coal exports have declined by about 35 percent. In particular, they declined sharply between 1998 and 2000, from over 70 million metric tons (over 77 million short tons) to nearly 55 million metric tons (nearly 61 million short tons). U.S. coal exports declined in almost every major world region. The decline in coal exports to Canada, Europe, and Asia was primarily attributed to competition from lower-priced coal from Australia, South Africa, Columbia, and Venezuela. Coal exports are projected to remain relatively stable, settling at slightly more than 50 million metric tons by 2020.

Natural Gas

Regulatory and legislative changes in the mid-1980s led to market pricing of natural gas. These changes heightened demand and boosted natural gas production, reversing the decline it had experienced in the 1970s and early 1980s. This increased production is projected to continue and even accelerate in the early decades of the 21st century. Nonetheless, growth in consumption is expected to outstrip that of production, leading to an increase in net imports, from the 1999 level of more than 85 billion cubic meters (3 trillion cubic feet) to a projected level of nearly 170 billion cubic meters (6 trillion cubic feet) in 2020.

Renewable Energy

Renewable sources currently constitute about 9 percent of U.S. energy production, and hydropower contributes 4 percent. Projected growth in renewable electricity generation is expected from biomass (currently at nearly 5 percent) and from solar, wind, and geothermal energy (currently at less than 1 percent). The largest increase in renewable electricity generation is projected for biomass, from more than 35 billion kilowatt hours in 1999 to over 65 billion in 2020.

Electricity Market Restructuring

The U.S. electric power generation industry is evolving from a regulated to a competitive industry. In many jurisdictions, wholesale markets have already become competitive, while retail markets have been slow to follow. Where power generation was once dominated by vertically integrated investor-owned utilities (IOUs) that owned most of the generation capacity, transmission, and distribution facilities, the electric power industry now has many new companies that generate and trade electricity. Although vertically integrated IOUs still produce most of the country's electrical power today, this situation is rapidly changing.

Competition in wholesale power sales received a boost from the Energy Policy Act of 1992 (EPAct), which expanded the Federal Energy Regulatory Commission's (FERC's) authority to order vertically integrated IOUs to allow nonutility power producers access to the transmission grid to sell power. In 1996, the FERC issued its Orders 888 and 889, which established a regime for nondiscriminatory access by all wholesale buyers and sellers to transmission facilities. More recently, in December 1999, FERC issued Order 2000, calling for the creation of regional transmission organizations (RTOs)—independent entities that will control and operate the transmission grid free of any discriminatory practices. Electric utilities were required to submit proposals to form RTOs by January 2001.

In addition to wholesale competition, for the first time in the history of the industry, retail customers in some states have been given a choice of electricity suppliers. As of July 1, 2000, 24 states and the District of Columbia had passed laws or regulatory orders to implement retail competition, and more are expected to follow. The introduction of wholesale and retail competition to the electric power industry has produced and will continue to produce significant changes in the industry.

In 2000, coal-fired power plants generated more than 50 percent of electricity produced in the United States, followed by nuclear power (nearly 20 percent), natural gas (a little over 15 percent), conventional hydropower (nearly 10 percent), petroleum (3 percent), and other fuels and renewables (2 percent). Over the past few years, and in near-term projections, natural gas has been the fuel of choice for new electricity-generating capacity. The restructuring of the electric power industry may accelerate this trend, due to the fact that natural gas generation is less capital-

The U.S. Energy Policy and Conservation Act

Several titles of the U.S. Energy Policy and Conservation Act of 1992 continue to be extremely important to the overall U.S. strategy of reducing greenhouse gas emissions. Important provisions of this Act were reauthorized in the Energy Conservation Reauthorization Act of 1998. Relevant titles of the original Act are summarized below.

Title I—Energy Efficiency

This title establishes energy efficiency standards, promotes electric utility energy management programs and dissemination of energy-saving information, and provides incentives to state and local authorities to promote energy efficiency.

Titles III, IV, V, and VI—Alternative Fuels and Vehicles

These titles provide monetary incentives, establish federal requirements, and support the research, design, and development of fuels and vehicles that can reduce oil use and, in some cases, carbon emissions as well.

Titles XII, XIX, XXI, and XXII—Renewable Energy, Revenue Provisions, Energy and Environment, and Energy and Economic Growth

These titles promote increased research, development, production, and use of renewable energy sources and more energy-efficient technologies.

Title XVI—Global Climate Change

This title provides for the collection, analysis, and reporting of information pertaining to global climate change, including a voluntary reporting program to recognize electric utility and industry efforts to reduce greenhouse gas emissions.

Title XXIV—Hydroelectric Facilities

This title facilitates efforts to increase the efficiency and electric power production of existing federal and nonfederal hydroelectric facilities.

Title XXVIII—Nuclear Plant Licensing

This title streamlines licensing for nuclear plants.

intensive than other technologies, and the cost of capital to the industry is expected to increase.

Consumption

On the consumption side, rapid economic and population growth, combined with the increasing energy demands of the transportation and buildings sectors, resulted in an 80 percent increase in energy demand from 1960 to 1979. Most of the increased demand was met by oil imports and by increased consumption of coal and natural gas. Total energy demand dampened during and after the international oil price shocks in 1973–74 and 1979–80, and overall energy consumption actually fell through the early 1980s. Energy consumption resumed its upward trend in the latter half of the 1980s, in response to declining oil and gas prices and renewed economic growth.

Another lingering effect of the oil price shocks was a shift in consumption away from oil. Power generation shifted toward natural gas, coal, and nuclear power, and space heating became more dependent on natural gas and electricity. Most of the shift away from oil to natural gas, however, occurred after the second oil price shock.

From 1949 to 2000, while the U.S. population expanded by nearly 90 percent, the amount of electricity sold by utilities grew by over 1,200 percent. Average per capita consumption of electricity in 2000 was seven times as high as in 1949. The growth in the economy, population, and distances traveled has contributed to increased U.S energy consumption. However, by 2000, energy use per dollar of GDP (or energy intensity) had decreased by nearly 45 percent from its peak in 1970. Most of these energy intensity improvements are due to an increase in the less energy-intensive industries and a decrease in the more energy-intensive industries. The household and the transportation sectors also experienced significant gains in efficiency. Today U.S. energy intensity is just slightly above OECD's average energy intensity (at 0.43 kg of oil equivalent per dollar of GDP, versus 0.41 kg for the OECD).

SECTORAL ACTIVITIES

In 2000, end users consumed about 75 quadrillion Btus (quads) of energy directly, including over 10 quads of electricity. In addition, about 25 quads of energy were used in the generation, transmission, and distribution of electricity. Industry and transportation consumed three-quarters of this direct energy, while the residential and commercial sectors used one-quarter. However, because most electricity is delivered to residential and commercial users, total primary energy consumption of nearly 100 quads is distributed fairly evenly among final users (Figure 2-7).

The remainder of this section discusses energy use by and emissions from industry, residential and commercial buildings, transportation, and the U.S. government, as well as waste. Agricultural and forest practices are addressed elsewhere in this chapter.

Industry

Comprised of manufacturing, construction, agriculture, and mining, the industrial sector accounted for more than 35 percent of total U.S. energy use in 2000 and slightly over 30 percent of total U.S. greenhouse gas emissions. Industry's energy consumption rose steadily until the early 1970s, and then dropped markedly, particularly in the early 1980s, following the second oil shock. Since the late 1980s, industrial energy consumption has resumed a gradual upward trend.

Similarly, from 1978 to 1999, industrial energy intensity (energy consumed by the individual sector per unit of

Figure 2-7 Energy Consumption by Sector: 1970–2000

Energy consumption is divided fairly evenly among the three sectors, with industrial being the largest and the buildings sector close behind. The rate of growth in energy consumption since 1970 has been highest in the buildings and transportation sectors.

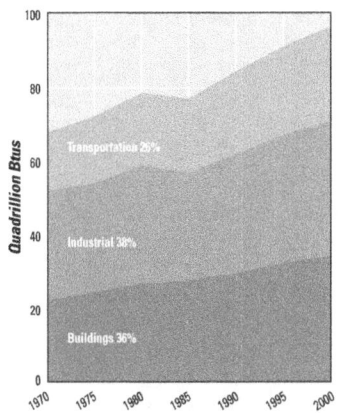

Notes: Sectoral share estimates correspond to 2000 data. Shares may not sum to 100 percent due to rounding.
Source: U.S. DOE/EIA 2000a.

industrial output) fell by about 25 percent. Approximately two-thirds of this decline is attributable to structural shifts, such as the changing array of products that industry produced during the period, while roughly one-third is attributable to efficiency improvements.

Over 80 percent of the energy consumed in the industrial sector is used for manufacturing (including feedstocks), with the remainder of the energy consumed by mining, construction, agriculture, fisheries, and forestry. In 1998, fuel consumption for manufacturing amounted to nearly 25 quadrillion Btus, an increase of nearly 10 percent since 1994. Of this, four subsectors accounted for nearly 80 percent of the total manufacturing fuel consumption: chemicals and allied products (25 percent), petroleum and coal products (over 30 percent), paper and allied products (over 10 percent), and primary metal industries (over 10 percent). Natural gas was the most commonly consumed energy source in manufacturing.

Natural gas and electricity together comprised nearly 45 percent of all energy sources (in terms of Btus). Over the past two decades energy intensity in the manufacturing sector has declined, although the rate of decline has slowed since energy prices fell in 1985. Of the 20 major energy-consuming industry groups in the manufacturing sector, most continued to reduce their energy intensity between 1985 and 1994.

Residential and Commercial Buildings

The number, size, and geographic distribution of residential and commercial buildings, as well as the market penetration of heating and cooling technologies and major appliances, all combine to influence the energy consumption and greenhouse gases associated with residential and commercial activities.

Residential and commercial buildings together account for roughly 35 percent of the U.S. carbon emissions associated with energy consumption. Commercial buildings—which encompass all nonresidential, privately owned, and public buildings—account for slightly over 15 percent of U.S. carbon emissions. Total energy use in the buildings sector has been increasing gradually, rising from more than 20 quadrillion Btus in 1970 to nearly 35 quadrillion Btus in 1998. The sector's share of total energy consumption relative to other end-use sectors has remained roughly stable over this period.

In 1997 the United States had more than 100 million households, approximately half of which lived in detached, single-family dwellings. Demographic changes have led to a steep decline in the average number of people per residence—from 3.3 in 1960 to 2.6 in 1990—and the sizes of houses have also increased. Since then, that number has remained fairly stable through 1996. The average heated space per person had increased to nearly 65 square meters (nearly 680 square feet) in 1990, compared to nearly 60 square meters (nearly 630 square feet) in 1980.

In addition, major energy-consuming

appliances and equipment came into widespread use during this period. By 1990, essentially all U.S. households had space and water heating, refrigeration and cooking appliances, and color television sets. In 1997, over 70 percent of the households had some form of air conditioning, over 75 percent had clothes washers, over 70 percent had clothes dryers, and about 50 percent had dishwashers (Figure 2-8).

New products have continued to penetrate the market. For example, in 1978, only 8 percent of U.S. households had a microwave oven; by 1997, nearly 85 percent had a microwave oven. Similarly, household survey data on personal computers were first collected in 1990, when slightly over 15 percent of households owned one or more PCs. By 1997 that share had more than doubled to 35 percent.

Despite this growth in appliances, products, and per capita heating and cooling space, large gains in the energy efficiency of appliances and building shells (e.g., through better insulation) have resulted in a modest decline in residential energy use per person and only modest increases in total U.S. energy demand in the residential sector. The increased use of nontraditional electrical appliances, such as computers and cordless (rechargeable) tools, is expected to drive a gradual (one half of 1 percent per year) rise in per-household residential energy consumption between 1990 and 2015.

The type of fuel used to heat U.S. homes has changed significantly over time. More than one-third of all U.S. housing units were warmed by coal in 1950, but by 1997 that share fell to less than one-half of 1 percent. During the same period, distillate fuel oil lost just over half of its share of the home-heating market, falling to 10 percent. Natural gas and electricity gained as home-heating sources. The share of natural gas rose from about a quarter of all homes in 1950 to over half in 1997, while electricity's share shot up from less than 1 percent in 1950 to nearly 30 percent in 1997.

In recent years, electricity and natural gas have been the most common sources

FIGURE 2-8 Energy Characteristics of U.S. Households

In 1997, household energy consumption was 10.25 quadrillion Btus. The primary energy source was natural gas, followed by electricity and oil. The graphic below depicts the percentage of households with a variety of energy-consuming appliances.

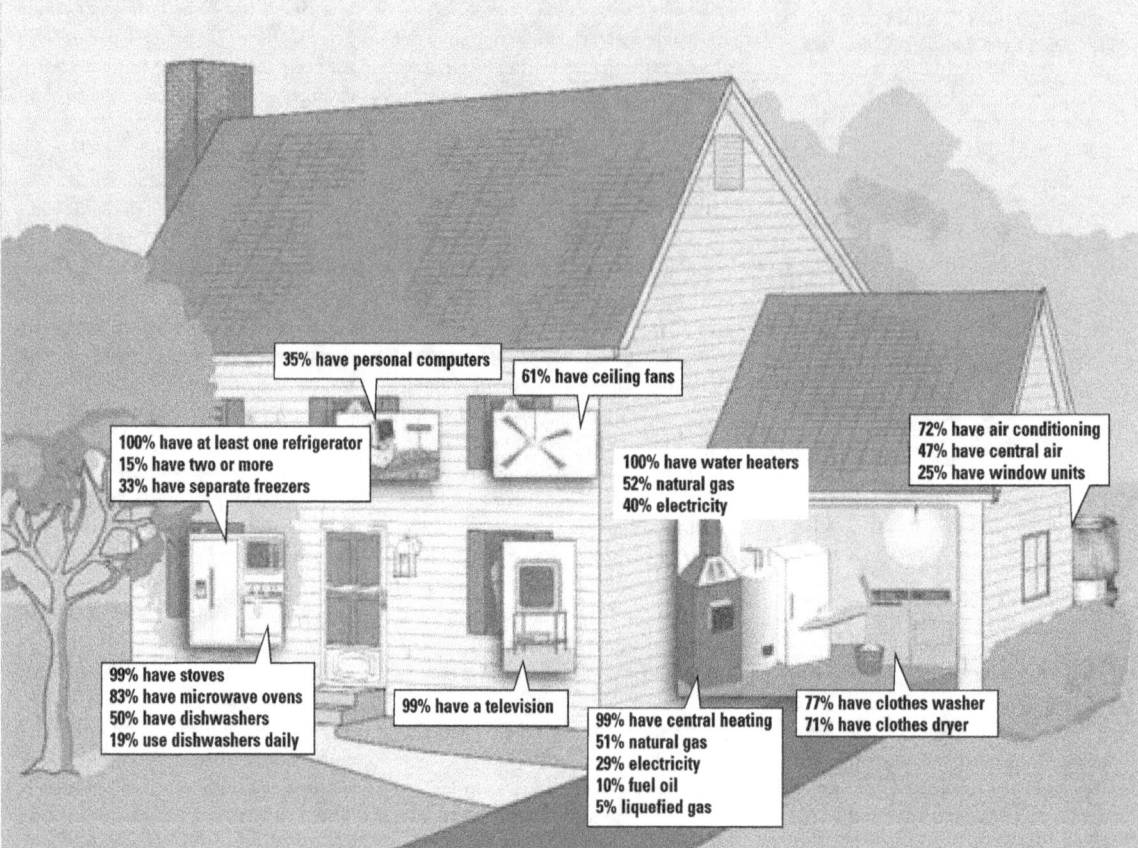

Source: U.S. DOE/EIA 2000a.

of energy used by commercial buildings as well. Commercial buildings house the rapidly growing financial and services sectors. Accordingly, their number and their total square footage have increased steadily. Over 85 percent of all commercial buildings are heated, and more than 75 percent are cooled. In addition, the past decade has seen a major increase in the use of computers and other energy-consuming office equipment, such as high-resolution printers, copiers, and scanners.

Rapid growth in the financial and services sectors has substantially increased the energy services required by commercial buildings. However, as in the residential sector, substantial efficiency gains have reduced the net increases in energy demand and carbon emissions. The widespread introduction of efficient lighting and more efficient office equipment, such as ENERGY STAR® labeled products, should help to continue this trend. The entry into the market of energy service companies, which contract with firms or government agencies to improve building energy efficiency and are paid out of the stream of energy savings, has aided the trend toward greater energy efficiency in the commercial buildings sector.

Transportation

Reflecting the nation's low population density, the U.S. transportation sector has evolved into a multimodal system that includes waterborne, highway, mass transit, air, rail, and pipeline transport, capable of moving large volumes of people and freight long distances. Automobiles and light trucks dominate the passenger transportation system. In 1997, the highway share of passenger miles traveled was nearly 90 percent, while air travel accounted for 10 percent. In contrast, transit and rail travel's combined share was only 1 percent (Figure 2-9).

FIGURE 2-9 U.S. Transportation: Characteristics and Trends

The U.S. transportation system relies heavily on private vehicles. Although fuel efficiency in automobiles has been rising steadily, there has also been a trend toward larger vehicles, such as light trucks and sport utility vehicles. Coupled with an increase in vehicle miles traveled, overall energy consumption has been increasing. Air travel has also experienced impressive growth, and the performance of freight modes has not off-set these increases in consumption.

Energy Use by Transportation Mode: 1998

In 1998 the transportation sector consumed nearly 26 quadrillion Btus. Highway vehicles accounted for about 80 percent of this consumption.

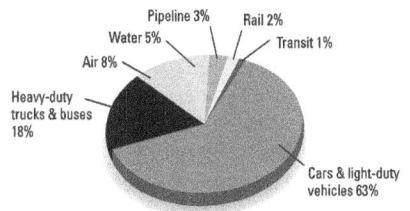

Passenger Miles Traveled: 1998

Of the nearly 5 trillion passenger miles traveled in 1998, passenger cars accounted for the single largest mode of transportation.

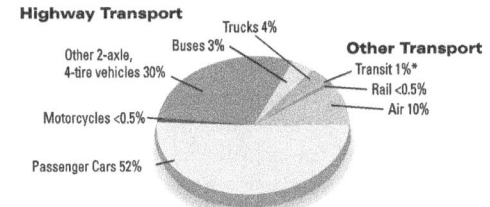

*Includes motor and trolley buses; light, heavy, and commuter rail; and ferry boats.

Passenger Car Use Index: 1980 = 1

Generally, although fuel efficiency has been improving as a result of CAFE standards, fuel consumption continues to rise due to increased U.S. vehicle miles traveled.

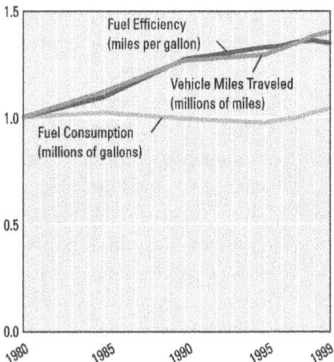

Gasoline Prices and Fuel Use Index: 1978 = 1

As real gasoline prices declined in the early 1990s, fuel consumption on our nation's highways increased.

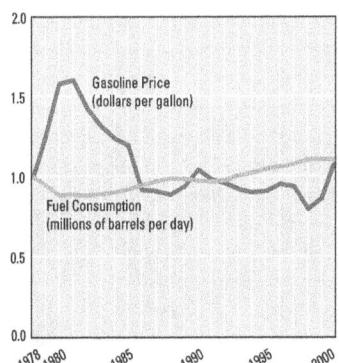

Air Transport (Billions of Miles)

Air transport has been rising over the past decade: revenue aircraft miles and available seat miles have been increasing at average annual rates of nearly 4 and 3 percent, respectively.

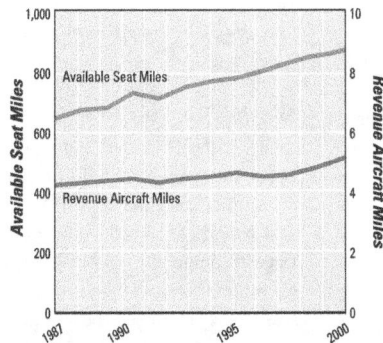

Note: Totals may not sum due to rounding.
Sources: U.S. DOE/EIA 2000a, U.S. DOT/BTS 2000a and 2000b, U.S. DOT/FAA 1998.

Efficiency of Freight Modes

The fuel efficiency of U.S. freight transportation is steadily improving. Most notably, the energy intensity of railroads decreased by nearly 45 percent during 1970-98.

Year	Trucks (mpg)	Class I Freight Railroads (Btus per Ton Mile)	Domestic Waterborne Commerce (Ton Miles per Barrel)
1970	5.5	645	4,820
1980	5.4	590	3,680
1990	6.0	420	3,370
1995	6.2	370	3,580
1996	6.2	365	3,580
1997	6.4	370	3,770
1998	6.1	360	3,660

Because of the dominance of motor vehicles in the U.S. transportation system, motor vehicle ownership rates, use, and efficiency drive energy consumption and greenhouse gas emissions in the transportation sector. Between 1960 and 1998, the number of cars and trucks registered in the United States almost tripled, from nearly 75 million to more than 210 million. Overall, the transportation sector consumed slightly over 25 quadrillion Btus in 1998, accounting for approximately one-third of U.S. greenhouse gas emissions. Rising incomes, population growth, and settlement patterns were the primary factors in this trend.

Both the number of vehicles on the road and the average distance they are driven have increased. In 1999, on average, passenger cars were driven over 19,000 kilometers (nearly 12,000 miles) per year, compared to approximately 16,000 kilometers (about 10,000 miles) in 1970. The distance traveled per car has increased steadily over the last two decades, interrupted only by the oil shocks in 1974 and 1979. Total U.S. vehicle miles traveled have increased by nearly 140 percent since 1970.

These increases have been significantly offset by enhanced efficiency. This can be attributed to a combination of factors, including the implementation of Corporate Average Fuel Economy (CAFE) standards for new cars, and improved average fuel consumption per kilometer—from a low of 18 liters per 100 kilometers (slightly over 13 miles per gallon) for the on-road passenger car fleet in 1973, to 11 liters per 100 kilometers (slightly over 21 miles per gallon) in 1999. Between 1998 and 1999, the fuel efficiency of passenger cars declined by about 1 percent, halting the growth trend in improvement of energy efficiency.

The fuel economy of light trucks and sport utility vehicles has also improved, although the increased share of light trucks in the total light-duty-vehicle fleet has diminished these overall gains. Thus, as in other sectors, efficiency improvements moderated the increase in motor fuel consumption (including air,

water, pipeline, and rail) in the transportation sector from nearly 8 million barrels per day in 1970 to about 12 million barrels per day in 1999.

The causes for the rapid rise in vehicle miles traveled are numerous, although their relative importance is unclear. In 1997, there was slightly over one vehicle per licensed driver—an increase of about 25 percent over 1970. This increase in ownership translates into a decrease in the use of carpools and public transportation, and an accompanying increase in personal vehicle use. Increased vehicle ownership and use are related to a host of factors, including changing patterns of land use, such as location of work and shopping centers; the changing composition of the work force, such as the growing number of women in the work force; and the reduced marginal costs of driving.

U.S. freight transportation, measured in ton-miles, grew at an average of 2 percent annually from 1970 to 1997, when it reached nearly three trillion ton-miles. In 1997, the predominant mode of freight transportation was trucks, followed closely by rail, then waterways, pipelines, and air.

- Heavy trucks account for most of the freight sector's energy use. From 1970 to 1997, their energy consumption more than doubled. While their fuel efficiency increased slightly, U.S. ton-miles of freight transported on intercity trucks nearly tripled between 1970 and 1997.
- Between 1970 and 1997, the number of railroad cars in use declined. However, they carried more freight for greater distances, resulting in nearly a 1 percent reduction in total fuel consumed for rail freight service since 1970, and nearly a 50 percent improvement in energy consumed (in terms of Btus) per freight ton-mile.
- Ton-miles shipped by air increased rapidly—by over 6 percent a year from 1970 to 1997.
- Water-transport and oil-pipeline shipments grew steadily over that same period.

Government

The U.S. government is the nation's single largest energy consumer. It uses energy in government buildings and operations widely dispersed across the entire nation and every climate zone, providing services to the U.S. population. Based on reports submitted to the Department of Energy by 28 federal agencies, the U.S. government consumed slightly over one quad of energy during fiscal year 1999 (about 1 percent of U.S. energy consumption), when measured in terms of energy actually delivered to the point of use. This total net energy consumption represented a 30 percent decrease from 1990. Based on these figures, the federal government was responsible for nearly 25 million metric tons of carbon emissions in 1999—a reduction of nearly 9 million metric tons, or over 25 percent, from 1990. The largest contribution to this reduction was from vehicle and equipment end-uses, which reduced their carbon emissions by nearly 35 percent.

The Department of Defense is the federal government's largest energy consumer, accounting for just over 80 percent of total federal energy use. The Postal Service is the second largest consumer of federal energy, and accounted for nearly 4 percent of total federal energy use. Overall in 1999, energy consumption by vehicles and equipment accounted for 60 percent of the total, buildings for 34 percent, and energy-intensive operations for 7 percent. In terms of energy use by fuel type, jet fuel accounted for nearly 55 percent; fuel oil, nearly 20 percent; electricity, more than 10 percent; natural gas, 10 percent; and other fuels, 6 percent.

Waste

In 1999, the United States generated approximately 230 million tons of municipal solid waste (MSW). Paper and paperboard products made up the largest component of MSW generated by weight (nearly 40 percent), and yard trimmings comprised the second largest material component (more than 10

Energy Savings in Federal Agencies

Initially in response to the energy crises of the 1970s, and later because it just made good financial sense, federal agencies have been steadily pursuing energy and cost savings in their buildings and operations. Under the Federal Energy Management Program, federal agencies have invested several billion dollars in energy efficiency over the past 20 years and have substantially reduced their energy consumption. In federal buildings, the primary focus of the program, 1999 energy consumption was down nearly 30 percent from 1985 levels and nearly 25 percent from 1990 levels. Within the same sector, carbon emissions have decreased by nearly 20 percent since 1990. This has been partly due to a 10 percent reduction in gross square footage since 1990 and about an 8 percent reduction in primary energy intensity (in terms of Btus per gross square footage).

The Energy Policy Act of 1992 and Executive Order 13123 further challenge federal energy managers to reduce energy use in federal buildings by 35 percent by 2010 from 1985 levels. With declining federal resources available, the Federal Energy Management Program is emphasizing the use of private-sector investment through energy-saving performance contracting and utility financing of energy efficiency to meet these goals. The combination of federal funding and the anticipated private-sector funding of about $4 billion through 2005 should make these goals attainable. In addition, agencies are making cost-effective investments in renewable-energy and water-conservation projects, and further savings are being pursued through an energy-efficient procurement initiative.

percent). Glass, metals, plastics, wood, and food each constituted between 5 and over 10 percent of the total MSW generated. Rubber, leather, and textiles combined made up about 7 percent of the MSW, while other miscellaneous wastes made up approximately 2 percent of the MSW generated in 1999.

Waste management practices include source reduction, recycling, and disposal (including waste combustion and landfilling). Management patterns changed dramatically in the late 1990s in response to changes in economic and regulatory conditions. The most significant change from a greenhouse gas perspective was the increase in the national average recycling rate, which rose from over 15 percent in 1990 to nearly 30 percent in 1999 (nearly 65 million tons). Of the remaining MSW generated, about 15 percent is combusted and nearly 60 percent is disposed of at landfills. The number of operating MSW landfills has decreased substantially over the last decade, from about 8,000 in 1988 to under 2,000 in 1999, while the average landfill size has increased.

Overall, waste management and treatment activities accounted for about 260 teragrams of carbon dioxide equivalent (Tg CO_2 Eq.), or nearly 4 percent of total U.S. greenhouse gas emissions in 1999. Of this, landfill emissions were over 210 Tg CO_2 Eq. Waste combustion, human sewage, and wastewater treatment constituted the rest of the emissions.

AGRICULTURE

Despite their decreased acreage, U.S. grazing lands are sustaining more animals, and agricultural lands are feeding more people. Enlightened land management policies and improved technologies are major contributors to their enhanced productivity.

Grazing Land

U.S. grazing lands—both grassland pasture and range and cropland used for pasture—are environmentally important. They include major recreational and scenic areas, serve as a principal source of wildlife habitat, and comprise a large area of the nation's watersheds. These ecosystems, like forest ecosystems, are vulnerable to rapid changes in climate, particularly shifts in temperature and moisture regimes. However, range ecosystems tend to be more resilient than forest ecosystems because of their ability to survive long-term droughts.

Grassland pasture and range ecosys-

tems can include a variety of different flora and fauna communities, usually denoted by the dominant vegetation. They are generally managed by varying grazing pressure, by using fire to shift species abundance, and by occasionally disturbing the soil surface to improve water infiltration.

In contrast, cropland used for pasture is a grazing ecosystem that relies on more intensive management inputs, such as fertilizer, chemical pest management, and introduced or domesticated species. U.S. cropland used for pasture includes native grasslands, savannas, alpine meadows, tundra, many wetlands, some deserts, and areas seeded by introduced and genetically improved species.

Grassland pasture and range accounted for nearly 240 million hectares (580 million acres), or over 25 percent of major land uses in 1997 (see Figure 2-2). However, the area of grassland pasture and range has declined since 1945, when it was nearly 270 million hectares (nearly 660 million acres). One reason for this decline is that farmers have improved the productivity of grazing lands. A second reason is that some of these land areas were also converted to cropland, rural residential, suburban, and urban land uses, as demand for grazing lands declined in recent years due to the decrease in the number of domestic animals—particularly sheep and draft animals—raised on grazing lands.

Agricultural Land

The United States enjoys a natural abundance of productive agricultural lands and a favorable climate for producing food crops, feed grains, and other agricultural commodities, such as oil seed crops. The area of the U.S. cropland used for crop production declined by 10 percent during the 16-year period between 1981 and 1997, from nearly 160 million hectares (nearly 390 million acres) to about 140 million hectares (nearly 350 million acres). During this same period, conservation programs for the most environmentally sensitive and highly erodible lands have removed nearly 15 million hectares

(35 million acres) from cropping systems.

Although the United States harvests about the same area as it did in 1910, it feeds a population that has grown two and one-half times since then, and its food exports have also expanded considerably. Agricultural productivity increases are due primarily to technological change in the food and agricultural sectors. In the absence of these improvements in productivity, substantially more land would need to be cultivated to achieve today's level of productivity.

The increase in no-till, low-till, and other erosion control practices reduced erosion on cropland and grazing land by 40 percent between 1982 and 1997. These practices also have helped to conserve carbon associated with those soils, protect soil productivity, and reduce other environmental impacts, such as pesticide and nutrient loadings in water bodies.

Although the number of cattle and sheep has been declining, greenhouse gas emissions from agricultural activities have been steadily rising, largely due to growth in emissions of nitrous oxides from agricultural soil management and methane emissions from manure management.

Forests

U.S. forests vary from the complex juniper forests of the arid interior West to the highly productive forests of the Pacific Coast and the Southeast. In 1997, forests covered about one-third (about 300 million hectares, or nearly 750 million acres) of the total U.S. land area. This includes both the forest-use lands and a portion of the special-use lands listed in Table 2-1 and Figure 2-2.

Excluding Alaska, U.S. forestland covers about 250 million hectares (620 million acres). Of this, nearly 200 million hectares are timberland, most of which is privately owned. However, much of the forested land is dedicated to special uses (i.e., parks, wilderness areas, and wildlife areas), which prohibits using the land for such activities as timber production. These areas increased from

about 9 million hectares (over 20 million acres) in 1945 to nearly 45 million hectares (about 100 million acres) in 1997. As a result, land defined as "forest-use land" declined consistently from the 1960s to 1997, while land defined as "special uses" increased.

Management inputs over the past several decades have been gradually increasing the production of marketable wood in U.S. forests. The United States currently grows more wood than it harvests, with a growth-to-harvest ratio of nearly 1.5. This ratio reflects substantial new forest growth; however, old-growth forests have continued to decline over the same period.

OTHER NATURAL RESOURCES

Climate change significantly affects other U.S. natural resources, including wetlands, wildlife, and water.

Wetlands

Wetland ecosystems are some of the more biologically important and ecologically significant systems on the planet. Because they represent a boundary condition ("ecotone") between land and aquatic ecosystems, wetlands have many functions. They provide habitats for many types of organisms, both plant and animal; serve as diverse ecological niches that promote preservation of biodiversity; are the source of economic products for food, clothing, and recreation; trap sediment, assimilate pollution, and recharge ground water; regulate water flow to protect against storms and flooding; and anchor shorelines and prevent erosion. The United States has a broad variety of wetland types, ranging from permafrost-underlain wetlands in Alaska to tropical rainforests in Hawaii.

Wetland ecosystems are highly dependent upon upland ecosystems.

Principles for Conservation

In September 2001, the U.S. Department of Agriculture presented its long-term view of the nation's agriculture and food system and a framework to foster strategic thinking and guiding principles for agricultural policies, including policies for environmental conservation. These Principles for Conservation were identified as key policy directives.

Sustain past environmental gains. Improvements in losses from soil erosion and wetlands benefit farmers and all Americans. These and other gains resulting from existing conservation programs should be maintained.

Accommodate new and emerging environmental concerns. Conservation policy should adapt to emerging environmental and community needs and incorporate the latest science. These new and emerging issues include the need for sources of renewable energy and the potential for reducing greenhouse gas emissions.

Design and adopt a portfolio approach to conservation policies. Targeted technical assistance, incentives for improved practices on working farms and forest lands, compensation for environmental achievements, and limited dedication of farmland and private forest lands to environmental use will provide a coordinated and flexible portfolio approach to agri-environmental goals.

Reaffirm market-oriented policies. Competition in the supply of environmental goods and services and targeted incentives ensure the maximum environmental benefits for each public dollar spent.

Ensure compatibility of conservation, farm, and trade policies. Producer compensation for conservation practices and environmental achievements should be consistent with "green box" criteria under World Trade Organization obligations.

Recognize the importance of collaboration. Nonfederal government agencies as well as private for-profit and not-for-profit organizations are playing an ever-increasing role in the delivery of technical assistance and in incentive programs for conservation.

Source: USDA 2001.

Therefore, they are vulnerable to changes in the health of upland ecosystems as well as to environmental change brought about by shifts in climate regimes. Wetlands, including riparian zones along waterways and areas of perennial wet soils or standing water, are both sources of and sinks for greenhouse gases.

Since the nation's settlement in the 18th century, the continental United States has lost about 40–45 million hectares (about 100–110 million acres) of approximately 90 million hectares (over 220 million acres) of its original wetlands. Most wetland conversion in the 19th century was originally for agricultural purposes, although converted land subsequently was often used for urban development. A significant additional share of wetlands was lost as a result of federal flood control and drainage projects.

The pace of wetland loss has slowed considerably in the past two decades. For example, while net wetland losses from the mid-1950s to the mid-1970s averaged 185,400 hectares (458,000 acres) a year, they fell to about 117,400 hectares (290,000 acres) a year from the mid-1970s to mid-1980s. Between 1982 and 1992, the net average rate of wetland conversion further dropped to about 32,000 hectares (80,000 acres) a year. During 1992–97, net wetland losses fell even further to roughly 13,000 hectares (32,600 acres) a year. Urban development accounted for nearly half of these losses, while agricultural conversion accounted for about one-quarter.

The reduced rate of wetland loss since the mid-1980s is attributable to a number of factors. Both government policies for protecting wetlands and low crop prices have decreased conversions of wetlands to agricultural uses. In addition, the majority of wetland restorations have occurred on agricultural land. Government programs, such as the Wetland Reserve Program, which provides funds and technical assistance to restore formerly drained wetlands, have aided such gains. Thus, agricultural land management has most likely contributed to overall gains in wetland areas, as losses to agricultural conversion are greatly reduced and previously drained areas are restored. Future losses are likely to be even smaller, because the United States has implemented a "no net loss" policy for wetlands.

Alaska's over 70 million hectares (175 million acres) of wetlands easily exceed the 45–50 million hectares (over 110–125 million acres) of wetlands in the continental United States. Many of these areas are federally owned. Total wetland losses in Alaska have been less than 1 percent since the mid-1800s, although in coastal areas, losses have been higher.

Wildlife

During the past 20 years, the United States has become more aware of the reduction in the diversity of life at all levels, both nationwide and worldwide. To better understand and catalog both previous and future changes, the United States is conducting a comprehensive, nationwide survey of its wildlife and biodiversity, referred to as the National Biological Survey.

Information on endangered species is already available through other sources. As of November 2000, over 960 species were listed as endangered, of which about 590 are plants and 370 are animals. In addition, over 140 plant and nearly 130 animal species were listed as threatened, for a total of nearly 1,240 threatened or endangered species. The United States continues to work to conserve species diversity through programs and laws like the Endangered Species Act.

Water

The development of water resources has been key to the nation's growth and prosperity. Abundant and reliable water systems have enabled urban and agricultural centers to flourish in arid and semi-arid regions of the country. For instance, between 1959 and 1997, irrigated agricultural land increased by nearly 70 percent, from less than 15 million hectares (nearly 35 million acres) to over 20 million hectares (55 million acres).

Currently, most of the nation's freshwater demands are met by diversions from streams, rivers, lakes, and reservoirs and by withdrawals from ground-water aquifers. Even though total withdrawals of surface water more than doubled from 1950 to 1980, withdrawals remained at about 20 percent of the renewable water supply in 1980. However, some areas of the country still experience intermittent water shortages during droughts.

There is increasing competition for water in the arid western sections of the country, not only to meet traditional agricultural and hydropower needs, but also for drinking water in growing urban areas; for American Indian water rights; and for industry, recreation, and natural ecosystems. The flows of many streams in the West are fully allocated to current users, limiting opportunities for expanded water use by major new facilities. Several states have adopted a market-based approach to water pricing and allocation, thus offering the potential to alleviate projected shortfalls. Also pertinent is the federal government's insistence that certain minimum-flow requirements be met to preserve threatened and endangered species.

These forces have contributed to a decline in per capita water use in the last two decades. After continual increases in the nation's total water withdrawals for off-stream use from 1950 to 1980, withdrawals declined from 1980 to 1995. The 1995 estimate of average withdrawals, which is over 400 million gallons a day, is 2 percent less than the 1990 estimate and nearly 10 percent less than the 1980 estimate, which was the peak year of water use. This decline in water withdrawals occurred even though population increased by over 15 percent from 1980 to 1995.

Chapter 3
Greenhouse
Gas Inventory

Central to any study of climate change is the development of an emissions inventory that identifies and quantifies a country's primary anthropogenic[1] sources and sinks of greenhouse gases. The *Inventory of U.S. Greenhouse Gas Emissions and Sinks: 1990–1999* (U.S. EPA 2001d) adheres to both (1) a comprehensive and detailed methodology for estimating sources and sinks of anthropogenic greenhouse gases, and (2) a common and consistent mechanism that enables signatory countries to the United Nations Framework Convention on Climate Change (UNFCCC) to compare the relative contribution of different emission sources and greenhouse gases to climate change. Moreover, systematically and consistently estimating national and

---

[1] In this context, the term "anthropogenic" refers to greenhouse gas emissions and removals that are a direct result of human activities or are the result of natural processes that have been affected by human activities (IPCC/UNEP/OECD/IEA 1997).

international emissions is a prerequisite for accounting for reductions and evaluating mitigation strategies.

In June 1992, the United States signed, and later ratified in October, the UNFCCC. The objective of the UNFCCC is "to achieve ... stabilization of greenhouse gas concentrations in the atmosphere at a level that would prevent dangerous anthropogenic interference with the climate system."[2] By signing the Convention, Parties make commitments "to develop, periodically update, publish and make available... national inventories of anthropogenic emissions by sources and removals by sinks of all greenhouse gases not controlled by the Montreal Protocol, using comparable methodologies...."[3] The United States views the *Inventory of U.S. Greenhouse Gas Emissions and Sinks* as an opportunity to fulfill this commitment.

This chapter summarizes information on U.S. anthropogenic greenhouse gas emission trends from 1990 through 1999. To ensure that the U.S. emissions inventory is comparable to those of other UNFCCC signatory countries, the emission estimates were calculated using methodologies consistent with those recommended in the *Revised 1996 IPCC Guidelines for National Greenhouse Gas Inventories* (IPCC/UNEP/OECD/IEA 1997). For most source categories, the IPCC default methodologies were expanded, resulting in a more comprehensive and detailed estimate of emissions.

Naturally occurring greenhouse gases include water vapor, carbon dioxide (CO_2), methane (CH_4), nitrous oxide (N_2O), and ozone (O_3). Several classes of halogenated substances that contain fluorine, chlorine, or bromine are also greenhouse gases, but they are, for the most part, solely a product of industrial activities. Chlorofluorocarbons (CFCs) and hydrochlorofluorocarbons (HCFCs) are halocarbons that contain chlorine, while halocarbons that contain bromine are referred to as bromofluorocarbons (i.e., halons). Because CFCs, HCFCs, and halons are stratospheric ozone-depleting substances, they are covered under the *Montreal Protocol on Substances That Deplete the Ozone Layer*. The UNFCCC defers to this earlier international treaty; consequently these gases are not included in national greenhouse gas inventories.[4] Some other fluorine-containing halogenated substances—hydrofluorocarbons (HFCs), perfluorocarbons (PFCs), and sulfur hexafluoride (SF_6)—do not deplete stratospheric ozone but are potent greenhouse gases. These latter substances are addressed by the UNFCCC and are accounted for in national greenhouse gas inventories.

There are also several gases that do not have a direct global warming effect but indirectly affect terrestrial radiation absorption by influencing the formation and destruction of tropospheric and stratospheric ozone. These gases include carbon monoxide (CO), nitrogen oxides (NO_x), and non-methane volatile organic compounds (NMVOCs).[5] Aerosols, which are extremely small particles or liquid droplets, such as those produced by sulfur dioxide (SO_2) or elemental carbon emissions, can also affect the absorptive characteristics of the atmosphere.

Although CO_2, CH_4, and N_2O occur naturally in the atmosphere, their atmospheric concentrations have been affected by human activities. Since pre-industrial time (i.e., since about 1750), concentrations of these greenhouse gases have increased by 31, 151, and 17 percent, respectively (IPCC 2001b). Because this build-up has altered the chemical composition of the Earth's atmosphere, it has affected the global climate system.

Beginning in the 1950s, the use of CFCs and other stratospheric ozone-depleting substances (ODSs) increased by nearly 10 percent per year until the mid-1980s, when international concern about ozone depletion led to the signing of the *Montreal Protocol*. Since then, the production of ODSs is being phased out. In recent years, use of ODS substitutes, such as HFCs and PFCs, has

Emission Reporting Nomenclature

The global warming potential (GWP)-weighted emissions of all direct greenhouse gases throughout this report are presented in terms of equivalent emissions of carbon dioxide (CO_2), using units of teragrams of CO_2 equivalents (Tg CO_2 Eq.). Previous years' inventories reported U.S. emissions in terms of carbon—versus CO_2-equivalent—emissions, using units of millions of metric tons of carbon equivalents (MMTCE). This change of units for reporting was implemented so that the U.S. inventory would be more consistent with international practices, which are to report emissions in units of CO_2 equivalents.

The following equation can be used to convert the emission estimates presented in this report to those provided previously:

$$\text{Tg } CO_2 \text{ Eq.} = \text{MMTCE (44/12)}$$

There are two elements to the conversion. The first element is simply nomenclature, since one teragram is equal to one million metric tons:

$$\text{Tg} = 10^9 \text{ kg} = 10^6 \text{ metric tons} = 1 \text{ million metric tons}$$

The second element is the conversion, by weight, from carbon to CO_2. The molecular weight of carbon is 12, and the molecular weight of oxygen is 16. Therefore, the molecular weight of CO_2 is 44 (i.e., 12+[16(2)], as compared to 12 for carbon alone. Thus, carbon comprises 12/44ths of CO_2 by weight.

---

[2] Article 2 of the Framework Convention on Climate Change published by the UNEP/WMO Information Unit on Climate Change. See http://www.unfccc.de.
[3] Article 4 of the Framework Convention on Climate Change published by the UNEP/WMO Information Unit on Climate Change (also identified in Article 12). See http://www.unfccc.de.
[4] Emission estimates of CFCs, HCFCs, halons, and other ozone-depleting substances are included in this chapter for informational purposes (see Table 3-12).
[5] Also referred to in the U.S. Clean Air Act as "criteria pollutants."

grown as they begin to be phased in as replacements for CFCs and HCFCs.

RECENT TRENDS IN U.S. GREENHOUSE GAS EMISSIONS

In 1999, total U.S. greenhouse gas emissions were 6,746 teragrams of CO_2 equivalents (Tg CO_2 Eq.),[6] 11.7 percent above emissions in 1990. The single-year increase in emissions from 1998 to 1999 was 0.9 percent (59.2 Tg CO_2 Eq.), which was less than the 1.2 percent average annual rate of increase for 1990 through 1999. The lower than average increase in emissions, especially given the robust economic growth in 1999, was primarily attributable to the following factors: (1) warmer than nor-mal summer and winter conditions, (2) significantly increased output from existing nuclear power plants, (3) reduced CH_4 emissions from coal mines, and (4) HFC-23 by-product emissions from the chemical manufacture of HCFC-22. Figures 3-1 through 3-3 illustrate the overall trends in total U.S. emissions by gas, annual changes, and absolute change since 1990. Table 3-1 provides a detailed summary of U.S. greenhouse gas emissions and sinks for 1990 through 1999.

Figure 3-4 illustrates the relative contribution of the direct greenhouse gases to total U.S. emissions in 1999. The primary greenhouse gas emitted by human activities was CO_2. The largest source of CO_2, and of overall greenhouse gas emissions in the United States, was fossil fuel combustion. Emissions of CH_4 resulted primarily from decomposition of wastes in landfills, enteric fermentation associated with domestic livestock, natural gas systems, and coal mining. Most N_2O emissions

---

[6] Estimates are presented in units of teragrams of carbon dioxide equivalents (Tg CO_2 Eq.), which weight each gas by its global warming potential, or GWP, value (see the following section).

FIGURE 3-1 U.S. Greenhouse Gas Emissions by Gas: 1990–1999 (Tg CO_2 Eq.)

In 1999, total U.S. greenhouse gas emissions rose to 6,746 teragrams of carbon dioxide equivalents (Tg CO_2 Eq.), which was 11.7 percent above 1990 emissions.

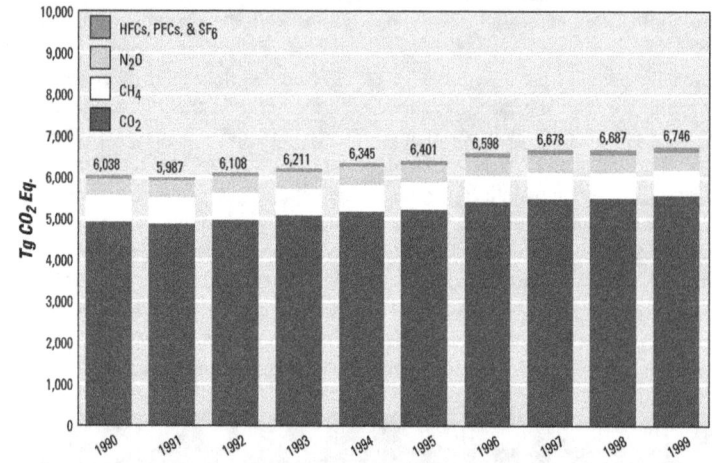

FIGURE 3-2 Annual Change in U.S. Greenhouse Gas Emissions Since 1990

The single-year increase in greenhouse gas emissions from 1998 to 1999 was 0.9 percent (59.2 Tg CO_2 Eq.), which was less than the 1.2 percent average annual rate of increase for 1990 through 1999.

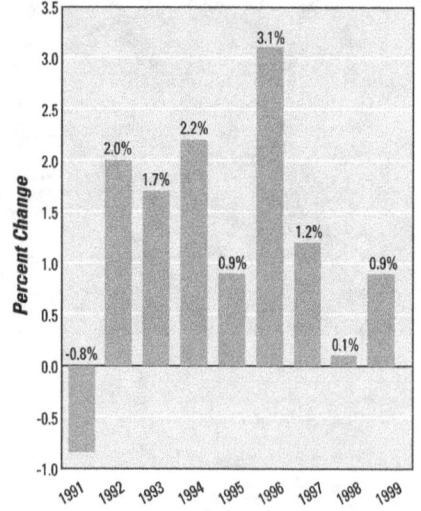

FIGURE 3-3 Absolute Change in U.S. Greenhouse Gas Emissions Since 1990

Greenhouse gas emissions increased a total of 707.9 Tg CO_2 Eq. between 1990 and 1999, or 11.7 percent since 1990.

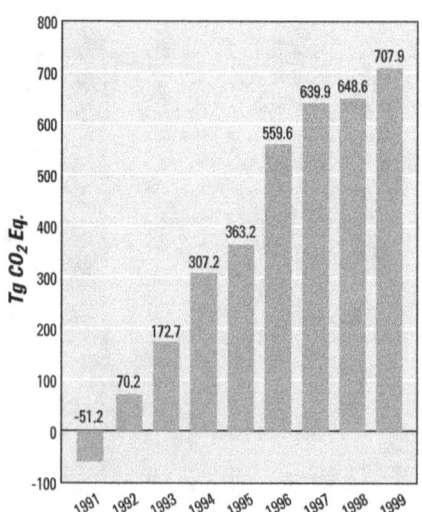

TABLE 3-1 Recent Trends in U.S. Greenhouse Gas Emissions and Sinks (Tg CO₂ Eq.)

From 1990 through 1999, total U.S. greenhouse gas emissions increased by 11.7 percent. Specifically, CO_2 emissions increased by 13.1 percent, CH_4 emissions decreased by 3.9 percent, N_2O emissions increased by 9.0 percent, and HFCs, PFCs, and SF_6 emissions increased by 61.7 percent.

Gas/Source	1990	1995	1996	1997	1998	1999
CO₂	4,913.0	5,219.8	5,403.2	5,478.7	5,489.7	5,558.1
Fossil Fuel Combustion	4,835.7	5,121.3	5,303.0	5,374.9	5,386.8	5,453.1
Cement Manufacture	33.3	36.8	37.1	38.3	39.2	39.9
Waste Combustion	17.6	23.1	24.0	25.7	25.1	26.0
Lime Manufacture	11.2	12.8	13.5	13.7	13.9	13.4
Natural Gas Flaring	5.1	13.6	13.0	12.0	10.8	11.7
Limestone and Dolomite Use	5.1	7.0	7.3	8.3	8.1	8.3
Soda Ash Manufacture and Consumption	4.1	4.3	4.3	4.4	4.3	4.2
Carbon Dioxide Consumption	0.8	1.0	1.1	1.3	1.4	1.6
Land-Use Change and Forestry (Sink)[a]	(1,059.9)	(1,019.1)	(1,021.6)	(981.9)	(983.3)	(990.4)
International Bunker Fuels[b]	114.0	101.0	102.2	109.8	112.8	107.3
CH₄	644.5	650.5	638.0	632.0	624.8	619.6
Landfills	217.3	222.9	219.1	217.8	213.6	214.6
Enteric Fermentation	129.5	136.3	132.2	129.6	127.5	127.2
Natural Gas Systems	121.2	124.2	125.8	122.7	122.1	121.8
Coal Mining	87.9	74.6	69.3	68.8	66.5	61.8
Manure Management	26.4	31.0	30.7	32.6	35.2	34.4
Petroleum Systems	27.2	24.5	24.0	24.0	23.3	21.9
Wastewater Treatment	11.2	11.8	11.9	12.0	12.1	12.2
Rice Cultivation	8.7	9.5	8.8	9.6	10.1	10.7
Stationary Combustion	8.5	8.9	9.0	8.1	7.6	8.1
Mobile Combustion	5.0	4.9	4.8	4.7	4.6	4.5
Petrochemical Production	1.2	1.5	1.6	1.6	1.6	1.7
Agricultural Residue Burning	0.5	0.5	0.6	0.6	0.6	0.6
Silicon Carbide Production	+	+	+	+	+	+
International Bunker Fuels[b]	+	+	+	+	+	+
N₂O	396.9	431.9	441.6	444.1	433.7	432.6
Agricultural Soil Management	269.0	285.4	294.6	299.8	300.3	298.3
Mobile Combustion	54.3	66.8	65.3	65.2	64.2	63.4
Nitric Acid	17.8	19.9	20.7	21.2	20.9	20.2
Manure Management	16.0	16.4	16.8	17.1	17.2	17.2
Stationary Combustion	13.6	14.3	14.9	15.0	15.1	15.7
Adipic Acid	18.3	20.3	20.8	17.1	7.3	9.0
Human Sewage	7.1	8.2	7.8	7.9	8.1	8.2
Agricultural Residue Burning	0.4	0.4	0.4	0.4	0.5	0.4
Waste Combustion	0.3	0.3	0.3	0.3	0.2	0.2
International Bunker Fuels[b]	1.0	0.9	0.9	1.0	1.0	1.0
HFCs, PFCs, and SF₆	83.9	99.0	115.1	123.3	138.6	135.7
Substitution of Ozone-Depleting Substances	0.9	24.0	34.0	42.1	49.6	56.7
HCFC-22 Production	34.8	27.1	31.2	30.1	40.0	30.4
Electrical Transmission and Distribution	20.5	25.7	25.7	25.7	25.7	25.7
Aluminum Production	19.3	11.2	11.6	10.8	10.1	10.0
Semiconductor Manufacture	2.9	5.5	7.0	7.0	6.8	6.8
Magnesium Production and Processing	5.5	5.5	5.6	7.5	6.3	6.1
Total Emissions	6,038.2	6,401.3	6,597.8	6,678.0	6,686.8	6,746.0
Net Emissions (Sources and Sinks)	4,978.3	5,382.3	5,576.2	5,696.2	5,703.5	5,755.7

+ Does not exceed 0.05 Tg CO₂ Eq.

[a] Sinks are only included in net emissions total, and are based partly on projected activity data.

[b] Emissions from international bunker fuels are not included in totals.

Notes: Totals may not sum due to independent rounding. Parentheses indicate negative values (or sequestration).

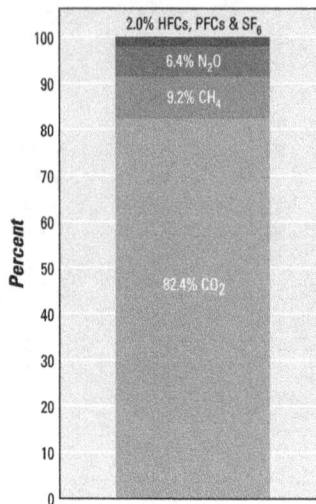

CO_2 was the principal greenhouse gas emitted by human activities, driven primarily by emissions from fossil fuel combustion.

were the result of agricultural soil management and mobile source fossil fuel combustion. The emissions of substitutes for ozone-depleting substances and emissions of HFC-23 during the production of HCFC-22 were the primary contributors to aggregate HFC emissions. Electrical transmission and distribution systems accounted for most SF_6 emissions, while the majority of PFC emissions were a by-product of primary aluminum production.

As the largest source of U.S. greenhouse gas emissions, CO_2 from fossil fuel combustion accounted for a nearly constant 80 percent of global warming potential (GWP)-weighted emissions in the 1990s.[7] Emissions from this source category grew by 13 percent (617.4 Tg CO_2 Eq.) from 1990 to 1999 and were responsible for most of the increase in national emissions during this period. The annual increase in CO_2 emissions from fossil fuel combustion was 1.2 percent in 1999, a figure close to the source's average annual rate of 1.4 per-

cent during the 1990s. Historically, changes in emissions from fossil fuel combustion have been the dominant factor affecting U.S. emission trends.

Changes in CO_2 emissions from fossil fuel combustion are influenced by many long-term and short-term factors, including population and economic growth, energy price fluctuations, technological changes, and seasonal temperatures. On an annual basis, the overall consumption of fossil fuels in the United States and other countries generally fluctuates in response to changes in general economic conditions, energy prices, weather, and the availability of non-fossil alternatives. For example, a year with increased consumption of goods and services, low fuel prices, severe summer and winter weather conditions, nuclear plant closures, and lower precipitation feeding hydroelectric output would be expected to have proportionally greater fossil fuel consumption than a year with poor economic performance, high fuel prices, mild temperatures, and increased output from nuclear and hydroelectric plants.

Longer-term changes in energy consumption patterns, however, tend to be more a function of changes that affect the scale of consumption (e.g., population, number of cars, and size of houses), the efficiency with which energy is used in equipment (e.g., cars, power plants, steel mills, and light bulbs), and consumer behavior (e.g., walking, bicycling, or telecommuting to work instead of driving).

Energy-related CO_2 emissions are also a function of the type of fuel or energy consumed and its carbon intensity. Producing heat or electricity using natural gas instead of coal, for example, can reduce the CO_2 emissions associated with energy consumption because of the lower carbon content of natural gas per unit of useful energy produced. Table 3-2 shows annual changes in emissions during the last few years of the 1990s for particular fuel types and sectors.

Emissions of CO_2 from fossil fuel combustion grew rapidly in 1996, due primarily to two factors: (1) fuel switching by electric utilities from natural gas to more carbon-intensive coal as colder winter conditions and the associated rise in demand for natural gas from residential, commercial, and industrial customers for heating caused gas prices to rise sharply; and (2) higher consumption of petroleum fuels for transportation. Milder weather conditions in summer and winter moderated the growth in emissions in 1997; however, the shutdown of several nuclear power plants led electric utilities to increase their consumption of coal and other fuels to offset the lost capacity. In 1998, weather conditions were again a dominant factor in slowing the growth in emissions. Warm winter temperatures resulted in a significant drop in residential, commercial, and industrial natural gas consumption. This drop in emissions from natural gas used for heating was primarily offset by two factors: (1) electric utility emissions, which increased in part due to a hot summer and its associated air conditioning demand; and (2) increased gasoline consumption for transportation.

In 1999, the increase in emissions from fossil fuel combustion was caused largely by growth in petroleum consumption for transportation. In addition, heating fuel demand partly recovered in the residential, commercial, and industrial sectors as winter temperatures dropped relative to 1998, although temperatures were still warmer than normal. These increases were offset, in part, by a decline in emissions from electric utilities due primarily to: (1) an increase in net generation of electricity by nuclear plants (8 percent) to record levels, which reduced demand from fossil fuel plants; and (2) moderated summer temperatures compared to the previous year, thereby reducing electricity demand for air conditioning. Utilization of existing nuclear power plants, measured by a

---

[7] If a full accounting of emissions from fossil fuel combustion is made by including emissions from the combustion of international bunker fuels and CH_4 and N_2O emissions associated with fuel combustion, then this percentage increases to a constant 82 percent during the 1990s.

[8] The capacity factor is defined as the ratio of the electrical energy produced by a generating unit for a given period of time to the electrical energy that could have been produced at continuous full-power operation during the same period (U.S. DOE/EIA 2000a).

plant's capacity factor,[8] increased from just over 70 percent in 1990 to over 85 percent in 1999.

Another factor that does not affect total emissions, but does affect the interpretation of emission trends, is the allocation of emissions from nonutility power producers. The Energy Information Administration (EIA) currently includes fuel consumption by nonutilities with the industrial end-use sector. In 1999, there was a large shift in generating capacity from regulated utilities to nonutilities, as restructuring legislation spurred the sale of 7 percent of utility generating capability (U.S. DOE/EIA 2000b). This shift is illustrated by the increase in industrial end-use sector emissions from coal and the associated decrease in electric utility emissions. However, emissions from the industrial end-use sector did not increase as much as would be expected, even though net generation by nonutilities increased from 11 to 15 percent of total U.S. electricity production (U.S. DOE/EIA 2000b).[9]

Overall, from 1990 to 1999, total emissions of CO_2 and N_2O increased by 645.2 (13 percent) and 35.7 Tg CO_2 Eq.

(9 percent), respectively, while CH_4 emissions decreased by 24.9 Tg CO_2 Eq. (4 percent). During the same period, aggregate weighted emissions of HFCs, PFCs, and SF_6 rose by 51.8 Tg CO_2 Eq. (62 percent). Despite being emitted in smaller quantities relative to the other principal greenhouse gases, emissions of HFCs, PFCs, and SF_6 are significant because many of them have extremely high global warming potentials and, in the cases of PFCs and SF_6, long atmospheric lifetimes. Conversely, U.S. greenhouse gas emissions were partly offset by carbon sequestration in forests and landfilled carbon, which were estimated to be 15 percent of total emissions in 1999.

Other significant trends in emissions from source categories over the nine-year period from 1990 through 1999 included the following:
* Aggregate HFC and PFC emissions resulting from the substitution of ozone-depleting substances (e.g., CFCs) increased by 55.8 Tg CO_2 Eq. This increase was partly offset, however, by reductions in PFC emissions from aluminum production (9.2 Tg CO_2 Eq. or 48 percent), and

reductions in emissions of HFC-23 from the production of HCFC-22 (4.4 Tg CO_2 Eq. or 13 percent). Reductions in PFC emissions from aluminum production were the result of both voluntary industry emission reduction efforts and lower domestic aluminum production. HFC-23 emissions from the production of HCFC-22 decreased due to a reduction in the intensity of emissions from that source, despite increased HCFC-22 production.
* Emissions of N_2O from mobile combustion rose by 9.1 Tg CO_2 Eq. (17 percent), primarily due to increased rates of N_2O generation in highway vehicles.
* CH_4 emissions from coal mining dropped by 26 Tg CO_2 Eq. (30 percent) as a result of the mining of less gassy coal from underground mines and the increased use of CH_4 from degasification systems.

---

[9] It is unclear whether reporting problems for electric utilities and the industrial end-use sector have increased with the dramatic growth in nonutilities and the opening of the electric power industry to increased competition.

TABLE 3-2 Annual Change in CO_2 Emissions from Fossil Fuel Combustion for Selected Fuels and Sectors

Changes in CO_2 emissions from fossil fuel combustion are influenced by many long- and short-term factors, including population and economic growth, energy price fluctuations, technological changes, and seasonal temperatures.

End-Use Sector /Fuel Type	1995–1996		1996–1997		1997–1998		1998–1999	
	Tg CO_2 Eq.	Percent	Tg CO_2 Eq.	Percent	Tg CO_2 Eq.	Percent	Tg CO_2 Eq.	Percent
Electric Utility								
Coal	89.9	5.7	52.0	3.1	14.3	0.8	-32.1	-1.8
Natural Gas	-25.3	-14.7	13.1	9.0	16.2	10.1	-7.8	-4.4
Petroleum	5.1	10.0	8.1	14.4	26.7	41.6	-17.4	-19.1
Transportation[a]								
Petroleum	38.8	2.5	7.6	0.5	34.1	2.1	57.6	3.6
Residential								
Natural Gas	21.4	8.1	-14.0	-4.9	-24.0	-8.9	8.5	3.4
Commercial								
Natural Gas	7.0	4.3	3.1	1.8	-11.1	-6.4	2.9	1.8
Industrial								
Coal	-7.3	-2.7	2.0	0.8	-1.1	-0.4	29.2	11.2
Natural Gas	17.8	3.4	-0.5	-0.1	-14.5	-2.7	1.6	0.3
All Sectors/All Fuels[b]	**181.7**	**3.5**	**71.9**	**1.4**	**11.9**	**0.2**	**66.4**	**1.2**

[a] Excludes emissions from international bunker fuels.
[b] Includes fuels and sectors not shown in table.

- N_2O emissions from agricultural soil management increased by 29.3 Tg CO_2 Eq. (11 percent), as fertilizer consumption and cultivation of nitrogen-fixing crops rose.
- By 1998, all of the three major adipic acid-producing plants had voluntarily implemented N_2O abatement technology. As a result, emissions fell by 9.3 Tg CO_2 Eq. (51 percent). The majority of this decline occurred from 1997 to 1998, despite increased production.

The following sections describe the concept of global warming potentials (GWPs), present the anthropogenic sources and sinks of greenhouse gas emissions in the United States, briefly discuss emission pathways, further summarize the emission estimates, and explain the relative importance of emissions from each source category.

GLOBAL WARMING POTENTIALS

Gases in the atmosphere can contribute to the greenhouse effect both directly and indirectly. Direct effects occur when the gas itself is a greenhouse gas. Indirect radiative forcing occurs when chemical transformations of the original gas produce a gas or gases that are greenhouse gases, when a gas influences the atmospheric lifetimes of other gases, and/or when a gas affects other atmospheric processes that alter the radiative balance of the Earth (e.g., affect cloud formation or albedo). The concept of a global warming potential (GWP) has been developed to compare the ability of each greenhouse gas to trap heat in the atmosphere relative to another gas. Carbon dioxide (CO_2) was chosen as the reference gas to be consistent with IPCC guidelines.

Global warming potentials are not provided for CO, NO_x, NMVOCs, SO_2, and aerosols (e.g., sulfate and elemental carbon) because there is no agreed-upon method to estimate the contribution of gases that are short-lived in the atmosphere and have only indirect effects on radiative forcing (IPCC 1996b).

Recent Trends in Various U.S. Greenhouse Gas Emissions-Related Data

There are several ways to assess a nation's greenhouse gas-emitting intensity. The basis for measures of intensity can be (1) per unit of aggregate energy consumption, because energy-related activities are the largest sources of emissions; (2) per unit of fossil fuel consumption, because almost all energy-related emissions involve the combustion of fossil fuels; (3) per unit of electricity consumption, because the electric power industry—utilities and nonutilities combined—was the largest source of U.S. greenhouse gas emissions in 1999; (4) per unit of total gross domestic product as a measure of national economic activity; or (5) on a per capita basis. Depending on the measure used, the United States could appear to have reduced or increased its national greenhouse gas intensity during the 1990s. Table 3-3 provides data on various statistics related to U.S. greenhouse gas emissions normalized to 1990 as a baseline year.

TABLE 3-3 AND FIGURE 3-5 Recent Trends in Various U.S. Data (Index: 1990 = 100)

Greenhouse gas emissions in the United States have grown at an average annual rate of 1.2 percent since 1990. This rate is slightly slower than that for total energy or fossil fuel consumption—indicating an improved or lower greenhouse gas-emitting intensity—and much slower than that for either electricity consumption or overall gross domestic product.

Variable	1991	1992	1993	1994	1995	1996	1997	1998	1999	Growth Rate[f]
GHG Emissions[a]	99	101	103	105	106	109	111	111	112	1.2%
Energy Consumption[b]	100	101	104	106	108	111	112	112	115	1.5%
Fossil Fuel Consumption[b]	99	101	103	105	107	110	112	112	113	1.4%
Electricity Consumption[b]	102	102	105	108	111	114	116	119	120	2.1%
Gross Domestic Product[c]	100	103	105	110	112	116	122	127	132	3.2%
Population[d]	101	103	104	105	106	108	109	110	112	1.2%
Atmospheric CO_2 Concentration[e]	100	101	101	101	102	102	103	104	104	0.4%

[a] GWP weighted values.
[b] Energy content weighted values (U.S. DOE/EIA 2000a).
[c] GDP in chained 1996 dollars (U.S. DOC/BEA 2000).
[d] U.S. DOC/Census 2000.
[e] Mauna Loa Observatory, Hawaii (Keeling and Whorf 2000).
[f] Average annual growth rate.

At the same time, total U.S. greenhouse gas emissions have grown at about the same rate as the national population during the last decade. Overall, global atmospheric CO_2 concentrations (a function of many complex anthropogenic and natural processes) are increasing at 0.4 percent per year.

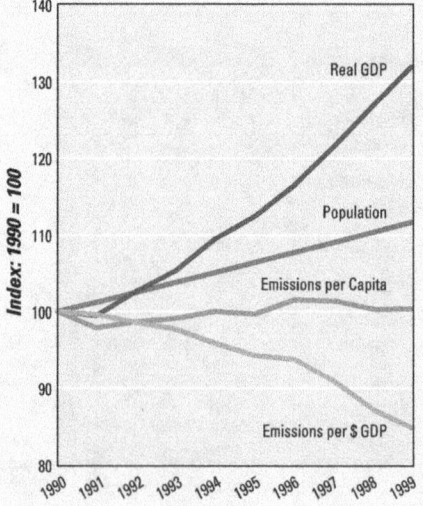

Sources: U.S. DOC/BEA 2000, U.S. DOC/Census 2001, and U.S. EPA 2001d.

Weather and Non-fossil Energy Adjustments to CO_2 from Fossil Fuel Combustion Trends

An analysis was performed using EIA's Short-Term Integrated Forecasting System (STIFS) model to examine the effects of variations in weather and output from nuclear and hydroelectric generating plants on U.S. energy-related CO_2 emissions.[10] Weather conditions affect energy demand because of the impact they have on residential, commercial, and industrial end-use sector heating and cooling demands. Warmer winters tend to reduce demand for heating fuels—especially natural gas—while cooler summers tend to reduce air conditioning-related electricity demand. Although changes in electricity output from hydroelectric and nuclear power plants do not necessarily affect final energy demand, increased output from these plants offsets electricity generation by fossil fuel power plants, and therefore leads to reduced CO_2 emissions.

FIGURE 3-6 Percent Difference in Adjusted and Actual Energy-Related CO_2 Emissions: 1997–1999

The results of this analysis show that CO_2 emissions from fossil fuel combustion would have been roughly 1.9 percent higher (102 Tg CO_2 Eq.) if weather conditions and hydroelectric and nuclear power generation had remained at normal levels.[11] Similarly, emissions in 1997 and 1998 would have been roughly 0.5 and 1.2 percent (7 and 17 Tg CO_2 Eq.) greater under normal conditions, respectively.

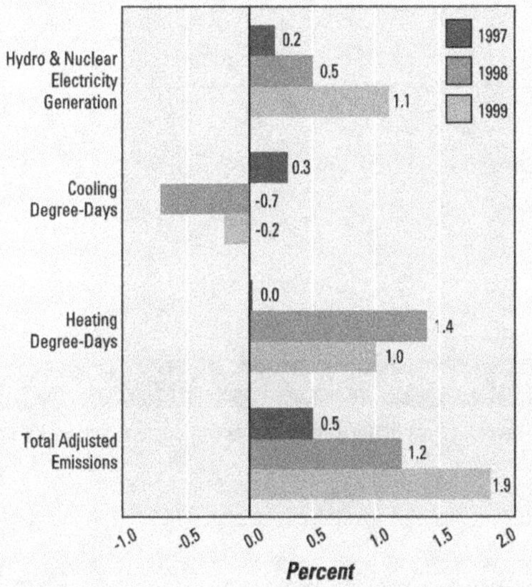

FIGURE 3-7 Recent Trends in Adjusted and Actual Energy-Related CO_2 Emissions: 1997–1999

In addition to the absolute level of emissions being greater, the growth rate in CO_2 emissions from fossil fuel combustion from 1998 to 1999 would have been 2.0 percent instead of the actual 1.2 percent if both weather conditions and non-fossil electricity generation had been normal. Similarly, emissions in 1998 would have increased by 0.9 percent under normal conditions versus the actual rate of 0.2 percent.

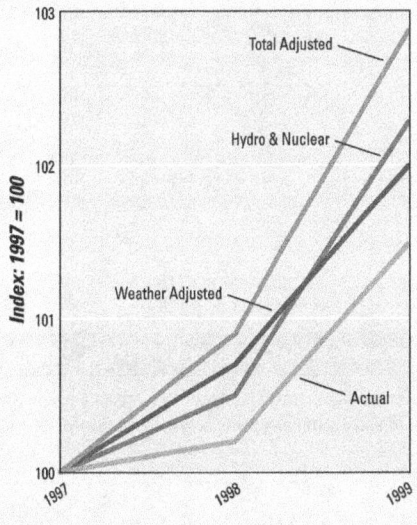

Warmer winter conditions in both 1998 and 1999 had a significant effect on U.S. CO_2 emissions by reducing demand for heating fuels. Heating degree-days in the United States in 1998 and 1999 were 14 and 7 percent below normal, respectively (see Figure 3-8).[12] These warm winters, however, were partly countered by increased electricity demand that resulted from hotter summers. Cooling degree-days in 1998 and 1999 were 18 and 3 percent above normal, respectively (see Figure 3-9).

Although no new U.S. nuclear power plants have been constructed in many years, the capacity factors[13] of existing plants reached record levels in 1998 and 1999, approaching 90 percent. This increase in utilization translated into increased electricity output by nuclear plants—slightly more than 7 percent in both years. Increased output, however, was partly offset by reduced electricity output by hydroelectric power plants, which declined by 10 and 4 percent in 1998 and 1999, respectively. Electricity generated by nuclear plants provides approximately twice as much of the energy consumed in the United States as hydroelectric plants. Figure 3-10 shows nuclear and hydroelectric power plant capacity factors since 1973 and 1989, respectively.

---

[10] The STIFS model is employed in producing EIA's Short-Term Energy Outlook (U.S. DOE/EIA 2000d). Complete model documentation can be found at http://www.eia.doe.gov/emeu/steo/pub/contents.html. Various other factors that influence energy-related CO_2 emissions were also examined, such as changes in output from energy-intensive manufacturing industries, and changes in fossil fuel prices. These additional factors, however, were not found to have a significant effect on emission trends.

[11] Normal levels are defined by decadal power generation trends.

[12] Degree-days are relative measurements of outdoor air temperature. Heating degree-days are deviations of the mean daily temperature below 65°F, while cooling degree-days are deviations of the mean daily temperature above 65°F. Excludes Alaska and Hawaii. Normals are based on data from 1961 through 1990. The variations in these normals during this time period were 10 percent and 14 percent for heating and cooling degree-days, respectively (99 percent confidence interval).

[13] The capacity factor is defined as the ratio of the electrical energy produced by a generating unit for a given period of time to the electrical energy that could have been produced at continuous full-power operation during the same period (U.S. DOE/EIA 2000a).

FIGURE 3-8 **Annual Deviations from Normal U.S. Heating Degree-Days: 1949–1999**

Warmer winter conditions in both 1998 and 1999 had a significant effect on U.S. CO_2 emissions by reducing demand for heating fuels. Heating degree-days in the United States in 1998 and 1999 were 14 and 7 percent below normal, respectively.

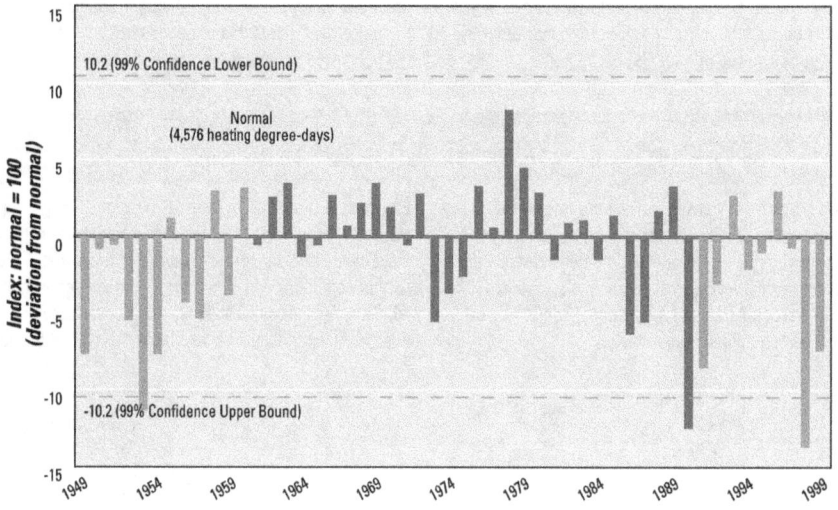

Note: Climatological normal data (1961–1990) are highlighted. Statistical confidence interval for "normal" climatology period of 1961 through 1990.

Sources: U.S. DOC/NOAA 1998a, b; 1999a, b; and 2001a, b.

FIGURE 3-9 **Annual Deviations from Normal U.S. Cooling Degree-Days: 1949–1999**

Warmer winters were partly countered by increased electricity demand that resulted from hotter summers. Cooling degree-days in 1998 and 1999 were 18 and 3 percent above normal, respectively.

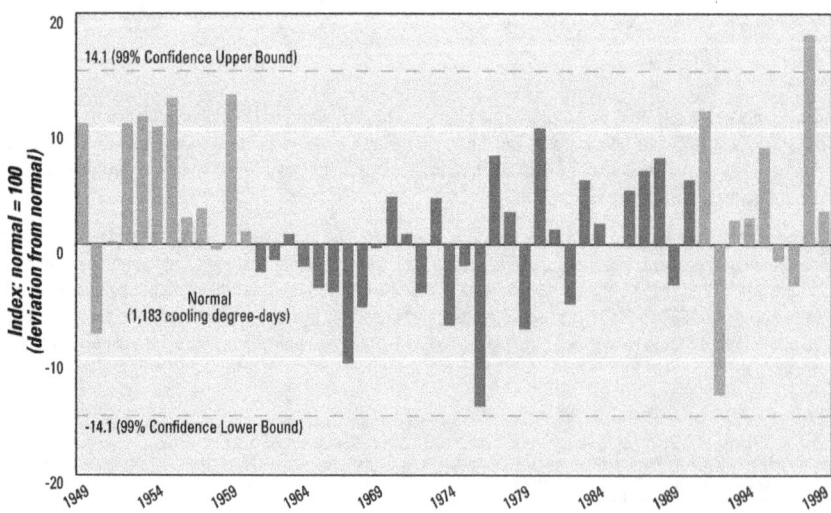

Note: Climatological normal data (1961–1990) are highlighted. Statistical confidence interval for "normal" climatology period of 1961 through 1990.

Sources: U.S. DOC/NOAA 1998a, b; 1999a, b; and 2001a, b.

FIGURE 3-10 U.S. Nuclear and Hydroelectric Power Plant Capacity Factors: 1973–1999

The utilization (i.e., capacity factors) of existing nuclear power plants reached record levels in 1998 and 1999, approaching 90 percent. This increase in utilization translated into an increase in electricity output by nuclear plants of slightly more than 7 percent in both years. However, it was partly offset by 10 and 14 percent respective declines in electricity output by hydroelectric power plants in 1998 and 1999.

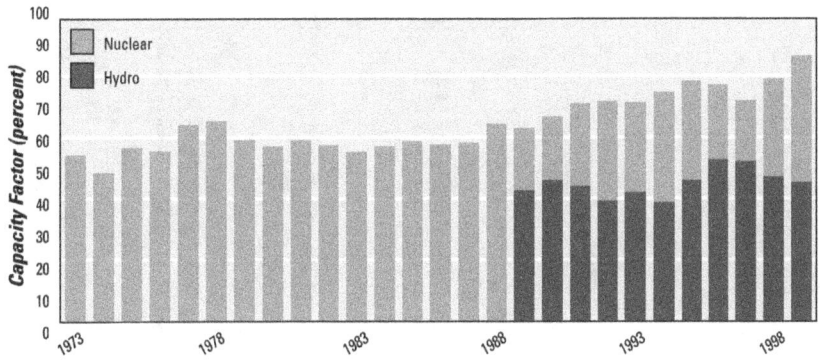

Greenhouse Gas Emissions from Transportation Activities

Motor vehicle use is increasing all over the world, including in the United States. Since the 1970s, the number of highway vehicles registered in the United States has increased faster than the overall population (U.S. DOT/FHWA 1999). Likewise, the number of miles driven—up 13 percent from 1990 to 1999 (U.S. DOT/FHWA 1999)—and gallons of gasoline consumed each year in the United States (U.S. DOC/EIA 2000a) have increased steadily since the 1980s. These increases in motor vehicle use are the result of a confluence of factors, including population growth, economic growth, urban sprawl, low fuel prices, and increasing popularity of sport utility vehicles and other light-duty trucks that tend to have lower fuel efficiency.[14] A similar set of social and economic trends led to a significant increase in air travel and freight transportation—by both air and road modes—during the 1990s.

Passenger cars, trucks, motorcycles, and buses emit significant quantities of air pollutants with local, regional, and global effects. Motor vehicles are major sources of CO, CO_2, CH_4, nonmethane volatile organic compounds (NMVOCs), NO_x, N_2O, and HFCs. They are also important contributors to many serious environmental pollution problems, including ground-level ozone (i.e., smog), acid rain, fine particulate matter, and global warming. Within the United States and abroad, government agencies have taken actions to reduce these emissions. Since the 1970s, the Environmental Protection Agency has required the reduction of lead in gasoline, developed strict emission standards for new passenger cars and trucks, directed states to enact comprehensive motor vehicle emission control programs, required inspection and maintenance programs, and, more recently, introduced the use of reformulated gasoline. New vehicles are now equipped with advanced emissions controls, which are designed to reduce emissions of NO_x, hydrocarbons, and CO.

Table 3-4 summarizes greenhouse gas emissions from all transportation-related activities. Overall, transportation activities, excluding international bunker fuels, accounted for an almost constant 26 percent of total U.S. greenhouse gas emissions from 1990 to 1999. These emissions were primarily CO_2 from fuel combustion, which increased by 16 percent from 1990 to 1999. However, because of larger increases in N_2O and HFC emissions during this period, overall emissions from transportation activities actually increased by 18 percent.

---

[14] The average miles per gallon achieved by the U.S. highway vehicle fleet decreased by slightly less than one percent in both 1998 and 1999.

TABLE 3-4 Transportation-Related Greenhouse Gas Emissions (Tg CO_2 Eq.)

Overall, transportation activities (excluding international bunker fuels) accounted for an almost constant 26 percent of total U.S. greenhouse gas emissions from 1990 to 1999. These emissions were primarily CO_2 from fuel combustion, which increased by 16 percent during that period. However, because of larger increases in N_2O and HFC emissions, overall emissions from transportation activities actually increased by 18 percent.

Gas/Vehicle Type	1990	1995	1996	1997	1998	1999
CO_2	1,474.4	1,581.8	1,621.2	1,631.4	1,659.0	1,716.4
Passenger Cars	620.0	641.9	654.1	660.2	674.5	688.9
Light-Duty Trucks	283.1	325.3	333.5	337.3	356.9	364.8
Other Trucks	206.0	235.9	248.1	257.0	257.9	269.7
Aircraft[a]	176.7	171.5	180.2	179.0	183.0	184.6
Boats and Vessels	59.4	66.9	63.8	50.2	47.9	65.6
Locomotives	28.4	31.5	33.4	34.4	33.6	35.1
Buses	10.7	13.5	11.3	12.0	12.3	12.9
Other[b]	90.1	95.3	96.7	101.4	93.0	94.9
International Bunker Fuels[c]	114.0	101.0	102.2	109.8	112.8	107.3
CH_4	5.0	4.9	4.8	4.7	4.6	4.5
Passenger Cars	2.4	2.0	2.0	2.0	2.0	1.9
Light-Duty Trucks	1.6	1.9	1.6	1.6	1.5	1.4
Other Trucks and Buses	0.4	0.5	0.7	0.7	0.7	0.7
Aircraft	0.2	0.1	0.1	0.2	0.1	0.2
Boats and Vessels	0.1	0.1	0.1	0.1	0.1	0.1
Locomotives	0.1	0.1	0.1	0.1	+	+
Other[d]	0.2	0.2	0.2	0.2	0.2	0.2
International Bunker Fuels[c]	+	+	+	+	+	+
N_2O	54.3	66.8	65.3	65.2	64.2	63.4
Passenger Cars	31.0	33.0	32.7	32.4	32.1	31.5
Light-Duty Trucks	17.8	27.1	23.9	24.0	23.3	22.7
Other Trucks and Buses	2.6	3.6	5.6	5.8	5.9	6.1
Aircraft[a]	1.7	1.7	1.8	1.7	1.8	1.8
Boats and Vessels	0.4	0.5	0.4	0.3	0.3	0.4
Locomotives	0.3	0.3	0.3	0.2	0.2	0.2
Other[d]	0.6	0.6	0.6	0.6	0.6	0.6
International Bunker Fuels[c]	1.0	0.9	0.9	1.0	1.0	1.0
HFCs	+	9.5	13.5	17.2	20.6	23.7
Mobile Air Conditioners[e]	+	9.5	13.5	17.2	20.6	23.7
Total[e]	**1,533.7**	**1,663.0**	**1,704.8**	**1,718.5**	**1,748.4**	**1,808.0**

+ Does not exceed 0.05 Tg CO_2 Eq.

[a] Aircraft emissions consist of emissions from all jet fuel (less bunker fuels) and aviation gas consumption.

[b] "Other" CO_2 emissions include motorcycles, construction equipment, agricultural machinery, pipelines, and lubricants.

[c] Emissions from international bunker fuels include emissions from both civilian and military activities, but are not included in totals.

[d] "Other" CH_4 and N_2O emissions include motorcycles; construction equipment; agricultural machinery; gasoline-powered recreational, industrial, lawn and garden, light commercial, logging, airport service, and other equipment; and diesel-powered recreational, industrial, lawn and garden, light construction, and airport service.

[e] Includes primarily HFC-134a.

Note: Totals may not sum due to independent rounding.

All gases in this report are presented in units of teragrams of carbon dioxide equivalents (Tg CO_2 Eq.). The relationship between gigagrams (Gg) of a gas and Tg CO_2 Eq. can be expressed as follows:

The GWP of a greenhouse gas is the ratio of global warming from one unit mass of a greenhouse gas to that of one unit mass of CO_2 over a specified period of time. While any time period can be selected, the 100-year GWPs recommended by the IPCC and employed by the United States for policymaking and reporting purposes were used in this report (IPCC 1996b). GWP values are listed in Table 3-5.

CARBON DIOXIDE EMISSIONS

The global carbon cycle is made up of large carbon flows and reservoirs. Billions of tons of carbon in the form of CO_2 are absorbed by oceans and living biomass (sinks) and are emitted to the atmosphere annually through natural processes (sources). When in equilibrium, carbon fluxes among these reservoirs are balanced.

Since the Industrial Revolution, this equilibrium of atmospheric carbon has been altered. Atmospheric concentrations of CO_2 have risen by about 31 percent (IPCC 2001b), principally because of fossil fuel combustion, which accounted for 98 percent of total U.S. CO_2 emissions in 1999. Changes in land use and forestry practices can also emit CO_2 (e.g., through conversion of forest land to agricultural or urban use) or can act as a sink for CO_2 (e.g., through net additions to forest biomass).

Figure 3-11 and Table 3-7 summarize U.S. sources and sinks of CO_2. The remainder of this section discusses CO_2 emission trends in greater detail.

Greenhouse Gas Emissions from Electric Utilities

Like transportation, activities related to the generation, transmission, and distribution of electricity in the United States resulted in a significant fraction of total U.S. greenhouse gas emissions. The electric power industry in the United States is composed of traditional electric utilities, as well as other entities, such as power marketers and nonutility power producers. Table 3-6 presents emissions from electric utility-related activities.

Aggregate emissions from electric utilities of all greenhouse gases increased by 11 percent from 1990 to 1999, and accounted for a relatively constant 29 percent of U.S. emissions during the same period. Emissions from nonutility generators are not included in these estimates. Nonutilities were estimated to have produced about 15 percent of the electricity generated in the United States in 1999, up from 11 percent in 1998 (U.S. DOE/EIA 2000b). Therefore, a more complete accounting of greenhouse gas emissions from the electric power industry (i.e., utilities and nonutilities combined) would account for roughly 40 percent of U.S. CO_2 emissions (U.S. U.S. DOE/EIA 2000c).

The majority of electric utility-related emissions resulted from the combustion of coal in boilers to produce steam that is passed through a turbine to generate electricity. Overall, the generation of electricity—especially when nonutility generators are included—results in a larger portion of total U.S. greenhouse gas emissions than any other activity.

TABLE 3-5 Global Warming Potentials (100-Year Time Horizon)

The concept of a global warming potential (GWP) has been developed to compare the ability of each greenhouse gas to trap heat in the atmosphere relative to another gas. Carbon dioxide was chosen as the reference gas to be consistent with IPCC guidelines.

Gas	GWP
Carbon Dioxide (CO_2)	1
Methane (CH_4)*	21
Nitrous Oxide (N_2O)	310
HFC-23	11,700
HFC-125	2,800
HFC-134a	1,300
HFC-143a	3,800
HFC-152a	140
HFC-227ea	2,900
HFC-236fa	6,300
HFC-4310mee	1,300
CF_4	6,500
C_2F_6	9,200
C_4F_{10}	7,000
C_6F_{14}	7,400
Sulfur Hexafluoride (SF_6)	23,900

* The methane GWP includes direct effects and those indirect effects due to the production of tropospheric ozone and stratospheric water vapor. The indirect effects due to the production of CO_2 are not included.
Source: IPCC 1996b.

TABLE 3-6 Electric Utility-Related Greenhouse Gas Emissions (Tg CO_2 Eq.)

Gas/Fuel Type or Source	1990	1995	1996	1997	1998	1999
CO_2	1,757.3	1,810.6	1,880.3	1,953.5	2,010.7	1,953.4
Coal	1,509.3	1,587.7	1,677.7	1,729.7	1,744.0	1,711.9
Natural Gas	151.1	171.8	146.5	159.6	175.8	168.0
Petroleum	96.8	51.0	56.0	64.1	90.8	73.4
Geothermal	0.2	0.1	0.1	0.1	0.1	+
CH_4	0.5	0.5	0.5	0.5	0.5	0.5
Stationary Combustion (Utilities)	0.5	0.5	0.5	0.5	0.5	0.5
N_2O	7.4	7.8	8.2	8.5	8.7	8.6
Stationary Combustion (Utilities)	7.4	7.8	8.2	8.5	8.7	8.6
SF_6	20.5	25.7	25.7	25.7	25.7	25.7
Electrical Transmission and Distribution	20.5	25.7	25.7	25.7	25.7	25.7
Total	**1,785.7**	**1,844.5**	**1,914.7**	**1,988.2**	**2,045.6**	**1,988.2**

+ Does not exceed 0.05 Tg CO_2 Eq.
Notes: Totals may not sum due to independent rounding. Excludes emissions from nonutilities, which are currently accounted for under the industrial end-use sector.

FIGURE 3-11 AND TABLE 3-7 U.S. Sources of CO_2 Emissions and Sinks (Tg CO_2 Eq.)

Carbon dioxide accounted for 82 percent of total U.S. greenhouse gas emissions in 1999, and fossil fuel combustion accounted for 98 percent of total CO_2 emissions. Changes in land use and forestry practices resulted in a net decrease of 990.4 Tg CO_2 Eq., or 18 percent, of CO_2 emissions.

Source or Sink	1990	1995	1996	1997	1998	1999
Fossil Fuel Combustion	4,835.7	5,121.3	5,303.0	5,374.9	5,386.8	5,453.1
Cement Manufacture	33.3	36.8	37.1	38.3	39.2	39.9
Waste Combustion	17.6	23.1	24.0	25.7	25.1	26.0
Lime Manufacture	11.2	12.8	13.5	13.7	13.9	13.4
Natural Gas Flaring	5.1	13.6	13.0	12.0	10.8	11.7
Limestone and Dolomite Use	5.1	7.0	7.3	8.3	8.1	8.3
Soda Ash Manufacture and Consumption	4.1	4.3	4.3	4.4	4.3	4.2
Carbon Dioxide Consumption	0.8	1.0	1.1	1.3	1.4	1.6
Land-Use Change and Forestry (Sink)[a]	(1,059.9)	(1,019.1)	(1,021.6)	(981.9)	(983.3)	(990.4)
International Bunker Fuels[b]	114.0	101.0	102.2	109.8	112.8	107.3
Total Emissions	**4,913.0**	**5,219.8**	**5,403.2**	**5,478.7**	**5,489.7**	**5,558.1**
Net Emissions (Sources and Sinks)	**3,853.0**	**4,200.8**	**4,381.6**	**4,496.8**	**4,506.4**	**4,567.8**

[a] Sinks are only included in net emissions total, and are based partly on projected activity data.
[b] Emissions from international bunker fuels are not included in totals.
Notes: Totals may not sum due to independent rounding. Parentheses indicate negative values (or sequestration).

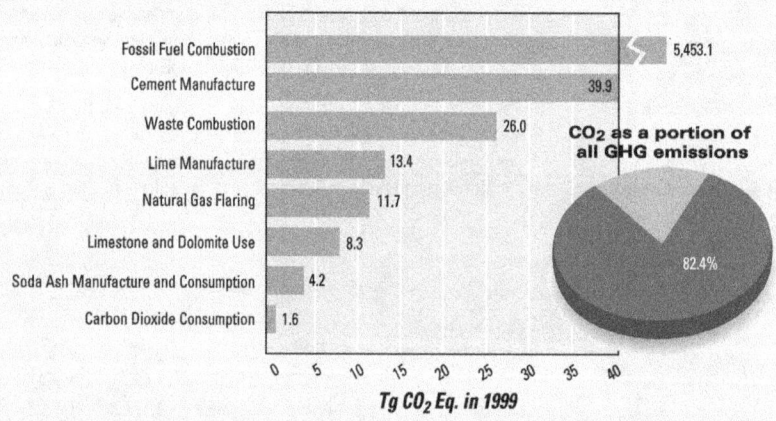

CO2 as a portion of all GHG emissions — 82.4%

Tg CO_2 Eq. in 1999

Energy

Energy-related activities accounted for the vast majority of U.S. CO_2 emissions from 1990 through 1999. Carbon dioxide from fossil fuel combustion was the dominant contributor. In 1999, approximately 84 percent of the energy consumed in the United States was produced through the combustion of fossil fuels. The remaining 16 percent came from other sources, such as hydropower, biomass, nuclear, wind, and solar energy (see Figures 3-12 and 3-13). This section discusses specific trends related to CO_2 emissions from energy consumption.

Fossil Fuel Combustion

As fossil fuels are combusted, the carbon stored in them is almost entirely emitted as CO_2. The amount of carbon in fuels per unit of energy content varies significantly by fuel type. For example, coal contains the highest amount of carbon per unit of energy, while petroleum has about 25 percent less carbon than coal, and natural gas about 45 percent less.

From 1990 through 1999, petroleum supplied the largest share of U.S. energy demands, accounting for an average of 39 percent of total energy consumption. Natural gas and coal followed in order of importance, accounting for an average of 24 and 23 percent of total energy consumption, respectively. Most petroleum was consumed in the transportation end-use sector, the vast majority of coal was used by electric utilities, and natural gas was consumed largely in the industrial and residential sectors.

Emissions of CO_2 from fossil fuel combustion increased at an average annual rate of 1.4 percent from 1990 to 1999. The fundamental factors behind this trend included (1) a robust domestic economy, (2) relatively low energy prices as compared to 1990, (3) fuel switching by electric utilities, and (4) heavier reliance on nuclear energy. Between 1990 and 1999, CO_2 emissions from fossil fuel combustion steadily increased from 4,835.7 to 5,453.1 Tg CO_2 Eq.—a 13 percent total increase over the ten-year period.

In 1999, fossil fuel emission trends were primarily driven by similar factors—a strong economy and an increased reliance on carbon-neutral nuclear power for electricity generation. Although the price of crude oil increased by over 40 percent between 1998 and 1999, and relatively mild weather conditions in 1999 moderated energy consumption for heating and cooling, emissions from fossil fuels still rose by 1.2 percent. Emissions from the combustion of petroleum products in 1999 grew the most (64 Tg CO_2 Eq., or about 3 percent), although emissions from the combustion of petroleum by electric utilities decreased by 19 percent. That decrease was offset by increased emissions from petroleum combustion in the residential, commer-

cial, industrial, and especially transportation end-use sectors. Emissions from the combustion of natural gas in 1999 increased slightly (5 Tg CO_2 Eq., or 0.4 percent), and emissions from coal consumption decreased slightly (3 Tg CO_2 Eq., or 0.1 percent) as the industrial end-use sector substituted more natural gas for coal in 1999.

Along with the four end-use sectors, electric utilities also emit CO_2, although these emissions are produced as they consume fossil fuel to provide electricity to one of the four end-use sectors. For the discussion in this chapter, electric utility emissions have been distributed to each end-use sector based upon their fraction of aggregate electricity consumption. This method of distributing emissions assumes that each end-use sector consumes electricity that is generated with the national average mix of fuels according to their carbon intensity. In reality, sources of electricity vary widely in carbon intensity. By assuming the same carbon intensity for each end-use sector's electricity consumption, for example, emissions attributed to the residential sector may be

overestimated, while emissions attributed to the industrial sector may be underestimated. Emissions from electric utilities are addressed separately after the end-use sectors have been discussed.

It is important to note, though, that all emissions resulting from the generation of electricity by the growing number of nonutility power plants are currently allocated to the industrial sector. Nonutilities supplied 15 percent of the electricity consumed in the United States in 1999. Emissions from U.S. territories are also calculated separately due to a lack of specific consumption data for the individual end-use sectors. Table 3-8, Figure 3-14, and Figure 3-15 summarize CO_2 emissions from fossil fuel combustion by end-use sector.

Industrial End-Use Sector. Industrial CO_2 emissions—resulting both directly from the combustion of fossil fuels and indirectly from the generation of electricity by utilities that is consumed by industry—accounted for 33 percent of CO_2 from fossil fuel combustion in 1999. About two-thirds of these

FIGURE 3-12 1999 U.S. Energy Consumption by Energy Source

Petroleum supplied the largest share of U.S. energy demands in 1999, accounting for 39 percent of total energy consumption. Natural gas and coal followed in order of importance, each accounting for 23 percent of total energy consumption.

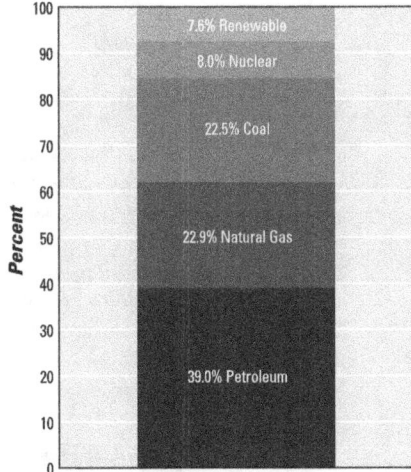

Source: U.S. DOE/EIA 2000a.

FIGURE 3-13 U.S. Energy Consumption: 1990–1999

In 1999, approximately 84 percent of the energy consumed in the United States was produced through the combustion of fossil fuels. The remaining 16 percent came from other energy sources, such as hydropower, biomass, nuclear, wind, and solar.

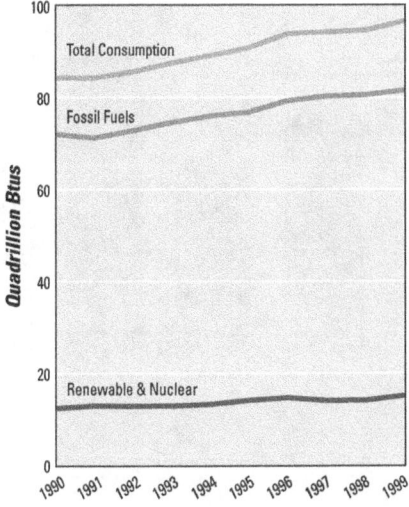

Source: U.S. DOE/EIA 2000a.

FIGURE 3-14 1999 CO$_2$ Emissions from Fossil Fuel Combustion by Sector and Fuel Type

Of the emissions from fossil fuel combustion in 1999, most petroleum was consumed in the transportation end-use sector. The vast majority of coal was consumed by electric utilities, and natural gas was consumed largely in the industrial and residential end-use sectors.

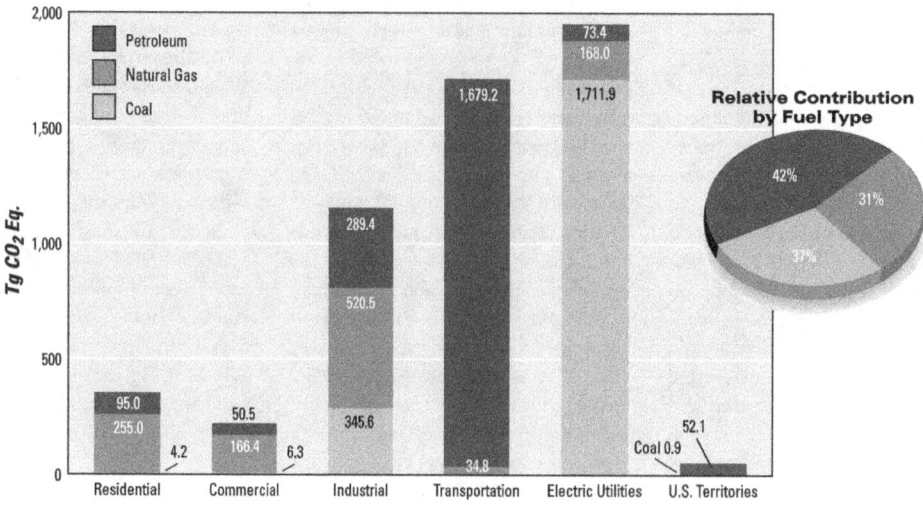

FIGURE 3-15 1999 End-Use Sector Emissions of CO$_2$ from Fossil Fuel Combustion

Electric utilities were responsible for 36 percent of the U.S. emissions of CO$_2$ from fossil fuel combustion in 1999. The remaining 64 percent of emissions resulted from the direct combustion of fuel for heat and other uses in the residential, commercial, industrial, and transportation end-use sectors.

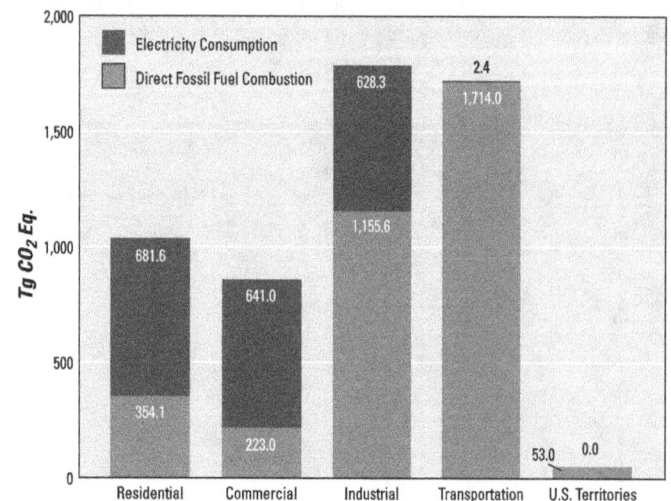

emissions resulted from direct fossil fuel combustion to produce steam and/or heat for industrial processes or by nonutility electricity generators that are classified as industrial, the latter of which are growing rapidly. The remaining third of emissions resulted from consuming electricity from electric utilities for motors, electric furnaces, ovens, lighting, and other applications.

Transportation End-Use Sector. Transportation activities (excluding international bunker fuels) accounted for 31 percent of CO$_2$ emissions from fossil fuel combustion in 1999.[15] Virtually all of the energy consumed in this end-use sector came from petroleum products. Slightly less than two-thirds of the emissions resulted from gasoline consumption in motor vehicles. The remaining emissions came from other transportation activities, including the

---

[15] If emissions from international bunker fuels are included, the transportation end-use sector accounted for 33 percent of U.S. emissions from fossil fuel combustion in 1999.

TABLE 3-8 CO_2 Emissions from Fossil Fuel Combustion by End-Use Sector (Tg CO_2 Eq.)

In 1999, industrial CO_2 emissions resulting from direct fossil fuel combustion and from the generation of electricity by utilities accounted for 33 percent of CO_2 from fossil fuel combustion. Transportation activities (excluding international bunker fuels) accounted for 31 percent of CO_2 emissions from fossil fuel combustion the same year, and the residential and commercial sectors accounted for 19 and 16 percent, respectively.

End-Use Sector*	1990	1995	1996	1997	1998	1999
Industrial	1,636.0	1,709.5	1,766.0	1,783.6	1,758.8	1,783.9
Transportation	1,474.4	1,581.8	1,621.2	1,631.4	1,659.0	1,716.4
Residential	930.7	988.7	1,047.5	1,044.2	1,040.9	1,035.8
Commercial	760.8	797.2	828.2	872.9	880.2	864.0
U.S. Territories	33.7	44.0	40.1	42.8	47.9	53.0
Total	**4,835.7**	**5,121.3**	**5,303.0**	**5,374.9**	**5,386.8**	**5,453.1**

* Emissions from electric utilities are allocated based on aggregate electricity consumption in each end-use sector.
Note: Totals may not sum due to independent rounding.

combustion of diesel fuel in heavy-duty vehicles and jet fuel in aircraft.

Residential and Commercial End-Use Sectors. The residential and commercial end-use sectors accounted for 19 and 16 percent, respectively, of CO_2 emissions from fossil fuel consumption in 1999. Both sectors relied heavily on electricity for meeting energy needs, with 66 and 74 percent, respectively, of their emissions attributable to electricity consumption for lighting, heating, cooling, and operating appliances. The remaining emissions were largely due to the consumption of natural gas and petroleum, primarily for meeting heating and cooking needs.

Electric Utilities. The United States relies on electricity to meet a significant portion of its energy demands, especially for lighting, electric motors, heating, and air conditioning. Electric utilities are responsible for consuming 27 percent of U.S. energy from fossil fuels and emitted 36 percent of the CO_2 from fossil fuel combustion in 1999. The type of fuel combusted by utilities significantly affects their emissions. For example, some electricity is generated with low CO_2-emitting energy technologies, particularly non-fossil fuel options, such as nuclear, hydroelectric, or geothermal energy. However, electric utilities rely

on coal for over half of their total energy requirements and accounted for 85 percent of all coal consumed in the United States in 1999. Consequently, changes in electricity demand have a significant impact on coal consumption and associated CO_2 emissions. Note, again, that all emissions resulting from the generation of electricity by nonutility plants are currently allocated to the industrial end-use sector.

Natural Gas Flaring

Carbon dioxide is produced when natural gas from oil wells is flared (i.e., combusted) to relieve rising pressure or to dispose of small quantities of gas that are not commercially marketable. In 1999, flaring activities emitted approximately 11.7 Tg CO_2 Eq., or about 0.2 percent of U.S. CO_2 emissions.

Biomass Combustion

Biomass in the form of fuel wood and wood waste was used primarily by the industrial end-use sector. The transportation end-use sector was the predominant user of biomass-based fuels, such as ethanol from corn and woody crops. Ethanol and ethanol blends, such as gasohol, are typically used to fuel public transport vehicles.

Although these fuels emit CO_2, in the long run the CO_2 emitted from biofuel consumption does not increase

atmospheric CO_2 concentrations if the biogenic carbon emitted is offset by the growth of new biomass. For example, fuel wood burned one year but regrown the next only recycles carbon, rather than creating a net increase in total atmospheric carbon. Net carbon fluxes from changes in biogenic carbon reservoirs in wooded areas or croplands are accounted for under the Land-Use Change and Forestry section of this chapter.

Gross CO_2 emissions from biomass combustion were 234.1 Tg CO_2 Eq. in 1999, with the industrial sector accounting for 81 percent and the residential sector 14 percent of the emissions. Ethanol consumption by the transportation sector accounted for only 3 percent of CO_2 emissions from biomass combustion.

Industrial Processes

Emissions are produced as a by-product of many nonenergy-related activities. For example, industrial processes can chemically transform raw materials. This transformation often releases such greenhouse gases as CO_2. The major production processes that emit CO_2 include cement manufacture, lime manufacture, limestone and dolomite use, soda ash manufacture and consumption, and CO_2 consumption. Total CO_2 emissions from these sources were approximately 67.4 Tg CO_2 Eq. in 1999, or about 1 percent of all CO_2 emissions. Between 1990 and 1999, emissions from most of these sources increased, except for emissions from soda ash manufacture and consumption, which have remained relatively constant.

Cement Manufacture
(39.9 Tg CO_2 Eq.)

Carbon dioxide is emitted primarily during the production of clinker, an intermediate product from which finished Portland and masonry cement are made. When calcium carbonate ($CaCO_3$) is heated in a cement kiln to form lime and CO_2, the lime combines with other materials to produce clinker, and the CO_2 is released to the atmosphere.

Lime Manufacture
(13.4 Tg CO_2 Eq.)

Lime is used in steel making, construction, pulp and paper manufacturing, and water and sewage treatment. It is manufactured by heating limestone (mostly calcium carbonate, $CaCO_3$) in a kiln, creating calcium oxide (quicklime) and CO_2, which is normally emitted to the atmosphere.

Limestone and Dolomite Use
(8.3 Tg CO_2 Eq.)

Limestone ($CaCO_3$) and dolomite ($Ca\,Mg(CO_3)_2$) are basic raw materials used by a wide variety of industries, including the construction, agriculture, chemical, and metallurgical industries. For example, limestone can be used as a purifier in refining metals. In the case of iron ore, limestone heated in a blast furnace reacts with impurities in the iron ore and fuels, generating CO_2 as a by-product. Limestone is also used in flue gas desulfurization systems to remove sulfur dioxide from the exhaust gases

Soda Ash Manufacture and
Consumption (4.2 Tg CO_2 Eq.)

Commercial soda ash (sodium carbonate, Na_2CO_3) is used in many consumer products, such as glass, soap and detergents, paper, textiles, and food.

During the manufacture of soda ash, some natural sources of sodium carbonate are heated and transformed into a crude soda ash, in which CO_2 is generated as a by-product. In addition, CO_2 is often released when the soda ash is consumed.

Carbon Dioxide Consumption
(1.6 Tg CO_2 Eq.)

Carbon dioxide is used directly in many segments of the economy, including food processing, beverage manufacturing, chemical processing, and a host of industrial and other miscellaneous applications. This CO_2 may be produced as a by-product from the production of certain chemicals (e.g., ammonia) from select natural gas wells, or by separating it from crude oil and natural gas. For the most part, the CO_2 used in these applications is eventually released to the atmosphere.

Land-Use Change and Forestry

(Sink) (990.4 Tg CO_2 Eq.)

When humans alter the terrestrial biosphere through land use, changes in land use, and forest management practices, they alter the natural carbon flux between biomass, soils, and the atmosphere. Forest management practices, the management of agricultural soils, and landfilling of yard trimmings have resulted in a net uptake (sequestration) of carbon in the United States that is equivalent to about 15 percent of total U.S. gross emissions.

Forests (including vegetation, soils, and harvested wood) accounted for approximately 91 percent of the total sequestration, agricultural soils (including mineral and organic soils and the application of lime) accounted for 8 percent, and landfilled yard trimmings accounted for less than 1 percent. The net forest sequestration is largely a result of improved forest management practices, the regeneration of previously cleared forest areas, and timber harvesting. Agricultural mineral soils account for a net carbon sink that is more than three times larger than the sum of emissions from organic soils and liming. Net

sequestration in these soils is largely due to improved management practices on cropland and grazing land, especially using conservation tillage (leaving residues on the field after harvest), and taking erodible lands out of production and planting them with grass or trees through the Conservation Reserve Program. Finally, the net sequestration from yard trimmings is due to their long-term accumulation in landfills.

Waste

Waste Combustion (26.0 Tg CO_2 Eq.)

Waste combustion involves the burning of garbage and nonhazardous solids, referred to as municipal solid waste (MSW), as well as the burning of hazardous waste. Carbon dioxide emissions arise from the organic (i.e., carbon) materials found in these wastes. Within MSW, many products contain carbon of biogenic origin, and the CO_2 emissions from their combustion are reported under the Land-Use Change and Forestry section. However, several components of MSW—plastics, synthetic rubber, synthetic fibers, and carbon black—are of fossil fuel origin, and are included as sources of CO_2 emissions.

METHANE EMISSIONS

Atmospheric methane (CH_4) is an integral component of the greenhouse effect, second only to CO_2 as a contributor to anthropogenic greenhouse gas emissions. The overall contribution of CH_4 to global warming is significant because it has been estimated to be 21 times more effective at trapping heat in the atmosphere than CO_2 (i.e., the GWP value of CH_4 is 21) (IPCC1996b). Over the last two centuries, the concentration of CH_4 in the atmosphere has more than doubled (IPCC 2001b). Experts believe these atmospheric increases were due largely to increasing emissions from anthropogenic sources, such as landfills, natural gas and petroleum systems, agricultural activities, coal mining, stationary and mobile combustion, wastewater treatment, and certain industrial processes (see Figure 3-16 and Table 3-9).

FIGURE 3-16 AND TABLE 3-9 U.S. Sources of Methane Emissions (Tg CO₂ Eq.)

Methane accounted for 9 percent of total U.S. greenhouse gas emissions in 1999. Landfills, enteric fermentation, and natural gas systems were the source of 75 percent of total CH₄ emissions.

Source	1990	1995	1996	1997	1998	1999
Landfills	217.3	222.9	219.1	217.8	213.6	214.6
Enteric Fermentation	129.5	136.3	132.2	129.6	127.5	127.2
Natural Gas Systems	121.2	124.2	125.8	122.7	122.1	121.8
Coal Mining	87.9	74.6	69.3	68.8	66.5	61.8
Manure Management	26.4	31.0	30.7	32.6	35.2	34.4
Petroleum Systems	27.2	24.5	24.0	24.0	23.3	21.9
Wastewater Treatment	11.2	11.8	11.9	12.0	12.1	12.2
Rice Cultivation	8.7	9.5	8.8	9.6	10.1	10.7
Stationary Combustion	8.5	8.9	9.0	8.1	7.6	8.1
Mobile Combustion	5.0	4.9	4.8	4.7	4.6	4.5
Petrochemical Production	1.2	1.5	1.6	1.6	1.6	1.7
Agricultural Residue Burning	0.5	0.5	0.6	0.6	0.6	0.6
Silicon Carbide Production	+	+	+	+	+	+
International Bunker Fuels*	+	+	+	+	+	+
Total*	**644.5**	**650.5**	**638.0**	**632.0**	**624.8**	**619.6**

+ Does not exceed 0.05 Tg CO₂ Eq.
* Emissions from international bunker fuels are not included in totals.
Note: Totals may not sum due to independent rounding.

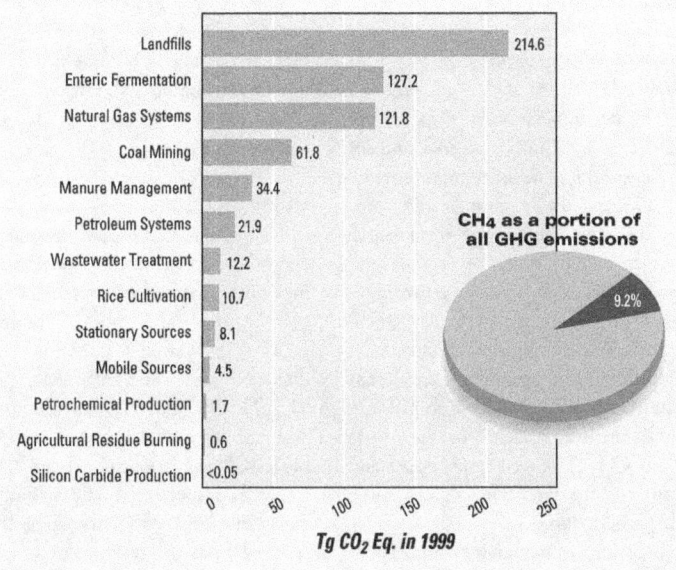

CH₄ as a portion of all GHG emissions

9.2%

Tg CO₂ Eq. in 1999

Landfills

Landfills are the largest source of anthropogenic CH₄ emissions in the United States. In an environment where the oxygen content is low or nonexistent, organic materials—such as yard waste, household waste, food waste, and paper—can be decomposed by bacteria, resulting in the generation of CH₄ and biogenic CO₂. Methane emissions from landfills are affected by site-specific factors, such as waste com-position, moisture, and landfill size.

In 1999, CH₄ emissions from U.S. landfills were 214.6 Tg CO₂ Eq., down by 1 percent since 1990. The relatively constant emission estimates are a result of two offsetting trends: (1) the amount of municipal solid waste in landfills contributing to CH₄ emissions has increased, thereby increasing the potential for emissions; and (2) the amount of landfill gas collected and combusted by landfill operators has also increased, thereby reducing emissions. Emissions from U.S. municipal solid waste landfills accounted for 94 percent of total landfill emissions, while industrial landfills accounted for the remainder. Approximately 28 percent of the CH₄ generated in U.S. landfills in 1999 was recovered and combusted, often for energy.

A regulation promulgated in March 1996 requires the largest U.S. landfills to collect and combust their landfill gas to reduce emissions of non-methane volatile organic compounds (NMVOCs). It is estimated that by the year 2000 this regulation will have reduced landfill CH₄ emissions by more than 50 percent.

Natural Gas and Petroleum Systems

Methane is the major component of natural gas. During the production, processing, transmission, and distribution of natural gas, fugitive emissions of CH₄ often occur. Because natural gas is often found in conjunction with petroleum deposits, leakage from petroleum systems is also a source of emissions. Emissions vary greatly from facility to facility and are largely a function of operation and maintenance procedures and equipment conditions. In 1999, CH₄ emissions from U.S. natural gas systems were estimated to be 121.8 Tg CO₂ Eq., accounting for approximately 20 percent of U.S. CH₄ emissions.

Petroleum is found in the same geological structures as natural gas, and the two are retrieved together. Methane is also saturated in crude oil, and volatilizes as the oil is exposed to the atmosphere at various points along the system. Emissions of CH_4 from the components of petroleum systems—including crude oil production, crude oil refining, transportation, and distribution—generally occur as a result of system leaks, disruptions, and routine maintenance. In 1999, emissions from petroleum systems were estimated to be 21.9 Tg CO_2 Eq., or just less than 4 percent of U.S. CH_4 emissions.

From 1990 to 1999, combined CH_4 emissions from natural gas and petroleum systems decreased by 3 percent. Emissions from natural gas systems have remained fairly constant, while emissions from petroleum systems have declined gradually since 1990, primarily due to production declines.

Coal Mining

Produced millions of years ago during the formation of coal, CH_4 trapped within coal seams and surrounding rock strata is released when the coal is mined. The quantity of CH_4 released to the atmosphere during coal mining operations depends primarily upon the depth and type of the coal that is mined.

Methane from surface mines is emitted directly to the atmosphere as the rock strata overlying the coal seam are removed. Because CH_4 in underground mines is explosive at concentrations of 5 to 15 percent in air, most active underground mines are required to vent this CH_4. At some mines, CH_4 recovery systems may supplement these ventilation systems. Recovery of CH_4 in the United States has increased in recent years. During 1999, coal mining activities emitted 61.8 Tg CO2 Eq. of CH_4, or 10 percent of U.S. CH_4 emissions. From 1990 to 1999, emissions from this source decreased by 30 percent due, in part, to increased use of the CH_4 collected by mine degasification systems.

Agriculture

Agriculture accounted for 28 percent of U.S. CH_4 emissions in 1999, with enteric fermentation in domestic livestock, manure management, and rice cultivation representing the majority. Agricultural waste burning also con-

tributed to CH_4 emissions from agricultural activities.

Enteric Fermentation
(127.2 Tg CO$_2$ Eq.)

During animal digestion, CH_4 is produced through the process of enteric fermentation, in which microbes residing in animal digestive systems break down the feed consumed by the animal. Ruminants, which include cattle, buffalo, sheep, and goats, have the highest CH_4 emissions among all animal types because they have a rumen, or large fore-stomach, in which CH_4-producing fermentation occurs. Nonruminant domestic animals, such as pigs and horses, have much lower CH_4 emissions.

In 1999, enteric fermentation was the source of about 21 percent of U.S. CH_4 emissions, and more than half of the CH_4 emissions from agriculture. From 1990 to 1999, emissions from this source decreased by 2 percent. Emissions from enteric fermentation have been generally decreasing since 1995, primarily due to declining dairy cow and beef cattle populations.

Manure Management
(34.4 Tg CO$_2$ Eq.)

The decomposition of organic animal waste in an anaerobic environment produces CH_4. The most important factor affecting the amount of CH_4 produced is how the manure is managed, because certain types of storage and treatment systems promote an oxygen-free environment. In particular, liquid systems tend to encourage anaerobic conditions and produce significant quantities of CH_4, whereas solid waste management approaches produce little or no CH_4. Higher temperatures and moist climate conditions also promote CH_4 production.

Emissions from manure management were about 6 percent of U.S. CH_4 emissions in 1999, and 20 percent of the CH_4 emissions from agriculture. From 1990 to 1999, emissions from this source increased by 8.0 Tg CO_2 Eq.—the largest absolute increase of all the CH_4 source categories. The bulk of this increase was from swine and dairy cow

manure, and is attributed to the shift in the composition of the swine and dairy industries toward larger facilities. Larger swine and dairy farms tend to use liquid management systems.

Rice Cultivation
(10.7 Tg CO$_2$ Eq.)

Most of the world's rice, and all of the rice in the United States, is grown on flooded fields. When fields are flooded, anaerobic conditions develop and the organic matter in the soil decomposes, releasing CH_4 to the atmosphere, primarily through the rice plants.

In 1999, rice cultivation was the source of 2 percent of U.S. CH_4 emissions, and about 6 percent of U.S. CH_4 emissions from agriculture. Emission estimates from this source have increased by about 23 percent since 1990, due to an increase in the area harvested.

Agricultural Residue Burning
(0.6 Tg CO$_2$ Eq.)

Burning crop residue releases a number of greenhouse gases, including CH_4. Because field burning is not common in the United States, it was responsible for only 0.1 percent of U.S. CH_4 emissions in 1999.

Other Sources

Methane is also produced from several other sources in the United States, including wastewater treatment, fuel combustion, and some industrial processes. Methane emissions from domestic wastewater treatment totaled 12.2 Tg CO_2 Eq. in 1999. Stationary and mobile combustion were responsible for CH_4 emissions of 8.1 and 4.5 Tg CO_2 Eq., respectively. The majority of emissions from stationary combustion resulted from the burning of wood in the residential end-use sector. The combustion of gasoline in highway vehicles was responsible for the majority of the CH_4 emitted from mobile combustion. Methane emissions from two industrial sources—petrochemical and silicon carbide production—were also estimated, totaling 1.7 Tg CO_2 Eq.

NITROUS OXIDE EMISSIONS

Nitrous oxide (N_2O) is a greenhouse gas that is produced both naturally, from a wide variety of biological sources in soil and water, and anthropogenically by a variety of agricultural, energy-related, industrial, and waste management activities. While total N_2O emissions are much smaller than CO_2 emissions, N_2O is approximately 310 times more powerful than CO_2 at trapping heat in the atmosphere (IPCC 1996b).

During the past two centuries, atmospheric concentrations of N_2O have risen by approximately 13 percent. The main anthropogenic activities producing N_2O in the United States are agricultural soil management, fuel combustion in motor vehicles, and adipic and nitric acid production processes (see Figure 3-17 and Table 3-10).

Agricultural Soil Management

Nitrous oxide is produced naturally in soils through microbial processes of nitrification and denitrification. A number of anthropogenic activities add to the amount of nitrogen available to be emitted as N_2O by these microbial processes. These activities may add nitrogen to soils either directly or indirectly. Direct additions occur through the application of synthetic and organic fertilizers; production of nitrogen-fixing crops; the application of livestock manure, crop residues, and sewage sludge; cultivation of high-organic-content soils; and direct excretion by animals onto soil. Indirect additions result from volatilization and subsequent atmospheric deposition, and from leaching and surface runoff of some of the nitrogen applied to soils as fertilizer, livestock manure, and sewage sludge.

In 1999, agricultural soil management accounted for 298.3 Tg CO_2 Eq., or 69 percent, of U.S. N_2O emissions. From 1990 to 1999, emissions from this source grew by 11 percent as fertilizer consumption, manure production, and crop production increased.

Fuel Combustion

Nitrous oxide is a product of the reaction that occurs between nitrogen and oxygen during fuel combustion. Both mobile and stationary combustion emit N_2O. The quantity emitted varies according to the type of fuel, technology, and pollution control device used, as well as maintenance and operating practices. For example, catalytic converters installed to reduce motor vehicle pollution can result in the formation of N_2O.

In 1999, N_2O emissions from mobile combustion totaled 63.4 Tg CO_2 Eq., or 15 percent of U.S. N_2O emissions. Emissions of N_2O from stationary combustion were 15.7 Tg CO_2 Eq., or 4 percent of U.S. N_2O emissions. From 1990 to 1999, combined N_2O emissions from stationary and mobile combustion increased by 16 percent, primarily due to increased rates of N_2O generation in motor vehicles.

Nitric Acid Production

Nitric acid production is an industrial source of N_2O emissions. Used primarily to make synthetic commercial fertilizer, this raw material is also a major component in the production of adipic acid and explosives.

Virtually all of the nitric acid manufactured in the United States is produced by the oxidation of ammonia, during which N_2O is formed and emitted to the atmosphere. In 1999, N_2O emissions

FIGURE 3-17 AND TABLE 3-10 U.S. Sources of Nitrous Oxide Emissions (Tg CO_2 Eq.)

Nitrous oxide accounted for 6 percent of total U.S. greenhouse gas emissions in 1999, and agricultural soil management represented 69 percent of total N_2O emissions.

Source	1990	1995	1996	1997	1998	1999
Agricultural Soil Management	269.0	285.4	294.6	299.8	300.3	298.3
Mobile Combustion	54.3	66.8	65.3	65.2	64.2	63.4
Nitric Acid	17.8	19.9	20.7	21.2	20.9	20.2
Manure Management	16.0	16.4	16.8	17.1	17.2	17.2
Stationary Combustion	13.6	14.3	14.9	15.0	15.1	15.7
Adipic Acid	18.3	20.3	20.8	17.1	7.3	9.0
Human Sewage	7.1	8.2	7.8	7.9	8.1	8.2
Agricultural Residue Burning	0.4	0.4	0.4	0.4	0.5	0.4
Waste Combustion	0.3	0.3	0.3	0.3	0.2	0.2
International Bunker Fuels*	1.0	0.9	0.9	1.0	1.0	1.0
Total*	**396.9**	**431.9**	**441.6**	**444.1**	**433.7**	**432.6**

* Emissions from international bunker fuels are not included in totals.

Note: Totals may not sum due to independent rounding.

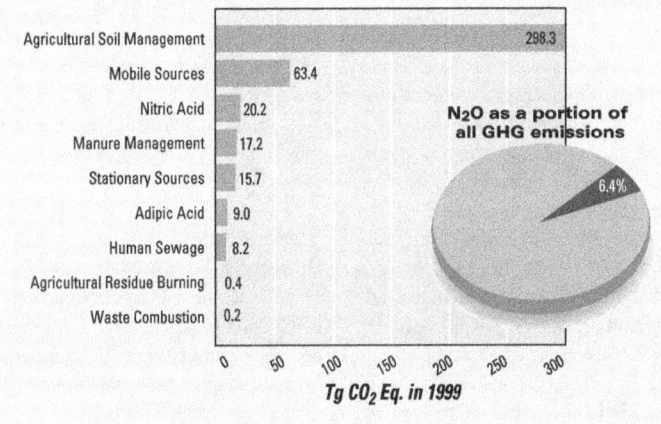

N2O as a portion of all GHG emissions

6.4%

Agricultural Soil Management — 298.3
Mobile Sources — 63.4
Nitric Acid — 20.2
Manure Management — 17.2
Stationary Sources — 15.7
Adipic Acid — 9.0
Human Sewage — 8.2
Agricultural Residue Burning — 0.4
Waste Combustion — 0.2

Tg CO_2 Eq. in 1999

from nitric acid production were 20.2 Tg CO_2 Eq., or 5 percent of U.S. N_2O emissions. From 1990 to 1999, emissions from this source category increased by 13 percent as nitric acid production grew.

Manure Management

Nitrous oxide is produced as part of microbial nitrification and denitrification processes in managed and unmanaged manure, the latter of which is addressed under agricultural soil management. Total N_2O emissions from managed manure systems in 1999 were 17.2 Tg CO_2 Eq., accounting for 4 percent of U.S. N_2O emissions. From 1990 to 1999, emissions from this source category increased by 7 percent, as poultry and swine populations have increased.

Adipic Acid Production

Most adipic acid produced in the United States is used to manufacture nylon 6,6. Adipic acid is also used to produce some low-temperature lubricants and to add a "tangy" flavor to foods. Nitrous oxide is emitted as a by-product of the chemical synthesis of adipic acid.

In 1999, U.S. adipic acid plants emitted 9.0 Tg CO_2 Eq. of N_2O, or 2 percent of U.S. N_2O emissions. Even though adipic acid production has increased, by 1998 all three major adipic acid plants in the United States had voluntarily implemented N_2O abatement technology. As a result, emissions have decreased by 51 percent since 1990.

Other Sources

Other sources of N_2O included agricultural residue burning, waste combustion, and human sewage in wastewater treatment systems. In 1999, agricultural residue burning and municipal solid waste combustion each emitted less than 1 Tg CO_2 Eq. of N_2O. The human sewage component of domestic wastewater resulted in emissions of 8.2 Tg CO_2 Eq. in 1999.

HFC, PFC, AND SF$_6$ EMISSIONS

Hydrofluorocarbons (HFCs) and perfluorocarbons (PFCs) are categories of synthetic chemicals that are being used as alternatives to the ozone-depleting substances (ODSs) being phased out under the *Montreal Protocol* and Clean Air Act Amendments of 1990. Because HFCs and PFCs do not directly deplete the stratospheric ozone layer, they are not controlled by the *Montreal Protocol*.

These compounds, however, along with sulfur hexafluoride (SF_6), are potent greenhouse gases. In addition to having high global warming potentials, SF_6 and PFCs have extremely long atmospheric lifetimes, resulting in their essentially irreversible accumulation in the atmosphere. Sulfur hexafluoride is the most potent greenhouse gas the IPCC has evaluated.

Other emissive sources of these gases include aluminum production, HCFC-22 production, semiconductor manufacturing, electrical transmission and distribution systems, and magnesium production and processing. Figure 3-18 and Table 3-11 present emission estimates for HFCs, PFCs, and SF_6, which totaled 135.7 Tg CO_2 Eq. in 1999.

Substitution of Ozone-Depleting Substances

The use and subsequent emissions of HFCs and PFCs as substitutes for ozone-depleting substances (ODSs) increased from small amounts in 1990 to 56.7 Tg CO_2 Eq. in 1999. This increase was the result of efforts to phase out CFCs and other ODSs in the United States, especially the introduction of HFC-134a as a CFC substitute in refrigeration applications. In the short term, this trend is expected to continue, and will most likely accelerate in the next decade as HCFCs, which are interim substitutes in many applications, are themselves phased out under the provisions of the Copenhagen Amendments to the *Montreal Protocol*. Improvements in the technologies associated with the use of these gases, however, may help to offset this anticipated increase in emissions.

Other Industrial Sources

HFCs, PFCs, and SF_6 are also emitted from a number of other industrial processes. During the production of primary aluminum, two PFCs—CF_4 and C_2F_6—are emitted as intermittent by-products of the smelting process. Emissions from aluminum production, which totaled 10.0 Tg CO_2 Eq., were estimated to have decreased by 48 percent between 1990 and 1999 due to voluntary emission reduction efforts by the industry and falling domestic aluminum production.

HFC-23 is a by-product emitted during the production of HCFC-22. Emissions from this source were 30.4 Tg CO_2 Eq. in 1999, and have decreased by 13 percent since 1990. The intensity of HFC-23 emissions (i.e., the amount of HFC-23 emitted per kilogram of HCFC-22 manufactured) has declined significantly since 1990, although production has been increasing.

The semiconductor industry uses combinations of HFCs, PFCs, SF_6, and other gases for plasma etching and to clean chemical vapor deposition tools. For 1999, it was estimated that the U.S. semiconductor industry emitted a total of 6.8 Tg CO_2 Eq. Emissions from this source category have increased with the growth in the semiconductor industry and the rising intricacy of chip designs.

The primary use of SF_6 is as a dielectric in electrical transmission and distribution systems. Fugitive emissions of SF_6 occur from leaks in and servicing of substations and circuit breakers, especially from older equipment. Estimated emissions from this source increased by 25 percent since 1990, to 25.7 Tg CO_2 Eq. in 1999.

Finally, SF_6 is also used as a protective cover gas for the casting of molten magnesium. Estimated emissions from primary magnesium production and magnesium casting were 6.1 Tg CO_2 Eq. in 1999, an increase of 11 percent since 1990.

Emissions of Ozone-Depleting Substances

Halogenated compounds were first emitted into the atmosphere during the 20th century. This family of manmade compounds includes chlorofluorocarbons (CFCs), halons, methyl chloroform, carbon tetrachloride, methyl

FIGURE 3-18 AND TABLE 3-11 U.S. Sources of HFC, PFC, and SF$_6$ Emissions (Tg CO$_2$ Eq.)

HFCs, PFCs, and SF$_6$ accounted for 2 percent of total U.S. greenhouse gas emissions in 1999, and substitutes for ozone-depleting substances comprised 42 percent of all HFC, PFC, and SF$_6$ emissions.

Source	1990	1995	1996	1997	1998	1999
Substitution of Ozone-Depleting Substances	0.9	24.0	34.0	42.1	49.6	56.7
HCFC-22 Production	34.8	27.1	31.2	30.1	40.0	30.4
Electrical Transmission and Distribution	20.5	25.7	25.7	25.7	25.7	25.7
Aluminum Production	19.3	11.2	11.6	10.8	10.1	10.0
Semiconductor Manufacture	2.9	5.5	7.0	7.0	6.8	6.8
Magnesium Production and Processing	5.5	5.5	5.6	7.5	6.3	6.1
Total	**83.9**	**99.0**	**115.1**	**123.3**	**138.6**	**135.7**

Note: Totals may not sum due to independent rounding.

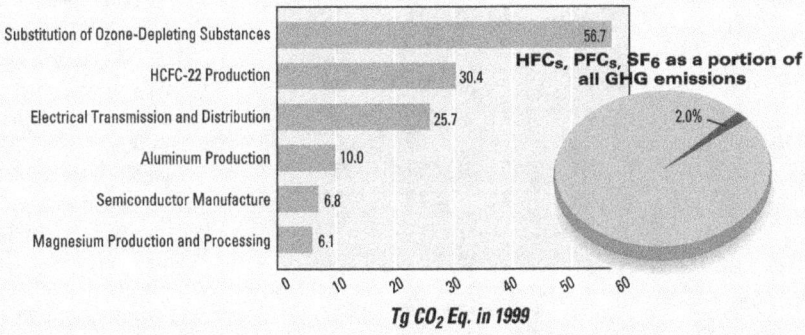

Tg CO$_2$ Eq. in 1999

bromide, and hydrochlorofluorocarbons (HCFCs). These substances have been used in a variety of industrial applications, including refrigeration, air conditioning, foam blowing, solvent cleaning, sterilization, fire extinguishing, coatings, paints, and aerosols.

Because these compounds have been shown to deplete stratospheric ozone, they are typically referred to as ozone-depleting substances (ODSs). However, they are also potent greenhouse gases.

Recognizing the harmful effects of these compounds on the ozone layer, 181 countries have ratified the *Montreal Protocol on Substances That Deplete the Ozone Layer* to limit the production and importation of a number of CFCs and other halogenated compounds. The United States furthered its commitment to phase out ODSs by signing and ratifying the Copenhagen Amendments to the *Montreal Protocol* in 1992. Under these amendments, the United States committed to ending the production and importation of halons by 1994, and CFCs by 1996.

The IPCC Guidelines and the United Nations Framework Convention on Climate Change do not include reporting instructions for estimating emissions of ODSs because their use is being phased out under the *Montreal Protocol*. The United States believes, however, that a greenhouse gas emissions inventory is incomplete without these emissions; therefore, estimates for several Class I and Class II ODSs are provided in Table 3-12. Compounds are grouped by class according to their ozone-depleting potential. Class I compounds are the primary ODSs; Class II compounds include partially halogenated chlorine compounds (i.e., HCFCs), some of which were developed as interim replacements for CFCs. Because these HCFC compounds are only partially halogenated, their hydrogen-carbon bonds are more vulnerable to oxidation in the troposphere and, therefore, pose only one-tenth to one-hundredth the threat to stratospheric ozone compared to CFCs.

It should be noted that the effects of these compounds on radiative forcing are not provided. Although many ODSs have relatively high direct global warming potentials, their indirect effects from ozone (also a greenhouse gas) destruction are believed to have negative radiative-forcing effects and, therefore, could significantly reduce the overall magnitude of their radiative-forcing effects. Given the uncertainties about the net effect of these gases, emissions are reported on an unweighted basis.

TABLE 3-12 Emissions of Ozone-Depleting Substances (Gigagrams)

Many ozone-depleting substances have relatively high direct global warming potentials. However, their indirect effects from ozone (also a greenhouse gas) destruction are believed to have negative radiative-forcing effects and, therefore, could significantly reduce the overall magnitude of their radiative-forcing effects. Given the uncertainties about the net effect of these gases, emissions are reported on an unweighted basis.

Compound	1990	1995	1996	1997	1998	1999
Class I						
CFC-11	52.4	19.1	11.7	10.7	9.8	9.2
CFC-12	226.9	71.1	72.2	63.6	54.9	64.4
CFC-113	39.0	7.6	+	+	+	+
CFC-114	0.7	0.8	0.8	0.8	0.6	+
CFC-115	2.2	1.6	1.6	1.4	1.1	1.1
Carbon Tetrachloride	25.1	5.5	+	+	+	+
Methyl Chloroform	27.9	8.7	1.6	+	+	+
Halon-1211	+	0.7	0.8	0.8	0.8	0.8
Halon-1301	1.0	1.8	1.9	1.9	1.9	1.9
Class II						
HCFC-22	33.9	46.2	48.8	50.6	52.3	83.0
HCFC-123	+	0.6	0.7	0.8	0.9	1.0
HCFC-124	+	5.6	5.9	6.2	6.4	6.5
HCFC-141b	+	20.6	25.4	25.1	26.7	28.7
HCFC-142b	+	7.3	8.3	8.7	9.0	9.5
HCFC-225ca/cb	+	+	+	+	+	+

+ Does not exceed 0.05 gigagrams.
Source: EPA estimates.

Sources and Effects of Sulfur Dioxide

Sulfur dioxide emitted into the atmosphere through natural and anthropogenic processes affects the Earth's radiative budget through its photochemical transformation into sulfate aerosols that can (1) scatter sunlight back to space, thereby reducing the radiation reaching the Earth's surface; (2) affect cloud formation; and (3) affect atmospheric chemical composition by providing surfaces for heterogeneous chemical reactions. The overall effect of SO_2-derived aerosols on radiative forcing is negative (IPCC 2001b). However, because SO_2 is short-lived and unevenly distributed in the atmosphere, its radiative-forcing impacts are highly uncertain.

Sulfur dioxide is also a contributor to the formation of urban smog, which can cause significant increases in acute and chronic respiratory diseases. Once SO_2 is emitted, it is chemically transformed in the atmosphere and returns to the Earth as the primary source of acid rain. Because of these harmful effects, the United States has regulated SO_2 emissions in the Clean Air Act.

Electric utilities are the largest source of SO_2 emissions in the United States, accounting for 67 percent in 1999. Coal combustion contributes nearly all of those emissions (approximately 93 percent). Emissions of SO_2 have decreased in recent years, primarily as a result of electric utilities switching from high-sulfur to low-sulfur coal and use of flue gas desulfurization.

CRITERIA POLLUTANT EMISSIONS

In the United States, carbon monoxide (CO), nitrogen oxides (NO_x), nonmethane volatile organic compounds (NMVOCs), and sulfur dioxide (SO_2) are commonly referred to as "criteria pollutants," as termed in the Clean Air Act. Criteria pollutants do not have a direct global warming effect, but indirectly affect terrestrial radiation absorption by influencing the formation and destruction of tropospheric and stratospheric ozone, or, in the case of SO_2, by affecting the absorptive characteristics of the atmosphere.

Carbon monoxide is produced when carbon-containing fuels are combusted incompletely. Nitrogen oxides (i.e., NO and NO_x) are created by lightning, fires, and fossil fuel combustion and are created in the stratosphere from nitrous oxide (N_2O). NMVOCs—which include such compounds as propane, butane, and ethane—are emitted primarily from transportation, industrial processes, and nonindustrial consumption of organic solvents. In the United States, SO_2 is primarily emitted from the combustion of coal by the electric power industry and by the metals industry.

In part because of their contribution to the formation of urban smog—and acid rain in the case of SO_2 and NO_x—criteria pollutants are regulated under the Clean Air Act. These gases also indirectly affect the global climate by reacting with other chemical compounds in the atmosphere to form compounds that are greenhouse gases. Unlike other criteria pollutants, SO_2 emitted into the atmosphere is believed to affect the Earth's radiative budget negatively; therefore, it is discussed separately.

One of the most important indirect climate change effects of NO_x and NMVOCs is their role as precursors for tropospheric ozone formation. They can also alter the atmospheric lifetimes of other greenhouse gases. For example, CO interacts with the hydroxyl radical—the major atmospheric sink for CH_4 emissions—to form CO_2.

TABLE 3-13 Emissions of NO$_x$, CO, NMVOCs, and SO$_2$ (Gg)

Fuel combustion accounts for the majority of emissions of criteria pollutants. Industrial processes—such as the manufacture of chemical and allied products, metals processing, and industrial uses of solvents—are also significant sources of CO, NO$_X$, and NMVOCs.

Gas/Activity	1990	1995	1996	1997	1998	1999
NO$_x$	21,955	22,755	23,663	23,934	23,613	23,042
Stationary Fossil Fuel Combustion	9,884	9,822	9,541	9,589	9,408	9,070
Mobile Fossil Fuel Combustion	10,900	11,870	12,893	13,095	13,021	12,794
Oil and Gas Activities	139	100	126	130	130	130
Industrial Processes	921	842	977	992	924	930
Solvent Use	1	3	3	3	3	3
Agricultural Burning	28	28	32	33	34	33
Waste	83	89	92	92	93	83
CO	85,846	80,678	87,196	87,012	82,496	82,982
Stationary Fossil Fuel Combustion	4,999	5,383	5,620	4,968	4,575	4,798
Mobile Fossil Fuel Combustion	69,523	68,072	72,390	71,225	70,288	68,179
Oil and Gas Activities	302	316	321	333	332	332
Industrial Processes	9,502	5,291	7,227	8,831	5,612	5,604
Solvent Use	4	5	1	1	1	1
Agricultural Burning	537	536	625	630	653	629
Waste	979	1,075	1,012	1,024	1,035	3,439
NMVOCs	18,843	18,663	17,353	17,586	16,554	16,128
Stationary Fossil Fuel Combustion	912	973	971	848	778	820
Mobile Fossil Fuel Combustion	8,154	7,725	8,251	8,023	7,928	7,736
Oil and Gas Activities	555	582	433	442	440	385
Industrial Processes	3,110	2,805	2,354	2,793	2,352	2,281
Solvent Use	5,217	5,609	4,963	5,098	4,668	4,376
Agricultural Burning	NA	NA	NA	NA	NA	NA
Waste	895	969	381	382	387	531
SO$_2$	21,481	17,408	17,109	17,565	17,682	17,115
Stationary Fossil Fuel Combustion	18,407	14,724	14,727	15,106	15,192	14,598
Mobile Fossil Fuel Combustion	1,339	1,189	1,081	1,116	1,145	1,178
Oil and Gas Activities	390	334	304	312	310	309
Industrial Processes	1,306	1,117	958	993	996	996
Solvent Use	0	1	1	1	1	1
Agricultural Burning	NA	NA	NA	NA	NA	NA
Waste	38	43	37	37	38	33

+ Does not exceed 0.5 gigagrams.
NA = Not Available.
Note: Totals may not sum due to independent rounding.
Source: EPA 2000, except for estimates from agricultural residue burning.

Therefore, increased atmospheric concentrations of CO limit the number of hydroxyl molecules (OH) available to destroy CH$_4$.

Since 1970, the United States has published estimates of annual emissions of criteria pollutants (U.S. EPA 2000).[16] Table 3-13 shows that fuel combustion accounts for the majority of emissions of these gases. Industrial processes, such as the manufacture of chemical and allied products, metals processing, and industrial uses of solvents, are also significant sources of CO, NO$_x$, and NMVOCs.

---

[16] NO$_x$ and CO emission estimates from agricultural residue burning were estimated separately and, therefore, not taken from U.S. EPA 2000.

Chapter 4
Policies
and Measures

In the past decade, the United States has made significant progress in reducing greenhouse gas emissions. In 2000 alone, U.S. climate change programs reduced the growth in greenhouse gas emissions by 242 teragrams of carbon dioxide equivalent[1] (Tg CO_2 Eq.) (see Table 4-1 at the end of this chapter). They have also significantly helped the United States reduce carbon intensity, which is the amount of CO_2 emitted per unit of gross domestic product .

While many policies and measures developed in the 1990s continue to achieve their goals, recent changes in the economy and in energy markets, coupled with the introduction of new science and technology, create a need to re-evaluate existing climate change programs to ensure they effectively meet future economic, climate, and other environmental

---

[1] Emissions are expressed in units of CO_2 equivalents for consistency in international reporting under the United Nations Framework Convention on Climate Change. One teragram is equal to one million metric tons.

goals. Our experience with greenhouse gas emissions highlights the importance of creating climate policy within the context of the overall economy, changing energy markets, technology development and deployment, and R&D priorities. Because global warming is a long-term problem, solutions need to be long lasting.

The U.S. government is currently pursuing a broad range of strategies to reduce net emissions of greenhouse gases. In addition, businesses, state and local governments, and nongovernmental organizations (NGOs) are addressing global climate change by improving the measurement and reporting of greenhouse gas emission reductions; by voluntarily reducing emissions, including using emission trading systems; and by sequestering carbon through tree planting and forest preservation, restoration, conversion of eroding cropland to permanent cover, and soil management.

NATIONAL POLICYMAKING PROCESS

Shortly after taking office in January 2001, President Bush directed a Cabinet-level review of U.S. climate change policy and programs. The President established working groups and requested them to develop innovative approaches that would:

- be consistent with the goal of stabilizing greenhouse gas concentrations in the atmosphere;
- be sufficiently flexible to allow for new findings;
- support continued economic growth and prosperity;
- provide market-based incentives;
- incorporate technological advances; and
- promote global participation.

Members of the Cabinet, the Vice President, and senior White House staff extensively reviewed and discussed climate science, existing technologies to reduce greenhouse gases and sequester carbon, current U.S. programs and policies, and innovative options for addressing concentrations of greenhouse gases in the atmosphere. They were assisted by a number of scientific, technical, and policy experts from the federal government, national laboratories, universities, NGOs, and the private sector. To obtain the most recent information and a balanced view of the current state of climate change science, the Cabinet group asked the National Academy of Sciences (NAS) to issue a report addressing areas of scientific consensus and significant gaps in our climate change knowledge (NRC 2001a). Appendix D of this report presents key questions posed by the Committee on the Science of Climate Change, along with the U.S. National Research Council's responses.

On June 11, 2001, the President issued the interim report of the Cabinet-level review (EOP 2001). Based on the NAS report (NRC 2001a) and the Cabinet's findings, President Bush directed the Department of Commerce, working with other federal agencies, to set priorities for additional investments in climate change research, to review such investments, and to maximize coordination among federal agencies to advance the science of climate change. The President is committed to fully funding all priority research areas that the review finds are underfunded or need to be accelerated relative to other research. Such areas could include the carbon and global water cycles and climate modeling.

The President further directed the Secretaries of Commerce and Energy, working with other federal agencies, to develop a National Climate Change Technology Initiative with the following major objectives:

- Evaluate the current state of U.S. climate change technology R&D and make recommendations for improvements.
- Develop opportunities to enhance private–public partnerships in applied R&D to expedite innovative

U.S. Strategies in Key Sectors to Reduce Net Emissions of Greenhouse Gases

The U.S. government is currently pursuing a broad range of strategies to reduce net emissions of greenhouse gases.

Electricity
Federal programs promote greenhouse gas reductions through the development of cleaner, more efficient technologies for electricity generation and transmission. The government also supports the development of renewable resources, such as solar energy, wind power, geothermal energy, hydropower, bioenergy, and hydrogen fuels.

Transportation
Federal programs promote development of fuel-efficient motor vehicles and trucks, research and development options for producing cleaner fuels, and implementation of programs to reduce the number of vehicle miles traveled.

Industry
Federal programs implement partnership programs with industry to reduce emissions of carbon dioxide (CO_2) and other greenhouse gases, promote source reduction and recycling, and increase the use of combined heat and power.

Buildings
Federal voluntary partnership programs promote energy efficiency in the nation's commercial, residential, and government buildings (including schools) by offering technical assistance as well as the labeling of efficient products, new homes, and office buildings.

Agriculture and Forestry
The U.S. government implements conservation programs that have the benefit of reducing agricultural emissions, sequestering carbon in soils, and offsetting overall greenhouse gas emissions.

Federal Government
The U.S. government has taken steps to reduce greenhouse gas emissions from energy use in federal buildings and in the federal transportation fleet.

Highlights of U.S. Climate Change Programs

Many U.S. climate change programs have been highly successful at stimulating participation and achieving measurable energy and cost savings, as well as reducing greenhouse gas emissions.*

- Minimum efficiency standards on residential appliances have saved consumers nearly $25 billion through 1999, avoiding cumulative emissions by an amount equal to almost 180 Tg CO_2 Eq. Four pending appliance standards (clothes washers, fluorescent light ballasts, water heaters, and central air conditioners) are projected to save consumers up to $10 billion and reduce cumulative emissions by as much as 80 Tg CO_2 Eq. through 2010.

- The ENERGY STAR® program promotes energy efficiency in U.S. homes and commercial buildings. It has reduced greenhouse gas emissions by more than 55 Tg CO_2 Eq. in 2000 alone, and is projected to increase this amount to about 160 Tg CO_2 Eq. a year by 2010.

- Public–private partnership programs have contributed to the decline in methane emissions since 1990, and are expected to hold emissions at or below 1990 levels through 2010 and beyond. Partners in the methane programs have reduced methane emissions by about 35 Tg CO_2 Eq. in 2000 and are projected to reduce emissions by 55 Tg CO_2 Eq. annually by 2010.

- Programs designed to halt the growth in emissions of the most potent greenhouse gases—the "high global warming potential (GWP) gases"—are achieving significant progress. These programs reduced high-GWP emissions by more than 70 Tg CO_2 Eq. in 2000 and are projected to reduce emissions by more than 280 Tg CO_2 Eq. annually by 2010.

- Federal, state, and local outreach has allocated nearly $10 million in grants and other awards since 1992 for 41 state greenhouse gas emission inventories, 27 state action plans, 16 demonstration projects, and 32 educational and outreach programs. Across the nation, 110 cities and counties, representing approximately 44 million people, are developing inventories and implementing climate change action plans.

* There is uncertainty in any attempt to project future emission levels and program impacts from what would have happened in the absence of these programs. These projections represent a best estimate. They are also based on the assumption that programs will continue to be funded at current funding levels.

and cost-effective approaches to reduce greenhouse gas emissions and the buildup of greenhouse gas concentrations in the atmosphere.

- Make recommendations for funding demonstration projects for cutting-edge technologies.
- Provide guidance on strengthening basic research at universities and national laboratories, including the development of the advanced mitigation technologies that offer the greatest promise for low-cost reductions of greenhouse gas emissions and global warming potential.
- Make recommendations to enhance coordination across federal agencies, and among the federal government, universities, and the private sector.
- Make recommendations for developing improved technologies for

measuring and monitoring gross and net greenhouse gas emissions.

Simultaneous with the President's climate change policy development is the implementation of the May 2001 *National Energy Policy* (NEPD Group 2001).[2] Developed under the leadership of Vice President Cheney, the *National Energy Policy* is a long-term, comprehensive strategy to advance the development of new, environmentally friendly technologies to increase energy supplies and encourage cleaner, more efficient energy use

The *National Energy Policy* identified a number of major energy challenges and contains 105 specific recommendations for dealing with them, many of which affect greenhouse gas emissions. For example, it promotes energy efficiency by calling for the intelligent use of

new technologies and information dissemination; confronts our increasing dependency on foreign sources of energy by calling for increased domestic production with advanced technologies; and addresses our increasing reliance on natural gas by paving the way for a greater balance among many energy sources, including renewable energy but also traditional sources, such as hydropower and nuclear energy. In addition, the *National Energy Policy* initiated a comprehensive technology review to re-prioritize energy R&D. The review, which is currently underway, is critically evaluating the research, development, demonstration, and deployment portfolio for energy efficiency, renewable energy, and alternative energy technologies as they apply to the buildings, transportation, industry, power generation, and government sectors.

FEDERAL POLICIES AND MEASURES

The United States recognizes that climate change is a serious problem, and has devoted significant resources to climate change programs and activities (Table 4-2). This section summarizes the progress of existing federal climate change programs, including new policies and measures not described in the 1997 *U.S. Climate Action Report* (CAR), and actions described in the 1997 CAR that are no longer ongoing (U.S. DOS 1997). Many of these programs have evolved substantially since the 1997 CAR, which is reflected in individual program descriptions. Although originally focused on important early reductions by the year 2000, the programs are building emission reductions that grow over time and provide even larger benefits in later years. In addition, the Cabinet-level climate change working group is continuing its review and is developing other approaches to reduce greenhouse gas emissions, including those that tap the power of the markets, help realize the promise of technology, and ensure the broadest possible global participation.

---

[2] Available at http://www.whitehouse.gov/energy

TABLE 4-2 Summary of Federal Climate Change Expenditures: 1999–2001
(Millions of Dollars)

Since the 1997 *U.S. Climate Action Report*, the United States has reassessed ongoing activities for their direct and indirect impacts on greenhouse emissions. This table summarizes the funding across a portion of these activities, which includes a range of research and development on energy efficiency and renewable energy, as well as setting efficiency standards.

Types of Programs	FY 1999	FY 2000	FY 2001 Estimate
Directly Related Programs & Policies	1,009	1,095	1,239
Other Climate-Related Programs	685	698	946
Total	**1,694**	**1,793**	**2,185**

Note: Funding for the U.S. Global Change Research Program, International Assistance, and the Global Environment Facility is described in later chapters and is not included in this total.

Source: OMB 2001.

Federal partnership programs promote improved energy efficiency and increased use of renewable energy technologies in the nation's commercial, residential, and government buildings (including schools) by offering technical assistance as well as the labeling of efficient products, efficient new homes, and efficient buildings. The U.S. government is implementing a number of partnership programs with industry to reduce CO_2 emissions, increase the use of combined heat and power, and promote the development of cleaner, more efficient technologies for electricity generation and transmission. The federal government is also supporting renewable resources, such as solar energy, wind power, geothermal energy, hydropower, bioenergy, and hydrogen fuels. In addition, the U.S. government's commitment to advanced research and development in the areas of energy efficiency, renewable energy, alternative energy technologies, and nuclear energy will play a central role in an effective long-term response to climate change.

Energy: Residential and Commercial

Residential and commercial buildings account for approximately 35 percent of U.S. CO_2 emissions from energy use. Electricity consumption for lighting, heating, cooling, and operating appliances accounts for the majority of these emissions. Many commercial buildings and new homes could effectively operate with 30 percent less energy if owners made investments in energy-efficient products, technologies, and best management practices. Federal partnership programs promote these investments through a market-based approach, using labeling to clearly identify which products, practices, new homes, and buildings are energy efficient. The United States also funds significant research on developing highly efficient building equipment and appliances. Following are descriptions of some of the key policies and measures in this area.

ENERGY STAR® for the Commercial Market

This program has evolved substantially since the last CAR. Its major focus now is on promoting high-performing (high-efficiency) buildings and providing decision makers throughout an organization with the information they need to undertake effective building improvement projects. While the partnership continues to work with more than 5,500 organizations across the country, this program also introduced a system in 1999 that allows the benchmarking of building energy performance against the national stock of buildings. As recommended in the *National Energy Policy*, this system is being expanded to represent additional major U.S. building types, such as schools (K–12), grocery stores, hotels,

hospitals, and warehouses. By the end of 2001, more than 75 percent of U.S. building stock could use this system. The national building energy performance rating system also allows for recognizing the highest-performing buildings, which can earn the ENERGY STAR® label. EPA estimates that ENERGY STAR® in the commercial building sector provided 23 Tg CO_2 Eq. reductions in 2000, and projects it will provide 62 Tg CO_2 Eq. reductions by 2010.

Commercial Buildings Integration

This program continues to work to realize energy-saving opportunities provided by the whole-building approach during the construction and major renovation of existing commercial buildings. The program is increasing its industry partnerships in design, construction, operation and maintenance, indoor environment, and control and diagnostics of heating, ventilation, air conditioning, lighting, and other building systems. Through these efforts, the Department of Energy (DOE) helps transfer the most energy-efficient building techniques and practices into commercial buildings through regulatory activities, such as supporting the upgrade of voluntary (model) building energy codes and promulgating upgraded federal commercial building energy codes. The program consists of *Updating State Building Codes* and *Partnerships for Commercial Buildings and Facilities*, and is supported by a number of DOE programs, such as *Commercial Building R&D*.

ENERGY STAR® for the Residential Market

This program has expanded significantly since the last CAR when it was focused on new home construction. It now also provides guidance for homeowners on designing efficiency into kitchen, additions, and whole-home improvement projects and works with major retailers and other organizations to help educate the public. In addition, it offers a Web-based audit tool and a home energy benchmark tool to help the homeowner implement a project and monitor progress. Builders have constructed more than 55,000 ENERGY STAR®-labeled new homes in the United States, at a pace that has doubled each year. These homes are averaging energy savings of about 35 percent better than the model energy code. The Environmental Protection Agency's (EPA's) ENERGY STAR®-labeled homes and home improvement effort are expected to provide about 20 Tg CO_2 Eq. in emission reductions in 2010.

Community Energy Program: Rebuild America

This program continues to help communities, towns, and cities save energy, create jobs, promote growth, and protect the environment through improved energy efficiency and sustainable building design and operation. The centerpiece of this newly consolidated program is *Rebuild America*—a program that assists states and communities in developing and implementing environmentally and economically sound activities through smarter energy use. The program provides one-stop shopping for information and assistance on how to plan, finance, implement, and manage retrofit projects to improve energy efficiency. As of May 2001, *Rebuild America* formed 340 partnerships committed to performing energy retrofits, which are complete or underway on approximately 550 million square feet of building space in the 50 states and two U.S. territories.

Residential Building Integration: Building America

This program represents the consolidation of a number of initiatives. It works with industry to jointly fund, develop, demonstrate, and deploy housing that integrates energy-efficiency technologies and practices. The *Energy Partnerships for Affordable Housing* consolidates the formerly separate systems engineering programs of *Building America, Industrialized Housing, Passive Solar Buildings, Indoor Air Quality,* and existing building research into a comprehensive program. Systems integration research and development activities analyze building com-

ponents and systems and integrate them so that the overall building performance is greater than the sum of its parts. *Building America* is a private–public partnership that provides energy solutions for production housing and combines the knowledge and resources of industry leaders with DOE's technical capabilities to act as a catalyst for change in the home building industry.

Energy Star®-Labeled Products

The strategy of this program has evolved substantially since the last CAR, not only with the addition of new products to the ENERGY STAR® family, but also with expanded outreach to consumers in partnership with their local utility or similar organization. The ENERGY STAR® label has been expanded to more than 30 product categories and, as recommended in the President's *National Energy Policy*, EPA and DOE are currently working to expand the program to additional products and appliances. ENERGY STAR® works in partnership with utilities representing about 50 percent of U.S. energy customers. The ENERGY STAR® label is now recognized by more than 40 percent of U.S. consumers, who have purchased over 600 million ENERGY STAR® products. Due to the increased penetration of these energy-efficient products, EPA estimates that 33 Tg CO_2 Eq. of emissions were avoided in 2000 and projects that 75 Tg CO_2 Eq. will be reduced in 2010.

Building Equipment, Materials, and Tools

This program conducts R&D on building components and design tools and issues standards and test procedures for a variety of appliances and equipment. Sample building components that increase the energy efficiency of buildings and improve building performance include innovative lighting, advanced space conditioning and refrigeration, and fuel cells. The program also conducts R&D on building envelope technologies, such as advanced windows, coatings, and insulation. It is improving analytical tools that effectively integrate all elements

affecting building energy use and help building designers, owners, and operators develop the best design strategies for new and existing buildings.

Additional Policies and Measures

Additional ongoing policies and measures in the residential and commercial sector include *Residential Appliance Standards, State and Community Assistance (State Energy Program, Weatherization Assistance Program, Community Energy Grants, Information Outreach), Heat Island Reduction Initiative,* and *Economic Incentives/Tax Credits.* Appendix B provides detailed descriptions of policies and measures.

Two policies and measures listed as new initiatives in the 1997 CAR no longer appear as separate programs. *Expand Markets for Next-Generation Lighting Products* and *Construction of Energy-Efficient Buildings* have been incorporated into other existing climate programs at DOE and EPA.

Energy: Industrial

About 27 percent of U.S. CO_2 emissions result from industrial activities. The primary source of these emissions is the burning of carbon-based fuels, either on site in manufacturing plants or through the purchase of generated electricity. Many manufacturing processes use more energy than is necessary. The following programs help to improve industrial productivity by lowering energy costs, providing innovative manufacturing methods, and reducing waste and emissions.

Industries of the Future

This program continues to work in partnership with the nation's most energy-intensive industries, enhancing their long-term competitiveness and accelerating research, development, and deployment of technologies that increase energy and resource efficiency. Led by DOE, the program's strategy is being implemented in nine energy- and waste-intensive industries. Two key elements of the strategy include: (1) an industry-driven report outlining each industry's vision for the future, and (2) a technology roadmap to identify the technologies that will be needed to reach that industry's goals.

Best Practices Program

This program offers industry the tools to improve plant energy efficiency, enhance environmental performance, and increase productivity. Selected best-of-class large demonstration plants are showcased across the country, while other program activities encourage the replication of those best practices in still greater numbers of large plants.

ENERGY STAR® for Industry

This new initiative integrates and builds upon the *Climate Wise* program and offers a more comprehensive partnership for industrial companies. ENERGY STAR® will enable industrial companies to evaluate and cost-effectively reduce their energy use. Through established energy performance benchmarks, strategies for improving energy performance, technical assistance, and recognition for accomplishing reductions in energy, the partnership will contribute to a reduction in energy use for the U.S. industrial sector. EPA estimates that awareness focused by *Climate Wise* reduced emissions by 11 Tg CO_2 Eq. in 2000, and projects that ENERGY STAR®'s industrial partnerships will provide 16 Tg CO_2 Eq. reductions in 2010.

Additional Policies and Measures

Additional ongoing policies and measures in the industrial sector include *Industrial Assessment Centers, Enabling Technologies*, and *Financial Assistance: NICE*[3]. Appendix B provides detailed descriptions of policies and measures.

Energy: Supply

Electricity generation is responsible for about 41 percent of CO_2 emissions in the United States. Federal programs promote greenhouse gas reductions through the development of cleaner, more efficient technologies for electricity generation and transmission. The U.S. government is also supporting renewable resources, such as solar

energy, wind power, geothermal energy, hydropower, bioenergy, and hydrogen fuels, as well as traditional nonemitting sources, such as nuclear energy. DOE's development programs have been very successful in reducing technology implementation costs. The cost of producing photovoltaic modules has decreased by 50 percent since 1991, and the cost of wind power has decreased by 85 percent since 1980. Commercial success has been achieved for both of these areas in certain applications.

Renewable Energy Commercialization

This program consists of several programs to develop clean, competitive renewable energy technologies, including wind, solar, geothermal, and biomass. Renewable technologies use naturally occurring energy sources to produce electricity, heat, fuel, or a combination of these energy types. The program also works to achieve tax incentives for renewable energy production and use. Some individual highlights follow.

Wind Energy. Use of wind energy is growing rapidly. Technologies under development by DOE and its partners can enable a twenty-fold or more expansion of usable wind resources and make wind energy viable without federal incentives. DOE will continue developing next-generation wind turbines able to produce power at 3.0 cents per kilowatt-hour in good wind regions, with the goal of having such turbines commercially available from U.S. manufacturers in 2004.

Solar Energy. Over the past 20 years, federal R&D has resulted in an 80 percent cost reduction in solar photovoltaics.

Geothermal Energy. The *Annual Energy Outlook* 2002 estimates geothermal energy will provide 5,300 megawatts of generating capacity by 2020 (U.S. DOE 2001a). However, geothermal could provide 25,000–50,000 megawatts from currently identified

hydrothermal resources if the technology existed to develop those resources at a reasonable cost. DOE's R&D program is working in partnership with U.S. industry to establish geothermal energy as an economically competitive contributor to the U.S. energy supply.

Biopower. DOE is testing and demonstrating biomass co-firing with coal, developing advanced technologies for biomass gasification, developing and demonstrating small modular systems, and developing and testing high-yield, low-cost biomass feedstocks.

Climate Challenge

This program is a joint, voluntary effort of the electric utility industry and DOE to reduce, avoid, or sequester greenhouse gases. Established as a Foundation Action under the 1993 *Climate Change Action Plan*, electric utilities developed Participation Accords with DOE to identify and implement cost-effective activities (EOP 1993). The program has now grown to include participation by over 650 utilities accounting for more than 70 percent of the sector's MWh production and CO_2 emissions. The Bush Administration and its industry partners are now considering successor efforts, building upon the experience and learning gained in the this program and in related industry-wide efforts.

Distributed Energy Resources

Distributed energy resources (DER) describe a variety of smaller electricity-generating options well suited for placement in homes, offices, and factories or near these facilities. The program focuses on technology development and the elimination of regulatory and institutional barriers to the use of DER, including interconnection to the utility grid and environmental siting and permitting. Distributed systems include combined cooling, heating, and power systems; biomass-based generators; combustion turbines; concentrating solar power and photovoltaic systems; fuel cells; microturbines; engines/generator sets; and wind turbine storage and control technologies. The program partners with industry to apply a wide array of technologies and integration strategies for on-site use, as well as for grid-enhancing systems. Successful deployment of DER technologies affects the industrial, commercial, institutional, and residential sectors of the U.S. economy—in effect, all aspects of the energy value chain.

High-Temperature Superconductivity

High-temperature superconductors conduct electricity with high efficiency when cooled to liquid nitrogen temperatures. This program supports industry-led projects to capitalize on recent breakthroughs in superconducting wire technology, aimed at developing such devices as advanced motors, power cables, and transformers. These technologies would allow more electricity to reach consumers and perform useful work with no increase in fossil CO_2 emissions.

Hydrogen Program

This program's mission is to advance and support the development of cost-competitive hydrogen technologies and systems that will reduce the environmental impacts of energy use and enable the penetration of renewable energy into the U.S. energy mix. The program has four strategies to carry out its objective: (1) expand the use of hydrogen fuels in the near term by working with industry, including hydrogen producers, to improve efficiency, lower emissions, and lower the cost of technologies that produce hydrogen from natural gas for distributed filling stations; (2) work with fuel cell manufacturers to develop hydrogen-based electricity storage and generation systems that will enhance the introduction and penetration of distributed, renewables-based utility systems; (3) coordinate with the Department of Defense and DOE's Office of Transportation Technologies to demonstrate safe and cost-effective fueling systems for hydrogen vehicles in urban nonattainment areas and to provide

onboard hydrogen storage systems; and (4) work with the national laboratories to lower the cost of technologies that produce hydrogen directly from sunlight and water.

Clean Energy Initiative

Through its new Clean Energy Initiative that has resulted from the President's *National Energy Policy*, EPA is promoting a variety of technologies, practices, and policies with the goal of reducing greenhouse gas emissions associated with the energy supply sector. The initiative has a three-pronged strategy: (1) expand markets for renewable energy; (2) work with state and local governments to develop policies that favor clean energy; and (3) facilitate combined heat and power and other clean "distributed generation" technologies in targeted sectors. Within this initiative, the United States has launched two new partnership programs—the *Green Power Partnership* and the *Combined Heat and Power Partnership*. EPA projects these efforts will spur new investments that will avoid about 30 Tg CO_2 Eq. emissions in 2010.

Nuclear Energy

The *Nuclear Energy Plant Optimization* program is working to further improve the efficiency and reliability of existing nuclear power plants, up to and beyond the end of their original operating licenses. It works to resolve open issues related to plant aging and applies new technologies to improve plant reliability, availability, and productivity, while maintaining high levels of safety. DOE also supports *Next-Generation Nuclear Energy Systems* through two programs: the *Nuclear Energy Research Initiative* (NERI) and the *Generation IV Initiative*. NERI funds small-scale research efforts on promising advanced nuclear energy system concepts, in areas that will promote novel next-generation, proliferation-resistant reactor designs, advanced nuclear fuel development, and fundamental nuclear science. The *Generation IV Initiative* is currently preparing a technology roadmap that will set forth a plan for large-scale research, development, and demonstration of promising advanced reactor con-

cepts. Research and development will be conducted to increase fuel lifetime, establish or improve material compatibility, improve safety performance, reduce system cost, effectively incorporate passive safety features, enhance system reliability, and achieve a high degree of proliferation resistance.

Carbon Sequestration

Carbon sequestration is one of the potentially lowest-cost approaches for reducing CO_2 emissions. This DOE program develops the applied science and demonstrates new technologies for addressing cost-effective, ecologically sound management of CO_2 emissions from the production and use of fossil fuels through capture, reuse, and sequestration. Its goal is to make sequestration options available by 2015. The program's technical objectives include reducing the cost of carbon sequestration and capture from energy production activities; establishing the technical, environmental, and economic feasibility of carbon sequestration using a variety of storage sites and fossil energy systems; determining the environmental acceptability of large-scale CO_2 storage; and developing technologies that produce valuable commodities from CO_2 reuse.

Additional Policies and Measures

Additional ongoing policies and measures in the energy supply sector include the *Hydropower Program, International Programs*, and *Economic Incentives/Tax Credits*. Appendix B provides detailed descriptions of policies and measures.

The program to *Promote Seasonal Gas Use for the Control of Nitrogen Oxides*, which was projected in the 1997 CAR to have no reductions in 2010 below baseline forecasts, is no longer included. ENERGY STAR® *Transformers* has been incorporated into ENERGY STAR®-labeled products.

Transportation

Cars, trucks, buses, aircraft, and other parts of the nation's transportation system are responsible for about one-third of U.S. CO_2 emissions. Emissions from transportation are growing rapidly as Americans drive more and use less fuel-efficient sport-utility and other large vehicles. The United States is currently promoting the development of fuel-efficient motor vehicles and trucks, researching options for producing cleaner fuels, and implementing programs to reduce the number of vehicle miles traveled. Furthermore, many communities are developing innovative ways to reduce congestion and transportation energy needs by improving highway designs and urban planning, and by encouraging mass transit.

FreedomCAR Research Partnership

This new public–private partnership with the nation's automobile manufacturers promotes the development of hydrogen as a primary fuel for cars and trucks. It will focus on the long-term research needed to develop hydrogen from domestic renewable sources and technologies that utilize hydrogen, such as fuel cells. *FreedomCAR* replaces and builds on the *Partnership for a New Generation of Vehicles* (PNGV) program. The transition of vehicles from gasoline to hydrogen is viewed as critical to reducing both CO_2 emissions and U.S. reliance on foreign oil. *FreedomCAR* will focus on technologies to enable mass production of affordable hydrogen-powered fuel cell vehicles and the hydrogen-supply infrastructure to support them. It also will support selected interim technologies that have the potential to dramatically reduce oil consumption and environmental impacts in the nearer term, and/or are applicable to fuel cell and hybrid approaches—e.g., batteries, electronics, and motors.

Innovative Vehicle Technologies and Alternative Fuels

DOE funds research, development, and deployment of technologies that can significantly alter current trends in oil consumption. Commercialization of innovative vehicle technologies and alternative fuels is the nation's best strategy for reducing its reliance on oil. These advanced technologies could also result in dramatic reductions of criteria pollutants and greenhouse gas emissions from the transportation sector. DOE's *Vehicle Systems R&D*

funds research and development for advanced power-train-technology (direct-injection) engines, hybrid-electric drive systems, advanced batteries, fuel cells, and lightweight materials for alternative fuels (including ethanol from biomass, natural gas, methanol, electricity, and biodiesel). The *Clean Cities* program works to deploy alternative-fuel vehicles and build supporting infrastructure, including community networks. And the *Biofuels Program* researches, develops, demonstrates, and facilitates the commercialization of biomass-based, environmentally sound, cost-competitive U.S. technologies to develop clean fuels for transportation.

EPA Voluntary Initiatives

EPA supports a number of voluntary initiatives designed to reduce emissions of greenhouse gases and criteria pollutants from the transportation sector.[3] Although many of these EPA initiatives generally fall under broader existing interagency transportation programs, EPA's efforts greatly increase the adoption in the market of the transportation strategies that have the potential to significantly reduce emissions of greenhouse gases. In addition to the initiatives and brief descriptions that follow, EPA is working with existing programs to further reduce greenhouse gas emissions and criteria pollutants in areas including congestion mitigation, transit demand-management strategies, and alternative transportation.

Commuter Options Programs.

Commuter Choice Leadership Initiative is a voluntary employer-adopted program that increases commuter flexibility by expanding mode options, using flexible scheduling, and increasing work location choices. *Parking Cash-Out* offers employees the option to receive taxable income in lieu of free or subsidized parking, and *Transit Check* offers nontaxable transit benefits, currently up to $100 monthly. EPA estimates emission reductions of 3.5 Tg CO_2 Eq. in 2000 and projects reductions of 14 Tg CO_2 Eq. in 2010 from these and other *Commuter Options* programs.

Smart Growth and Brownfields Policies. These programs, such as the *Air-Brownfields Pilot Program*, demonstrate the extent to which brownfields redevelopment and local land use policies can reduce the growth rate of vehicle miles traveled. EPA estimates reductions of 2.7 Tg CO_2 Eq. in 2000 and projects almost 11 Tg CO_2 Eq. will be avoided in 2010 from these policies.

Ground Freight Transportation Initiative. This voluntary program is aimed at reducing emissions from the freight sector through the implementation of advanced management practices and efficient technologies. EPA projects this program will reduce emissions by 18 Tg CO_2 Eq. in 2010.

Clean Automotive Technology. This program includes research activities and partnerships with the automotive industry to develop advanced clean, fuel-efficient automotive technology. EPA is collaborating with its industry partners to transfer the unique efficient hybrid engine and power-train components, originally developed for passenger car applications, to meet the more demanding size, performance, durability and towing requirements of sport-utility and urban-delivery vehicle applications, while being practical and affordable with ultra-low emissions and ultra-high fuel efficiency. The successful technology development under this program has laid the foundation for cost-effective commercialization of high-fuel-economy/ low-emission vehicles for delivery to market, with an aim toward putting a pilot fleet of vehicles on the road by the end of the decade.

DOT Emission-Reducing Initiatives

The U.S. Department of Transportation (DOT) provides funding for and oversees transportation projects and programs that are implemented by the states and metropolitan areas across the country. Although these activities are not designed specifically as climate

---

[3] These initiatives replace the *Transportation Partners Programs*.

programs, highway funds are used for projects that may have significant ancillary benefits for reducing greenhouse gas emissions, such as transit and pedestrian improvements, bikeways, ride-sharing programs, and other transportation demand-management projects, as well as system improvements on the road network. It is very difficult to estimate the amount of greenhouse gas emission reductions from these programs, since project selection is left to the individual states and metropolitan areas, and reductions will vary among projects. Some significant DOT programs that are likely to have ancillary greenhouse gas benefits follow.

Transit Programs. Funded at about $6.8 billion per year, these programs will likely reduce greenhouse gas emissions by carrying more people per gallon of fuel consumed than those driving alone in their automobiles.

Congestion Mitigation and Air Quality Improvement. This program is targeted at reducing criteria pollutants and provides about $1.35 billion per year to the states to fund new transit services, bicycle and pedestrian improvements, alternative fuel projects, and traffic flow improvements that will likely reduce greenhouse gases as well.

Additional Policies and Measures

Appendix B describes *Transportation Enhancements*, the *Transportation and Community System Preservation Pilot Program*, and *Corporate Average Fuel Economy Standards*. The *Fuel Economy Labels for Tires* program, which was listed in the 1997 CAR, was never implemented and is no longer included.

Industry (Non-CO_2)

Although CO_2 accounts for the largest share of U.S. greenhouse gas emissions, non-CO_2 greenhouse gases have significantly higher global warming potentials. For example, over a 100-year time horizon, methane is more than 20 times more effective than CO_2 at trapping heat in the atmosphere, nitrous oxide is about 300 times more effective,

and hydrofluorocarbons (HFCs) are 100 to 12,000 times more effective. In addition, perfluorocarbons (PFCs) and sulfur hexafluoride (SF_6) also have extremely long atmospheric lifetimes.

Methane and Industry

U.S. industry works in concert with the federal government through a variety of voluntary partnerships that are directed toward eliminating market barriers to the profitable collection and use of methane that otherwise would be released to the atmosphere. Collectively, EPA projects these programs will hold methane emissions below 1990 levels through and beyond 2010.

Natural Gas STAR. Since its launch in 1993, *Natural Gas STAR* has been a successful means of limiting methane emissions. In 2000, it was expanded to the processing sector and included companies representing 40 percent of U.S. natural gas production, 72 percent of transmission company pipeline miles, 49 percent of distribution company service connections, and 23 percent of processing throughput. EPA estimates the program reduced methane emissions by 15 Tg CO_2 Eq. in 2000. Because of the program's expanded reach, EPA projects the estimated reduction for 2010 reported in the 1997 CAR will increase from 15 to 22 Tg CO_2 Eq.

Coalbed Methane Outreach Program. The fraction of coal mine methane from degasification systems that is captured and used grew from 25 percent in 1990 to more than 85 percent in 1999. Begun in 1994, the *Coalbed Methane Outreach Program* (CMOP) is working to demonstrate technologies that can eliminate the remaining emissions from degasification systems, and is addressing methane emissions in ventilation air. EPA estimates that CMOP reduced 7 Tg CO_2 Eq. in 2000. Due to unanticipated mine closures, EPA projects that the program's reduction in 2010 will be reduced slightly from that reported in the 1997 submission, from 11 to 10 Tg CO_2 Eq. However, CMOP's anticipated success in reducing ventilation air

methane over the next few years may lead to an upward revision in the projected reductions for 2010 and beyond.

HFC, PFC, SF_6 Environmental Stewardship

The United States is one of the first nations to develop and implement a national strategy to control emissions of high-GWP gases. The strategy is a combination of industry partnerships and regulatory mechanisms to minimize atmospheric releases of HFCs, PFCs and SF_6, which contribute to global warming, while ensuring a safe, rapid, and cost-effective transition away from chlorofluorocarbons (CFCs), hydrochlorofluorocarbons (HCFCs), halons, and other ozone-depleting substances across multiple industry sectors.

Significant New Alternatives Program. This program continued to facilitate the smooth transition away from ozone-depleting chemicals in major industrial and consumer sectors, while minimizing risks to human health and the environment. Hundreds of alternatives determined to reduce overall risks have been listed as substitutes for ozone-depleting chemicals. By limiting use of global warming gases in specific applications where safe alternatives are available, the program reduced emissions by an estimated 50 Tg CO_2 Eq. in 2000 and is projected to reduce emissions by 162 Tg CO_2 Eq. in 2010.

HFC-23 Partnership. This partnership continued to encourage companies to develop and implement technically feasible, cost-effective processing practices or technologies to reduce HFC-23 emissions from the manufacture of HCFC-22. Despite a 35 percent increase in production since 1990, EPA estimates that total emissions are below 1990 levels—a reduction of 17 Tg CO_2 Eq., compared to business as usual. EPA projects reductions of 27 Tg CO_2 Eq. for 2010.

Partnership with Aluminum Producers. This partnership continued to reduce CF_4 and C_2F_6 where cost-effective technologies and practices are technically feasible. It met its overall goal for 2000, with emissions reduced by about 50 percent relative to 1990 levels on an emissions per unit of product basis. EPA estimates that the partnership reduced emissions by 8 Tg CO_2 Eq. in 2000 and projects reductions of 10 Tg CO_2 Eq. in 2010.

Environmental Stewardship Initiative. This initiative was a new action proposed as part of the 1997 CAR, based on new opportunities to reduce emissions of high-GWP gases. Its initial objective was to limit emissions of HFCs, PFCs, and SF_6 in three industrial applications: semiconductor production, electric power distribution, and magnesium production. Additional sectors are being assessed for the availability of cost-effective emission reduction opportunities and are being added to this initiative. EPA's current projections are that the programs will reduce emissions by 94 Tg CO_2 Eq. in 2010. Because resource constraints delayed implementation of the electric power system and magnesium partnerships, EPA's estimate of the total 2000 reduction is 3 Tg CO_2 Eq. less than was expected in 1997.

Agriculture

The U.S. government maintains a broad portfolio of research and outreach programs aimed at enhancing the overall environmental performance of U.S. agriculture, including reducing greenhouse gas emissions and increasing carbon sinks.

AgSTAR and RLEP

The U.S. government also implements programs targeting greenhouse gas emissions from agriculture. Specific practices aimed at directly reducing greenhouse gas emissions are developed, tested, and promoted through outreach programs. These programs, including *AgSTAR* and the *Ruminant Livestock Efficiency Program* (RLEP), have focused on reducing methane emissions. Although the overall impact of *AgSTAR* and RLEP on greenhouse gas

emissions has been small on a national scale, program stakeholders in the agricultural community have demonstrated that the practices can reduce greenhouse gas emissions and increase productivity. The practices being tested under *AgSTAR* and RLEP can be incorporated into U.S. Department of Agriculture (USDA) broad conservation programs.

Nutrient Management Tools

Efforts to reduce nitrous oxide emissions focus on improving the efficiency of fertilizer use. For example, in 1996 USDA's Natural Resources Conservation Service began collaborating with partners on two nutrient management tools that can improve the efficiency of farm-level fertilizer use. The project's goal is to construct a database of such information and make it available to producers. These tools will enable farmers to develop nutrient management plans and detailed crop nutrient budgets, and to assess the impact of management practices on nitrous oxide emissions.

Conservation Programs

Several conservation programs are providing significant benefits in reducing greenhouse gas emissions and increasing carbon sequestration in agricultural soils.

Conservation Reserve Program. This USDA program cost-effectively assists farm owners and operators in conserving and improving soil, water, air, and wildlife resources by removing environmentally sensitive land from agricultural production and keeping it under long-term, resource-conserving cover. Currently, USDA estimates that the program removes 34 million acres of environmentally sensitive cropland from production and generates long-term environmental benefits, including the offset of about 56 Tg CO_2 Eq. each year. Projections indicate that total enrollment in the program will reach the maximum authorized level of slightly over 36 million acres by the end of 2002.

Changing Management Practices. USDA also offers conservation programs that are aimed at changing management practices rather than removing land from production. For example, the *Environmental Quality Incentive Program* provides technical, educational, and financial assistance to landowners who face serious natural resource challenges. It helps producers make beneficial and cost-effective changes to cropping and grazing systems; improve manure, nutrient, and pest management; and implement conservation measures to improve soil, water, and related natural resources. Similarly, *Conservation and Technical Assistance* supports locally led, voluntary conservation through unique partnerships. The program provides technical assistance to farmers for planning and implementing soil- and water-conservation practices.

Conservation Compliance Plans. In addition to direct assistance programs, USDA farm program "conservation compliance" eligibility policy protects existing wetlands on agricultural land and requires that excess erosion on highly erodible agricultural land be controlled through implementation of a conservation plan. The ancillary benefits of this policy to greenhouse gas mitigation include increased soil carbon sequestration on working agricultural land and preservation of soil carbon associated with wetlands.

Bio-based Products and Bioenergy

The goal of this USDA–DOE collaborative research program is to triple the nation's use of bio-based products and bioenergy. One of the objectives is to use renewable agricultural and forestry biomass for a range of products, including biofuels, as an offset to CO_2 emissions.

Additional Policies and Measures

Appendix B describes two additional programs: the *USDA Commodity Credit Corporation's Bioenergy Program* and the *Conservation Reserve Program Biomass Project*.

Forestry

The U.S. government supports efforts to sequester carbon in both forests and harvested wood products to minimize unintended carbon emissions from forests by reducing the catastrophic risk of wildfires.

Forest Stewardship

USDA's *Forest Stewardship Program* and *Stewardship Incentive Program* provide technical and financial assistance to nonindustrial, private forest owners. About 147 million hectares (363 million acres) of U.S. forests are nonindustrial, private forestlands and provide many ecological and economic benefits and values. These forests provide about 60 percent of our nation's timber supply, with increases expected in the future. The acceleration of tree planting on nonindustrial, private forestlands and marginal agricultural lands can help meet resource needs and provide important ancillary benefits that improve environmental quality—e.g., wildlife habitat, soil conservation, water quality protection and improvement, and recreation. Additionally, tree planting and forest management increase the uptake of CO_2 and the storage of carbon in living biomass, soils, litter, and long-life wood products. Both programs are managed by USDA's Forest Service in cooperation with state forestry agencies.

National Fire Plan

The recently completed National Fire Plan will improve fire management on forested lands, especially in the western parts of the United States. The effort is designed to foster a proactive, collaborative, and community-based approach to reducing risks from wildland fires, using hazardous fuels reduction, integrated vegetation management, and traditional firefighting strategies. While the initiative recognizes that fire is part of natural ecosystems, it will have long-term benefits in reducing greenhouse gas emissions because the risks of catastrophic forest fires will be lower. In addition, the initiative will generate a great volume of small-diameter, woody materials as part of hazardous fuel-reduction activities. Some of these materials have the potential to be used for biomass electric power and composite structural building products.

Waste Management

The U.S. government's waste management programs work to reduce municipal solid waste and greenhouse gas emissions through energy savings, increased carbon sequestration, and avoided methane emissions from landfill gas—the largest contributor to U.S. anthropogenic methane emissions.

Climate and Waste Program

This program was introduced to encourage recycling and source reduction for the purpose of reducing greenhouse gas emissions. EPA is implementing a number of targeted efforts within this program to achieve its goals. *WasteWise* continues to work with organizations to reduce solid waste. New initiatives, including extended product responsibility and biomass waste, further waste reduction efforts through voluntary or negotiated agreements with product manufacturers, and market development activities for recycled-content and bio-based products. Since the last CAR, the *Pay-As-You-Throw Initiative* was launched to provide information and education on community-based programs that provide cost incentives for residential waste reduction. EPA is also continuing to conduct supporting outreach, technical assistance, and research efforts on the linkages between climate change and waste management to complement these activities. Reductions in 2000 are estimated by EPA at 8 Tg CO_2 Eq. and are projected to increase to 20 Tg CO_2 Eq. in 2010.

Stringent Landfill Rule

Promulgated under the Clean Air Act in March 1996, the New Source Performance Standards and Emissions Guidelines (*Landfill Rule*) require large landfills to capture and combust their landfill gas emissions. Since the last CAR, implementation of the rule began at the state level in 1998. Preliminary data on the rule's impact indicate that increasing its stringency has significantly increased the number of landfills that must collect and combust their landfill gas. Methane reductions in 2000 are estimated by EPA at 15 Tg CO_2 Eq. The current EPA projection for 2010 is 33 Tg CO_2 Eq., although the preliminary data suggest that reductions from the more stringent rule may be even greater over the next decade.

Agriculture and Forestry: Opportunities and Challenges

The array of conservation issues has grown with changes in the structure of agriculture and in farm and forest management practices, and with greater public concern about a wider range of issues, including greenhouse gas emissions and carbon sequestration, and energy production and conservation. The agriculture and forestry sectors have been responsive to this concern, and progress has been made in each of these areas.

Today, U.S. forests and forest products are sequestering a significant quantity of carbon every year, equivalent to roughly 15 percent of overall U.S. emissions. Carbon sequestration in agricultural soils is offsetting an additional 2 percent of U.S. greenhouse gas emissions. Given appropriate economic incentives, much of the vast landscape managed by farmers and forest landowners could be managed to store additional carbon, produce biomass and biofuels to replace fossil fuels, and reduce energy use. The challenge is to identify and implement low-cost opportunities to increase carbon storage in soils, provide low-cost tools for enhanced farm and forest management, and ensure that the production of energy raw materials is environmentally beneficial. Realizing these opportunities will take a number of efforts, including an adequate system for measuring the carbon storage and greenhouse gas emissions from agriculture and forests.

For more information about the Administration's effort to formulate a longer-term view of the nation's agriculture and food system, see *Food and Agricultural Policy: Taking Stock for the New Century*, which is available at www.usda.gov (USDA/NRCS 2001).

More comprehensive data will be available by the next CAR submission.

Landfill Methane Outreach Program

This program continues to encourage landfills not regulated by the *Landfill Rule* to capture and use their landfill gas emissions. Capturing and using landfill gas reduces methane emissions directly and reduces CO_2 emissions indirectly through the utilization of landfill gas as a source of energy, thereby displacing the use of fossil fuels. Since the last CAR, the *Landfill Methane Outreach Program* (LMOP) continues to work with landfill owners, state energy and environmental agencies, utilities and other energy suppliers, industry, and other stakeholders to lower the barriers to landfill gas-to-energy projects. LMOP has developed a range of tools to help landfill operators overcome barriers to project development, including feasibility analyses, software for evaluation project economics, profiles of hundreds of candidate landfills across the country, a project development handbook, and energy end user analyses. Due to these efforts, the number of landfill gas-to-energy projects has grown from less than 100 in the early 1990s to almost 320 projects by the end of 2000. EPA estimates that LMOP reduced greenhouse gas emissions from landfills by about 11 Tg CO_2 Eq. in 2000 and projects reductions of 22 Tg CO_2 Eq. in 2010.

Cross-sectoral

The federal government has taken the lead to reduce greenhouse gas emissions from energy use in federal buildings and transportation fleets by:

- requiring federal agencies, through Executive Order 13221, to purchase products that use no more than one watt in standby mode;
- directing the heads of executive departments and agencies to take appropriate actions to conserve energy use at their facilities, review existing operating and administrative processes and conservation programs, and identify and implement ways to reduce energy use;

- requiring all federal agencies to take steps to cut greenhouse gas emissions from energy use in buildings by 30 percent below 1990 levels by 2010;
- directing federal agencies in Washington, D.C., to offer their employees up to $100 a month in transit and van pool benefits; and
- requiring federal agencies to implement strategies to reduce their fleets' annual petroleum consumption by 20 percent relative to 1990 consumption levels and to use alternative fuels a majority of the time.

Federal Energy Management Program

This program reduces energy use in federal buildings, facilities, and operations by advancing energy efficiency and water conservation, promoting the use of renewable energy, and managing utility choices of federal agencies. The program accomplishes its mission by leveraging both federal and private resources to provide federal agencies the technical and financial assistance they need to achieve their goals.

State and Local Climate Change Outreach Program

This EPA program continues to provide technical and financial assistance to states and localities to conduct greenhouse gas inventories, to develop state and city action plans to reduce greenhouse gas emissions, to study the impacts of climate change, and to demonstrate innovative mitigation policies or outreach programs. New or developing projects include estimates of forest carbon storage for each state, a spreadsheet tool to facilitate state inventory updates, a software tool to examine the air quality benefits of greenhouse gas mitigation, a study of the health benefits of greenhouse gas mitigation, and a working group on voluntary state greenhouse gas registries. To date, 41 states and Puerto Rico have initiated or completed inventories, and 27 states and Puerto Rico have completed or are developing action plans. While the program's primary purpose is to build climate change capacity

and expertise at the state and local levels, EPA estimates that the program reduced greenhouse gas emissions by about 6 Tg CO_2 Eq. in 2000.

NONFEDERAL POLICIES AND MEASURES

All federal climate initiatives are conducted in cooperation with private-sector parties. The private sector's support is essential for the success of emission reduction policies. Businesses, state and local governments, and NGOs are also moving forward to address global climate change—through programs to improve the measurement and reporting of emission reductions; through voluntary programs, including emissions trading programs; and through sequestration programs.

State Initiatives

In 2000, the National Governors Association reaffirmed its position on global climate change policy. At the domestic level, the governors recommended that the federal government continue its climate research, including regional climate research, to improve scientific understanding of global climate change. The governors also recommended taking steps that are cost-effective and offer other social and economic benefits beyond reducing greenhouse gas emissions. In particular, the governors supported voluntary partnerships to reduce greenhouse gas emissions while achieving other economic and environmental goals.

NEG-ECP 2001 Climate Change Action Plan

The New England Governors and Eastern Canadian Premiers (NEG–ECP) adopted a Resolution accepting the goals of the *NEG–ECP 2001 Climate Change Action Plan*. The plan sets an overall goal for reducing greenhouse gases in New England States and Eastern Canadian Provinces to 1990 emission levels by 2010, and to 10 percent below 1990 emissions by 2020. The plan's long-term goal is to reduce regional greenhouse gas emissions sufficiently to eliminate any dangerous threat to the climate (75–85 percent below current levels).

Massachusetts Regulation of Electric Utility Emissions

In April 2001, the governor of Massachusetts released a regulation requiring additional controls on Massachusetts electric utility sources, making the state the first in the nation to adopt binding reduction requirements for CO_2. The new regulation sets a cap on total emissions and creates an emission standard that will require CO_2 reductions of about 10 percent below the current average emission rate. The regulation allows companies to buy carbon credits to meet their reduction requirements.

New York and Maryland Executive Orders

The governors of New York and Maryland issued Executive Orders requiring state facilities to: (1) purchase a percentage of energy from green sources; (2) evaluate energy efficiency in state building design and maintenance; and (3) purchase ENERGY STAR®-labeled products when available. Both states are developing comprehensive action plans to reduce greenhouse gas emissions.

New Jersey Executive Order

The State of New Jersey issued an Executive Order to reduce the state's annual greenhouse gas emissions to 3.5 percent below 1990 levels by 2005, using "no regrets" measures that are readily available and that pay for themselves within the short term. The potential emission reductions are based on policies and technologies identified in the *New Jersey Sustainability Greenhouse Gas Action Plan* (NJ 2000). Approximately two-thirds of the reductions will be achieved through energy efficiency and innovative energy technologies in residential, commercial, and industrial buildings. The remainder will come from energy conservation and innovative technologies in the transportation sector, waste management improvements, and natural resource conservation.

Other State Initiatives

California, Maine, New Hampshire, New Jersey, Wisconsin, and Texas are developing *voluntary registries for greenhouse gas emissions*. In addition, 12 states have established *renewable portfolio standards*, and 19 out of 24 states have included *public benefit charges* (also called *system benefit charges*) as a component of their electricity restructuring policy to support continued investment in energy efficiency and renewable energy. Approximately $11 billion, for the period 1998–2012, is expected to be available nationwide through public benefit fund programs. Greenhouse gas *emission inventories* have been completed in 37 states, with four more in progress; and 19 states completed *action plans* to reduce greenhouse gas emissions, with 8 more in progress.

Local Initiatives

A total of 110 U.S. cities and counties, representing nearly 44 million people, are participating in the International Council for Local Environmental Initiatives' *Cities for Climate Protection Campaign*. This program offers training and technical assistance to cities, towns, and counties for projects focusing on reducing emissions. Actions implemented through the campaign are reducing emissions by over 7 Tg CO_2 Eq. each year. Also, in June 2000, the U.S. Conference of Mayors passed a resolution recognizing the seriousness of global warming and calling for increased cooperation between cities and the federal government in taking action to address the challenge.

Private-Sector and NGO Initiatives

Following are some highlights of private-sector and NGO efforts that are demonstrative of leadership by example.

Green Power Market Development Group

In May 2000 a number of private corporations not directly involved with the electric utility industry organized the *Green Power Market Development Group* to support the development of green U.S. energy markets. Together, the Group's 11 members—Alcoa, Cargill-Dow, Delphi, DuPont, General Motors, IBM, Interface, Johnson & Johnson, Kinko's, Oracle, and Pitney Bowes—account for about 7 percent of U.S. industrial energy use. They are working with the World Resources Institute and Business for Social Responsibility to purchase 1,000 megawatts of new green energy capacity and otherwise provide support to the development of green energy markets. The Group believes that such markets are essential to provide competitively priced energy that also protects the Earth's climate and reduces conventional air pollutants. The members are exploring a variety of green energy purchase opportunities to identify those that are cost-competitive. This is a long-term process, with companies hoping to support market development over a 10-year period.

Business Environmental Leadership Council

The U.S. business community, many times in partnership with environmental NGOs, is moving forward on climate change in many other ways. For example, the Pew Center on Climate Change launched a $5 million campaign in 1998 to build support for taking action on climate change. Boeing, DuPont, Shell, Weyerhaeuser, and 32 other major corporations joined the Center's *Business Environmental Leadership Council*, agreeing that "enough is known about the science and environmental impacts of climate change for us to take actions to address its consequences."

Climate Savers

Johnson & Johnson, IBM, Polaroid, and Nike have joined this new partnership to help business voluntarily lower energy consumption and reduce greenhouse gas emissions. In joining *Climate Savers*, partners make specific commitments to reduce their emissions and participate in an independent verification process.

Partnership for Climate Action

Seven companies, including BP, DuPont, and Shell International joined Environmental Defense in the creation of the *Partnership for Climate Action*. Each company has set a firm target for greenhouse gas reductions and has agreed to measure and publicly report its emissions.

Voluntary Reporting of Greenhouse Gases

Under this program, provided by section 1605(b) of the Energy Policy Act of 1992, more than 200 companies have voluntarily reported to DOE more than 1,715 voluntary projects to reduce, avoid, or sequester greenhouse gas emissions.

Auto Manufacturers' Initiatives

U.S. auto manufacturers have announced production plans for hybrid gas and electric vehicles in 2003 or 2004 and have pledged to increase their sport-utility vehicles' fuel economy by 25 percent by 2005.

TABLE 4-1 Summary of Actions to Reduce Greenhouse Gas Emissions

Name of Policy or Measure	Objective and/or Activity Affected	GHG Affected	Type of Instrument	Status	Implementing Entity/Entities	Estimated Mitigation Impact for 2000 (Tg CO$_2$ Eq.)
ENERGY: COMMERCIAL AND RESIDENTIAL						56.8
ENERGY STAR® for the Commercial Market	Promotes the improvement of energy performance in commercial buildings.	CO$_2$	Voluntary	Implemented	EPA	
Commercial Buildings Integration: Updating State Buildings Codes; Partnerships for Commercial Buildings and Facilities	Realizes energy-saving opportunities provided by whole-building approach during construction and major renovation of existing commercial buildings.	CO$_2$	Research, regulatory	Implemented	DOE	
ENERGY STAR® for the Residential Market	Promotes the improvement of energy performance in residential buildings beyond the labeling of products.	CO$_2$	Voluntary, outreach	Implemented	EPA	
Community Energy Program: Rebuild America	Helps communities, towns and cities save energy, create jobs, promote growth, and protect the environment through improved energy efficiency and sustainable building design and operation.	CO$_2$	Voluntary, information, education	Implemented	DOE	
Residential Building Integration: Building America	Funds, develops, demonstrates, and deploys housing that integrates energy-efficiency technologies and practices.	CO$_2$	Voluntary, research, education	Implemented	DOE	
ENERGY STAR®-Labeled Products	Label distinguishes energy-efficient products in the marketplace.	CO$_2$	Voluntary, outreach	Implemented	EPA/DOE	
Building Equipment, Materials, and Tools: Superwindow Collaborative; Lighting Partnerships; Partnerships for Commercial Buildings and Facilities; Collaborative Research and Development	Conducts R&D on building components and design tools and issues standards and test procedures for a variety of appliances and equipment.	CO$_2$	Information, research	Implemented	DOE	
Residential Appliance Standards	Reviews and updates efficiency standards for most major household appliances.	CO$_2$	Regulatory	Implemented	DOE	
State and Community Assistance: State Energy Program; Weatherization Assistance Program; Community Energy Grants; Information Outreach	Provides funding for state and communities to provide local energy-efficiency programs, including services to low-income families; to implement sustainable building design and operation; and to adopt a systematic approach to marketing and communication objectives.	CO$_2$	Economic, information	Implemented	DOE	
Heat Island Reduction Initiative	Reverses the effects of urban heat islands by encouraging the use of mitigation strategies.	CO$_2$	Voluntary, information, research	Implemented	EPA	
Economic Incentives/ Tax Credits	Provides tax credits to residential solar energy systems.	CO$_2$	Economic	Proposed		

TABLE 4-1 (continued) Summary of Actions to Reduce Greenhouse Gas Emissions

Name of Policy or Measure	Objective and/or Activity Affected	GHG Affected	Type of Instrument	Status	Implementing Entity/Entities	Estimated Mitigation Impact for 2000 (Tg CO$_2$ Eq.)
ENERGY: INDUSTRIAL						27.9
Industries of the Future	Helps nine key energy-intensive industries reduce their energy consumption while remaining competitive and economically strong.	All	Voluntary, information	Implemented	DOE	
Best Practices Program	Offers industry tools to improve plant energy efficiency, enhance environmental performance, and increase productivity.	All	Voluntary, information	Implemented	DOE	
ENERGY STAR® for Industry (Climate Wise)	Enables industrial companies to evaluate and cost-effectively reduce their energy use.	CO$_2$	Voluntary	Implemented	EPA	
Industrial Assessment Centers	Assesses and provides recommendations to manufacturers in identifying opportunities to improve productivity, reduce waste, and save energy.	All	Information, research	Implemented	DOE	
Enabling Technologies	Addresses the critical technology challenges partners face for developing materials and production processes.	All	Information, research	Implemented	DOE	
Financial Assistance: NICE[3]	Provides funding to state and industry partnerships for projects that develop and demonstrate advances in energy efficiency and clean production technologies.	All	Research	Implemented	DOE	
ENERGY: SUPPLY						14.7
Renewable Energy Commercialization: Wind; Solar; Geothermal; Biopower	Develops clean, competitive power technologies using renewable resources.	All	Research, regulatory	Implemented	DOE	
Climate Challenge	Promotes efforts to reduce, avoid, or sequester greenhouse gases from electric utilities.	All	Voluntary	Implemented	DOE	
Distributed Energy Resources (DER)	Focuses on technology development and the elimination of regulatory and institutional barriers to the use of DER.	All	Information, research, education, regulatory	Implemented	DOE	
High-Temperature Superconductivity	Advances R&D of high-temperature superconducting power equipment for energy transmission, distribution, and industrial use.	All	Research	Implemented	DOE	
Hydrogen Program	Enhances and supports the development of cost-competitive hydrogen technologies and systems to reduce the environmental impacts of their use.	All	Research, education	Implemented	DOE	

TABLE 4-1 (continued) Summary of Actions to Reduce Greenhouse Gas Emissions

Name of Policy or Measure	Objective and/or Activity Affected	GHG Affected	Type of Instrument	Status	Implementing Entity/Entities	Estimated Mitigation Impact for 2000 (Tg CO_2 Eq.)
Clean Energy Initiative: Green Power Partnership; Combined Heat and Power Partnership	Removes market barriers to increased penetration of cleaner, more efficient energy supply.	CO_2	Voluntary, education, technical assistance	Implemented	EPA	
Nuclear Energy Plant Optimization	Recognizes the importance of existing nuclear plants in reducing greenhouse gas emissions.	CO_2	Information, technical assistance	Implemented	DOE	
Development of Next-Generation Nuclear Energy Systems: Nuclear Energy Research Initiative; Generation IV Initiative	Supports research, development, and demonstration of an advanced nuclear energy system concept.	CO_2	Research, technical assistance	Implemented	DOE	
Support Deployment of New Nuclear Power Plants in the United States	Ensures the availability of near-term nuclear energy options that can be in operation in the U.S. by 2010.	CO_2	Information	Implemented	DOE	
Carbon Sequestration	Develops new technologies for addressing cost-effective management of CO_2 emissions from the production and use of fossil fuels.	CO_2	Research	Implemented	DOE	
Hydropower Program	Improves the technical, societal, and environmental benefits of hydropower.	All	Information, research	Implemented	DOE	
International Programs	Accelerates the international development and deployment of clean energy technologies.	All	Information, technical assistance	Implemented	DOE	
Economic Incentives/ Tax Credits	Provides tax credits to electricity generated from wind- and biomass-based generators.	CO_2	Economic	Proposed		

TABLE 4-1 (continued) Summary of Actions to Reduce Greenhouse Gas Emissions

Name of Policy or Measure	Objective and/or Activity Affected	GHG Affected	Type of Instrument	Status	Implementing Entity/Entities	Estimated Mitigation Impact for 2000 (Tg CO_2 Eq.)
TRANSPORTATION						8.4
FreedomCAR Research Partnership	Promotes the development of hydrogen as a primary fuel for cars and trucks.	CO_2	Research, information	Implemented	DOE	
Vehicle Systems R&D	Promotes the development of cleaner, more efficient passenger vehicles.	CO_2	Research, information	Implemented	DOE	
Clean Cities	Supports public–private partnerships to deploy alternative-fuel vehicles and builds supporting infrastructure, including community networks.	All	Voluntary, information	Implemented	DOE	
Biofuels Program	Researches, develops, demonstrates, and facilitates the commercialization of biomass-based, environmentally sound fuels for transportation.	All	Information, research	Implemented	DOE	
Commuter Options Programs	Reduces single-occupant-vehicle commuting by providing incentives and alternative modes, timing, and locations for work.	CO_2	Voluntary agreements, tax incentives, information, education, outreach	Implemented	EPA/DOT	
Smart Growth and Brownfields Policies	Reduces motorized trips and trip distance by promoting more efficient location choice.	CO_2	Technical assistance, outreach	Implemented	EPA	
Ground Freight Transportation Initiative	Increases efficient management practices for ground freight.	CO_2	Voluntary/ negotiated agreements	Adopted	EPA	
Clean Automotive Technology	Develops advanced clean and fuel-efficient automotive technology.	CO_2	Voluntary, research	Implemented	EPA	
DOT Emission-Reducing Initiatives	Provides funding mechanisms for alternative modes to personal motorized vehicles.	CO_2	Funding mechanisms	Implemented	DOT	

TABLE 4-1 (continued) Summary of Actions to Reduce Greenhouse Gas Emissions

Name of Policy or Measure	Objective and/or Activity Affected	GHG Affected	Type of Instrument	Status	Implementing Entity/Entities	Estimated Mitigation Impact for 2000 (Tg CO$_2$ Eq.)
INDUSTRY (NON-CO$_2$)						88.7
Natural Gas STAR Program	Reduces methane emissions from U.S. natural gas systems through the widespread adoption of industry best management practices.	CH$_4$	Voluntary agreement	Implemented	EPA	
Coalbed Methane Outreach Program	Reduces methane emissions from U.S. coal mining operations through cost-effective means.	CH$_4$	Information, education, outreach	Implemented	EPA	
Significant New Alternatives Program	Facilitates smooth transition away from ozone-depleting chemicals in industrial and consumer sectors.	High GWP	Regulatory, information	Implemented	EPA	
HFC-23 Partnership	Encourages reduction of HFC-23 emissions through cost-effective practices or technologies.	High GWP	Voluntary agreement	Implemented	EPA	
Partnership with Aluminum Producers	Encourages reduction of CF$_4$ and C$_2$F$_6$ where technically feasible and cost-effective.	PFCs	Voluntary agreement	Implemented	EPA	
Environmental Stewardship Initiative	Limits emissions of HFCs, PFCs, and SF$_6$ in industrial applications.	High GWP	Voluntary agreement	Implemented	EPA	
AGRICULTURE						
Agricultural Outreach Programs: AgSTAR; RLEP	Promotes practices to reduce GHG emissions at U.S. farms.	CH$_4$	Information, education, outreach	Implemented	EPA/ USDA	
Nutrient Management Tools	Aims to reduce nitrous oxide emissions through improving by efficiency of fertilizer nitrogen.	N$_2$O	Technical assistance, information	Implemented	EPA/ USDA	
USDA CCC Bioenergy Program	Encourages bioenergy production through economic incentives to commodity producers.	CO$_2$	Economic	Implemented	USDA	
Conservation Reserve Program: Biomass Project	Encourages land-use changes to increase the amount of feedstock available for biomass projects.	CO$_2$ N$_2$O	Economic	Implemented (pilot phase)	USDA	
FORESTRY						
Forest Stewardship	Sequesters carbon in trees, forest soils, forest litter, and understudy plants.	CO$_2$	Technical/ financial assistance	Implemented	USDA	

TABLE 4-1 (continued) Summary of Actions to Reduce Greenhouse Gas Emissions

Name of Policy or Measure	Objective and/or Activity Affected	GHG Affected	Type of Instrument	Status	Implementing Entity/Entities	Estimated Mitigation Impact for 2000 (Tg CO_2 Eq.)
WASTE MANAGEMENT						39.2
Climate and Waste Program	Encourages recycling, source reduction, and other progressive integrated waste management activities to reduce GHG emissions.	All	Voluntary agreements, technical assistance, information, research	Implemented	EPA	
Stringent Landfill Rule	Reduces methane/landfill gas emissions from U.S. landfills.	CH_4	Regulatory	Implemented	EPA	
Landfill Methane Outreach Program	Reduces methane emissions from U.S. landfills through cost-effective means.	CH_4	Voluntary agreements, information, education, outreach	Implemented	EPA	
CROSS-SECTORAL						6.2
Federal Energy Management Program	Promotes energy efficiency and renewable energy use in federal buildings, facilities, and operations.	All	Economic, information, education	Implemented	DOE	
State and Local Climate Change Outreach Program	Assists key state and local decision makers in maintaining and improving economic and environmental assets given climate change.	All	Information, education, research	Implemented	EPA	
TOTAL						241.9

Chapter 5
Projected Greenhouse Gas Emissions

This chapter provides estimates for national emissions under many of the implemented policies and measures for reducing emissions through technology improvements and dissemination, demand-side efficiency gains of many specific types, more efficient regulatory procedures, and shifts to cleaner fuels. The anticipated expansion of the U.S. economy under the impetus of population and output growth at projected rates contributes to rising greenhouse gas emissions. These emissions are partly offset by reductions from ongoing efforts to decrease energy use and from implemented policies and measures. Even with projected growth in absolute emissions, there are near-term and continuing reductions in emissions per unit of gross domestic product (GDP). These projections do not include the impact of the President's climate change initiative announced in February 2002, nor do they include the effects of policies in the *National Energy*

Policy that have not yet been implemented (NEPD Group 2001).

THE NEMS MODEL AND POLICIES COVERAGE

The U.S. Department of Energy's (DOE's) *Annual Energy Outlook 2002* (AEO 2002) presents mid-term forecasts of energy supply, demand, and prices through 2020 based on results from the Energy Information Administration's National Energy Modeling System (NEMS) (U.S. DOE/EIA 2001a). This integrated model looks at all determinants of carbon emissions simultaneously, accounting for interaction and feedback effects. But in some cases, it uses assumptions about diffusion and adoption rates that are different from the assumptions used for the independent policies and measures estimates in Chapter 4 of this report.

The NEMS uses a market-based approach that balances supply and demand with price competition between fuels and sectors. It is a comprehensive, but simplified, representation of the energy economy. Rather than explicitly including and replicating every transaction, the NEMS measures aggregate impacts using empirically developed statistical proxies. Its strength lies in the consistency it brings in representing and accounting for the large number of concurrent, interrelated, and competing energy transactions, investment transactions, and production and consumption decisions that occur in the national energy sector.

The AEO 2002 projections are based on the assumption that the trend in funding levels for policies continues to follow historical patterns. Policies or programs adopted since July 2001—such as the Green Power Partnership, the Combined Heat and Power Partnership, and the Ground Freight Transportation Initiative—are not included in these emission estimates. The methods used to create the projections are regularly updated as new information and methods emerge. However, there is a time lag in the representation of the future effects of some of the adopted measures when using an economic model based on history, such as the NEMS. Consequently, actual growth in energy use and emissions may be different from the projected levels, and the AEO 2002 projections should not be interpreted as reflecting the ultimate impact of policies and measures over the 20-year horizon.

The reported impacts of the individual policies and measures in Chapter 4 of this report are based on specific assumptions for the impacts and adoption of each measure. However, those impacts recognize fewer interaction and competitive effects within and among the economic sectors in which the individual measures are applied. A precise mapping of the emission reductions from individual policies and measures against the aggregate estimates of the NEMS used in the AEO modeling exercise is not possible. Readers are cautioned not to interpret the difference between the estimates in Chapter 4 and this chapter as the numeric difference between the "with measures" and "without measures" cases. The direct impact measures of Chapter 4 compare the effects of provisions that avoid large interaction effects between each other or broadly competitive alternatives. The NEMS results, which address interaction effects and potentially nonmarginal changes, reflect integrated responses to a comprehensive set of economic variables.

Assumptions Used to Estimate Future CO_2 Emissions

This projection of emissions for distant future years is always subject to certain assumptions and uncertainties. These assumptions relate to the prospective implementation and funding of policies and measures adopted but not yet funded; to the actual discovery, adoption, and efficacy of technologies not yet tested in the marketplace; and to the pace of future economic growth.

The AEO 2002 projects a declining ratio of emissions to GDP by incorporating the impacts—including costs—of legislation and regulations adopted as of July 1, 2001. These provisions include, for example, rising appliance efficiencies driven by upgraded ENERGY STAR® specifications for products and homes, progressive upgrades to commercial lighting, and adoption of electric and alternative-fuel vehicles in accord with federal and state requirements. Utility Climate Challenge plans are represented in large measure, with the exception of tree-planting programs and purchases of emission offsets. Renewable-fuels power generation is included, consistent with announced utility building plans through 2020. A description of the policies and measures and technology assumptions embodied in the AEO projections is provided in Appendix G of the AEO 2002.

The assumptions under which the AEO 2002 estimates were prepared include real GDP growth of 3 percent annually over the 20-year period, without specific regard to interim business cycles. The degree of technology improvement reflected in the projections is internally generated in the modeling process based on the Energy Information Administration's judgment about the market readiness, cost, and performance of available technologies, their rates of adoption, and their potential for efficiency improvement. Based on the AEO 2002 estimates, real oil prices are expected to average just over $21 a barrel in 2002, and then rise gradually to $24–$25 a barrel by 2020. Natural gas supplies are assumed to be adequate to support the projected growth in demand. Natural gas prices are projected to rise from just over $2 per thousand cubic feet in 2002, to $3.26 in real terms per thousand cubic feet in 2020. The projection exercise assumes that current laws and regulations will continue in force, but does not anticipate measures not yet enacted or implemented.

Table 5-1 presents several measures of the U.S. economy that generate energy consumption and related carbon emissions, and compares the values used in the 1997 *U.S. Climate Action Report* (CAR) to those relied upon for this report. In this 2001 CAR, 2020 real GDP is notably higher, energy intensity per dollar of GDP is notably lower, natural gas prices are higher, and gasoline prices are lower compared to the levels assumed in the 1997 CAR.

U.S. GREENHOUSE GAS EMISSIONS: 2000–2020

This report contains reported levels of greenhouse gas emissions for the year 2000 and estimates to 2020. The projections of U.S. greenhouse gas emissions described here reflect national estimates of net greenhouse gas emissions considering population growth, long-term economic growth potential, historical rates of technology improvement, normal weather patterns, and many of the implemented policies and measures. The covered gases include carbon dioxide (CO_2), methane, nitrous oxide, hydrofluorocarbons, perfluorocarbons, and sulfur hexafluoride.

DOE's Energy Information Administration computed the energy-related CO_2 projections and the estimated adjustments for bunker fuel use (U.S. DOE/EIA 2001a). The U.S. Environmental Protection Agency (EPA) prepared the emission projections for source categories other than CO_2 emissions resulting from fossil fuel consumption (U.S. EPA 1999; 2001a, b, d). And the U.S. Department of Agriculture (USDA) prepared the estimates of carbon sequestration rates (USDA 2000). The projections reflect long-run trends and do not attempt to mirror short-run departures from those trends.

Rather than the carbon tonnages often used in the United States, emission projections in this report are converted to metric tons of carbon dioxide equivalents, in keeping with the reporting guidelines of the United Nations Framework Convention on Climate Change (UNFCCC). The conversions of non-CO_2 gases to CO_2 equivalents are based on the 100-year global warming potentials (GWPs) listed in the Intergovernmental Panel on Climate Change's (IPCC's) second assessment report (IPCC 1996b).

U.S. greenhouse gas emissions from energy consumption, industrial and agricultural activities, and other anthropogenic sources continued to grow from levels reported in the 1997 *U.S. Climate Action Report* (Figure 5-1). However, emissions of a few of the non-CO_2 gases— e.g., methane and industrial gases associated with the production of aluminum and HCFC-22—have declined from 1990 levels and are projected to remain below 1990 levels out to 2020 (Figure 5-2).

As shown in Figure 5-1, while carbon sequestration partly offsets gross emissions of greenhouse gases, net emissions are projected to rise nonetheless under the impetus of population and economic growth. Increased efforts to use cleaner fuels, improved technologies, and better management methods for agriculture, forestry, mines, and landfills have kept the growth of greenhouse gas emissions below the concurrent growth of the U.S. economy. The policies and measures described in Chapter 4 of this report are expected to further decouple economic growth and greenhouse gas emissions.

The most recent historical measures of greenhouse gas emissions are for 2000, but these measures are still preliminary and are thus subject to possible revision after this report's publication. Nevertheless, the projections use the report's preliminary 2000 data as a point of departure for estimating greenhouse

---

TABLE 5-1 Comparison of 1997 and 2001 CAR Assumptions and Model Results

Several sectors of the U.S. economy involve energy consumption and related carbon emissions. This table compares the values used in the 1997 *U.S. Climate Action Report* (CAR) to those relied upon for this report. In this 2001 CAR, 2020 real GDP is notably higher, energy intensity per dollar of GDP is notably lower, natural gas prices are higher, and gasoline prices are lower compared to the levels assumed in the 1997 CAR.

Factors	2000		2010		2020	
	1997 CAR	*2001 CAR*	*1997 CAR*	*2001 CAR*	*1997 CAR*	*2001 CAR*
Real GDP *(billions of 1996 dollars)*	8,152	9,224	9,925	12,312	11,467	16,525
Population *(millions)*	276	276	299	300	324	325
Residential Housing Stock *(millions)*	103.0	105.2	114.7	116.0	125.4	127.1
Commercial Floor Space *(billion sq. ft.)*	72.3	64.5	78.5	77.5	85.3	89.6
Energy Intensity *(Btus per 1996 dollar GDP)*	11,903	10,770	10,572	9,400	9,631	7,920
Light-Duty Vehicle Miles Traveled *(billions)*	2,373	2,340	2,885	2,981	3,368	3,631
Energy Commodity Prices						
World Oil Price *(2000 dollars/barrel)*	19.86	27.72	22.16	23.36	24.18	24.68
Wellhead Natural Gas *(2000 dollars/1,000 cu. ft.)*	2.02	3.60	2.16	2.85	2.61	3.26
Minemouth Coal *(2000 dollars/ton)*	19.60	16.45	18.00	14.11	16.70	12.79
Average Price Electricity *(2000 cents/kWh)*	7.31	6.90	6.98	6.30	6.66	6.50
Average Price Gasoline *(2000 dollars/gallon)*	1.49	1.53	1.52	1.40	1.56	1.40

FIGURE 5-1 **Gross and Net U.S. Greenhouse Gas Emissions: 2000–2020**

Although carbon sequestration partly offsets gross greenhouse gas emissions, net emissions are projected to increase nonetheless under the impetus of population and economic growth.

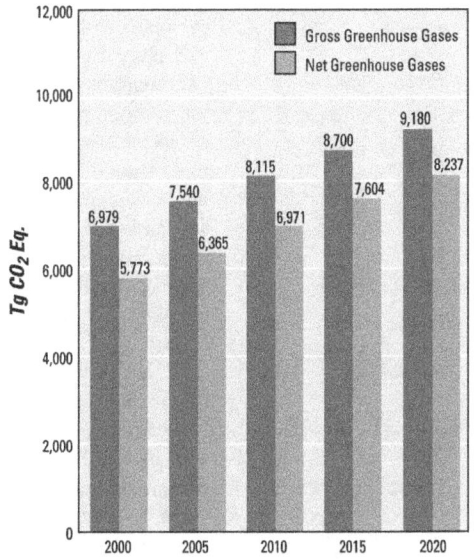

FIGURE 5-2 **U.S. Greenhouse Gas Emissions by Gas: 2000–2020**

A few of the non-CO_2 gases—e.g., methane and industrial gases associated with the production of aluminum and HFC-22—have declined from 1990 levels and are projected to remain below 1990 levels out to 2020.

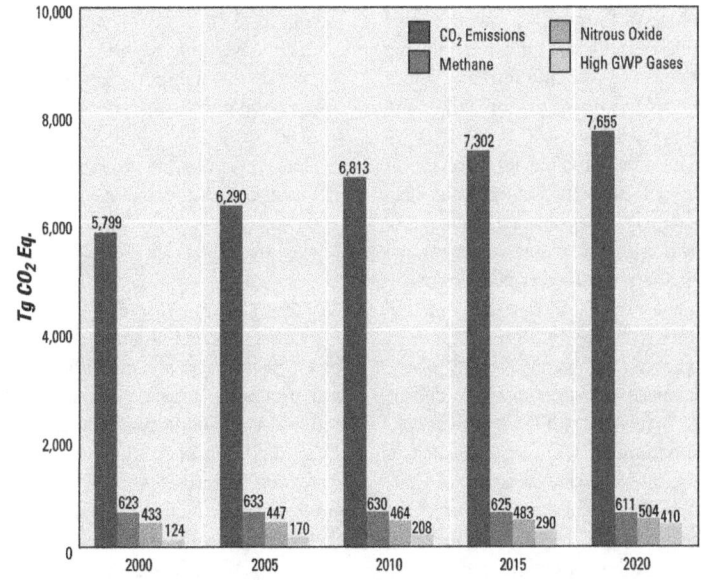

Note: CO_2 emissions reported are net of adjustments.

gas emissions at the 5-year interval benchmarks of 2005, 2010, 2015, and 2020. The text that follows describes changes in emission levels and intensities to the end-point year 2020.

Net U.S. Greenhouse Gas Emissions: 2000–2020

The total projected levels of U.S. greenhouse gas emissions are tallied by (1) combining the CO_2 contributions of energy and nonenergy activities and the non-CO_2 greenhouse gas emissions of methane, nitrous oxide (including forestry and agriculture), and the high GWP gases; (2) subtracting for projected levels of carbon sequestration; and (3) making noted adjustments. Because some of the individual greenhouse gas emissions apart from energy-related portions are not attributed to particular economic sectors, the totals are reported in aggregate.

Total net U.S. greenhouse gas emissions are projected to rise by 42.7 percent, from 5,773 teragrams of CO_2 equivalent (Tg CO_2 Eq.)[1] as the (preliminary) actual level for 2000, to 8,237 Tg CO_2 Eq. projected for 2020 (Table 5-2). However, when examined by 5-year intervals, the rate of increase in U.S. greenhouse gas emissions is expected to diminish over the 20-year projection period. The declining 5-year growth rates reflect the influence of development and implementation of cleaner, more efficient technologies that reduce the ratio of greenhouse gas emissions to GDP over the period; the substitution of fuels that emit lower volumes of greenhouse gases; and changes in the composition of GDP to goods and services with fewer fuel inputs. Some of the mitigating factors are also the subject of implemented policies and measures that reduce emissions relative to a hypothetical "business as usual" path. In addition, there are adopted policies and measures, not yet fully implemented, and the possibility of additional policies and measures prior to 2020 that are not yet defined, which together may further reduce the

---

[1] One teragram equals one million metric tons.

TABLE 5-2 Projected U.S. Greenhouse Gas Emissions from All Sources: 2000–2020 (Tg CO₂ Eq.)

Between 2000 and 2020, total net U.S. greenhouse gas emissions are projected to rise by 42.7 percent. However, the rate of increase in emissions is projected to diminish over the same period, reflecting the development and implementation of cleaner, more efficient technologies; the substitution of fuels that emit lower volumes of greenhouse gases; and changes in the composition of GDP to goods and services with fewer fuel inputs.

All Covered Sources	2000	2005	2010	2015	2020
Energy-Related CO₂	5,726	6,210	6,727	7,206	7,655
Non-energy CO₂	132	138	145	153	161
Methane	623	633	630	625	611
Nitrous Oxide	433	447	464	483	504
High GWP Gases	124	170	208	290	410
Sequestration Removals	-1,205	-1,175	-1,144	-1,096	-1,053
Adjustments	-59	-58	-59	-57	-51
Total	**5,773**	**6,366**	**6,972**	**7,604**	**8,237**

GWP = global warming potential.

Notes: These total U.S. CO₂ equivalent emissions correspond to carbon weights of 1,574 teragrams (Tg) for year 2000; 1,901 Tg for 2010; and 2,246 Tg for 2020. Totals may not sum due to independent rounding.

20-year greenhouse gas path below the aggregate and sectoral levels projected in this report.

The projected emission levels of this report for the years 2010 and 2020 are higher than the levels projected for those same years in the 1997 *U.S. Climate Action Report*, and the preliminary actual level of emissions reported for 2000 is lower than the 1997 projected value. The sections that follow present more detailed projections of specific categories of total U.S. greenhouse gas emissions.

CO₂ Emissions

From 2000 to 2020, energy-related CO₂ emissions are projected to increase by 33.6 percent, compared to cumulative projected economic growth of 80 percent (Table 5-3). The nation's carbon intensity has declined from 721 grams of CO₂ per dollar of GDP in 1990 to 621 grams per dollar in 2000, and is projected to decline further to 463 grams per dollar of GDP by 2020.

In the first 5-year interval, CO₂ emissions are projected to grow by 1.6 percent annually, but by the final 5-year period growth in emissions will have diminished to 1.2 percent annually. The

estimated level of U.S. CO₂ emissions from energy-related activities for the year 2020 is 7,655 Tg CO₂. This level of emissions results from the projected long-term economic, technological, and demographic path, and from the impacts of implemented policies and measures. Additional policies and measures, adopted but not yet implemented—including both new recommendations of the *National Energy Policy* and expanded emphasis on some measures already implemented—could further reduce U.S. CO₂ emissions for 2020 and interim years.

The rising absolute levels of greenhouse gas emissions for the entire U.S. economy occur against a background of growth assumptions for population and GDP. Over the 20-year period, population and personal income are projected to rise respectively by 18 and 79 percent.

- The CO₂ emission intensity of the *residential sector* is expected to decline by 30 percent, while the sector's contributions of CO₂ are estimated to rise by 25 percent, to a total of 1,397 Tg CO₂ annually by 2020. Over the same period, the sector is expected to contribute a diminishing share of total U.S. CO₂ emissions.

- The projected CO₂ emission intensity of the *commercial sector* is expected to decline by 16 percent over the 20-year interval, as measured against the projected 79 percent increase in GDP. The sector's absolute emission contributions are estimated to rise by 42.5 percent to a total of 1,363 Tg CO₂ annually by 2020. Over the 20-year projection period, the commercial sector is expected to contribute a rising share of total U.S. CO₂ emissions.

- The projected CO₂ emission intensity of the *industrial sector* is expected to decline by 27 percent over the 20-year interval, as measured against the projected 79 percent increase in GDP. The sector's absolute emission contributions are estimated to rise by 22 percent to a total of 2,135 Tg CO₂ annually by 2020. Over the 20-year projection period, the industrial sector is expected to contribute a diminishing share of total U.S. CO₂ emissions.

- The projected CO₂ emission intensity of the *transportation sector* is expected to decline by 19 percent over the 20-year interval, as measured against the projected 79 percent increase in GDP. The sector's absolute emission contributions are estimated to rise by 46 percent to a total of 2,760 Tg CO₂ annually by 2020. Over the 20-year projection period, the transportation sector is expected to contribute a rising share of estimated total U.S. CO₂ emissions, reflecting the growth of travel demand and the relatively limited projected use of low-emission fuels even by 2020.

Nonenergy CO₂ Emissions

Other, nonfuel, sources that emit CO₂ include natural gas production and processing, the cement industry, and waste handling and combustion. These CO₂ emissions are subject to increasing voluntary control and are using recapture technologies to reduce their emission levels. Because the underlying sources are so varied, no clear projection method, other than historical extrapolation, is available. These sources are projected to grow by 1 percent annually, well below the 79 percent GDP growth rate assumed in the fuel emission projections. These nonfuel emissions are projected to grow from 132 Tg CO_2 in 2000 to 161 Tg CO_2 in 2020.

CO_2 Emissions from the Electricity Sector

Electricity generation typically produces significant CO_2 emissions, with the important exceptions of electricity generated from nuclear power and from renewable sources, such as hydropower, geothermal, wind, biomass and biomass conversion, and solar power applications. While electricity producers differ greatly in their reliance on various primary fuel inputs, their overall CO_2 contributions are attributable to the nationwide electricity purchases of customers in all economic sectors.

The electricity sector's CO_2 emission intensity is projected to decline by 6 percent over the 20-year interval, as measured against a 43 percent projected increase in total sales of electric energy. Absolute emissions contributions from the electricity sector are estimated to rise by 35 percent during the same period to a total of 2,897 Tg CO_2, reflecting rising electric power sales from 2000 to 2020. The sector's share of total U.S. CO_2 emissions is expected to rise as well, due to the growing role of electricity in powering activities in all economic sectors.

By 2020, the mix of primary fuels in electricity production is expected to be significantly different from the mix during 2000. The expanding role of natural gas, with its relatively low greenhouse gas impact, and the growing dominance of highly efficient generation technologies are projected to reduce the sector's greenhouse gas emissions to a level far below what they would have been without these changes. As noted above, the emission intensity of electricity production is estimated to decline significantly over the projection period. By fuel type, the 2020 CO_2 emissions from electricity generation are 22 Tg CO_2 for energy generated from petroleum, 554 Tg CO_2 for energy generated from natural gas, and 2,322 Tg CO_2 for generation from coal (Table 5-4). Greenhouse gas emissions from nuclear and renewable sources are essentially zero.

Sectoral CO_2 Emissions from Electricity Use

Customers in all sectors use electricity. In that sense, the greenhouse gas emissions that result from electricity production and distribution can be attributed to the end-use sectors (Table 5-5).

- Electricity demand by the **residential sector** is projected to rise by 40 percent from 2000 to 2020, while the CO_2 emissions from the sector's electricity consumption are projected to rise by 31.7 percent. The absolute level of projected CO_2 emissions attributable to the residential sector from electricity use in 2020 is 985 Tg CO_2.
- Electricity demand by the **commercial sector** is projected to rise by 49 percent from 2000 to 2020, while the CO_2 emissions from the sector's electricity consumption are

TABLE 5-3 U.S. CO_2 Emissions by Sector and Source: 2000–2020 (Tg CO_2)

Improvements in CO_2 emission intensity and the absolute levels of future CO_2 emissions vary among economic sectors. The projected 1 percent annual growth in CO_2 emissions from nonenergy sources is well below the 79 percent GDP growth rate assumed in the fuel emission projections.

Sector/Source	2000	2005	2010	2015	2020
Residential	1,122	1,223	1,269	1,325	1,397
Petroleum	101	95	90	86	83
Natural Gas	268	292	300	311	325
Coal	4	4	5	5	5
Electricity	748	832	874	924	985
Commercial	957	1,057	1,163	1,264	1,363
Petroleum	52	48	50	51	51
Natural Gas	181	199	213	228	245
Coal	7	6	7	7	7
Electricity	717	803	893	979	1,059
Industrial	1,753	1,818	1,951	2,049	2,135
Petroleum	344	362	393	414	432
Natural Gas	499	541	581	612	632
Coal	239	231	232	234	237
Electricity	671	684	745	790	834
Transportation	1,895	2,112	2,345	2,568	2,760
Petroleum	1,843	2,055	2,280	2,495	2,679
Natural Gas	42	45	50	57	61
Other	0	*	*	*	*
Electricity	11	13	14	16	19
Total Energy Uses	5,726	6,210	6,727	7,206	7,655
Petroleum	2,339	2,560	2,813	3,045	3,245
Natural Gas	990	1,077	1,145	1,208	1,263
Coal	250	242	244	245	249
Other	0	*	*	*	*
Electricity	2,147	2,331	2,526	2,709	2,897
Nonenergy CO_2 Emissions	132	138	145	153	161
Natural Gas Production	39	40	41	43	44
Industrial Processes	92	98	104	110	117
Total CO_2 Emissions	**5,858**	**6,348**	**6,872**	**7,359**	**7,816**

Note: Totals may not sum due to independent rounding.

* = less than 0.5 Tg.

projected to rise by 47.7 percent. The absolute level of projected CO_2 emissions attributable to the commercial sector from electricity use in 2020 is 1,059 Tg CO_2.

- Electricity demand by the **industrial sector** is projected to rise by 32 percent from 2000 to 2020, while the CO_2 emissions from the sector's electricity consumption are projected to rise by 24.2 percent. The absolute level of projected CO_2 emissions attributable to the industrial sector from electricity use in 2020 is 834 Tg CO_2.
- Emissions of CO_2 from the **transportation sector's** electricity use are projected to rise by 71 percent from 2000 to 2020. However, this sector's overall electricity use is expected to remain small, constituting less than 1 percent of total U.S. electricity demand in 2020.
- For all sectors, demand for **electricity** is projected to grow more rapidly than direct fuel use in other sectors, as electricity assumes an expanding role in meeting the energy demands

of the U.S. economy. Emissions of CO_2 from the electricity sector are projected to rise by 34.9 percent over the 20-year projection period. Efficient production and use of electricity, as well as development of clean fuels, will be a continuing policy focus for the United States.

U.S. CO_2 Emissions from Energy Activities

Total CO_2 emissions are projected to increase by 33.4 percent from 2000 to 2020, to an absolute level of 7,816 Tg CO_2 (Table 5-6). By contrast, cumulative GDP growth over the same period is projected at 79 percent. Consolidating end-use sectors and the electricity industry to examine the projected levels of CO_2 emissions by principal primary fuels shows a growing relative share for natural gas emissions, reflecting rising natural gas use. This share growth for natural gas is an important cause of the declining ratio of greenhouse gas—particularly CO_2—emissions to U.S. economic output.

Emissions of CO_2 from primary fuels

are projected to rise as follows: petroleum, 35.4 percent; natural gas, 49.7 percent; and coal, 22.4 percent. Emissions of CO_2 from the ancillary power needs for electricity generation from non-fossil fuels—primarily nuclear and hydro-power, but also including other renewable sources—remain at negligible levels (less than 0.4 Tg CO_2), even though the utilization of low-emission energy sources is expected to double by 2020. Natural gas is projected to meet a growing share of U.S. energy demand; coal, a reduced share; and petroleum fuels, approximately the same share. The impact of the changing shares of primary fuels is to reduce the intensity of the GDP's greenhouse gas emissions. Nonenergy CO_2 emissions are expected to grow by 22 percent over the projection period.

Non-CO_2 Greenhouse Gas Emissions

Emissions other than CO_2 include methane emissions from natural gas production and transmission, coal mine operation, landfills, and livestock operations; nitrous oxide emissions from agriculture and, to a lesser degree, transportation; and hydrofluorocarbon (HFC), perfluorocarbon (PFC), and sulfur hexafluoride (SF_6) gases from industrial activities and, in some cases, the life cycles of the resulting products (Table 5-7).

Methane

Methane emissions are estimated for 1990 and 2000, and over the 5-year benchmarks of 2005, 2010, 2015, and 2020 (U.S. EPA 1999, 2001a). Over this period, total methane emissions are estimated to decline by 5.2 percent, primarily due to reductions in methane emissions from coal mines and landfills. However, this decline is expected to be offset in part by rising methane emissions from livestock operations. Projected methane emissions from natural gas production, transportation, and use remain nearly unchanged, as the rising natural gas volumes produced and transported are governed by policies and practices that

TABLE 5-4 **U.S. CO_2 Emissions from Electricity Generators: 2000–2020** (Tg CO_2)

By 2020, the mix of primary fuels in electricity production is expected to be significantly different from the mix during 2000.

Primary Fuel	2000	2005	2010	2015	2020
Petroleum	73	25	116	19	22
Natural Gas	224	295	369	479	554
Coal	1,850	2,011	2,141	2,211	2,322
Total	**2,147**	**2,331**	**2,526**	**2,709**	**2,898**

Note: Totals may not sum due to independent rounding.

TABLE 5-5 **Sectoral U.S. CO_2 Emissions from Electricity Use: 2000–2020** (Tg CO_2)

For all sectors, demand for electricity is projected to grow more rapidly than direct fuel use in other sectors, as electricity assumes an expanding role in meeting the energy demands of the U.S. economy. Emissions of greenhouse gases from the electricity sector are projected to rise by 34.9 percent over the 20-year projection period.

Sector	2000	2005	2010	2015	2020
Residential	748	832	874	924	985
Commercial	717	803	893	979	1,059
Industrial	671	684	745	790	834
Transportation	11	13	14	16	19
Total	**2,147**	**2,331**	**2,526**	**2,709**	**2,898**

Note: Totals may not sum due to independent rounding.

will curtail methane releases with increasing effectiveness over the projection period.

Natural Gas Operations. Methane emissions from natural gas operations are projected to increase from 116 Tg CO_2 Eq. in 2000 to 119 Tg CO_2 Eq. in 2020—an increase of only 2.5 percent, despite the more than 60 percent projected increase in natural gas use over the 20-year period.

Coal Mine Operations. Methane emissions from coal mine operations are projected to decline from 70 Tg CO_2 Eq. in 2000 to 66 Tg CO_2 Eq. in 2020—a decrease of 6 percent, primarily due to the closure of very gassy mines and to a projected shift in coal production from underground to surface mines. Coal mine methane is subject to continually improving management practices. This decline in coal-related methane emissions is expected, despite the more than 20 percent increase in coal production projected over the 20-year period.

Landfills. Landfill methane emissions are projected to decrease from 214 Tg CO_2 Eq. in 2000 to 186 Tg CO_2 Eq. in 2020—a decrease of 13 percent, despite growing volumes of municipal waste in place over the period. Landfill sites are assumed to be subject to continually improving methane recovery practices over the 20-year period.

Livestock Operations and Other Activities. Methane emissions from livestock operations, manure management, and other activities not separately listed are expected to rise from 224 Tg CO_2 Eq. in 2000 to 240 Tg CO_2 Eq. in 2020—an increase of 10.3 percent. Anticipated emission management practices for the agricultural and other categories do not fully offset projected agricultural growth over the 20-year period.

Total Methane Emissions. Total U.S. methane emissions from all sources are projected to decline from 623 Tg CO_2 Eq. in 2000 to 611 Tg CO_2 Eq. in 2020—a decrease of 2.1 percent.

HFCs, PFCs, and SF_6

Emissions of HFCs, PFCs, and SF_6 are estimated by EPA for 1990, 2000, and over the 5-year interval benchmarks of 2005, 2010, 2015, and 2020 (U.S. EPA 2001e). While total emissions are projected to rise from 124 Tg CO_2 Eq. in 2000 to 410 Tg CO_2 Eq. in 2020, this increase is expected to be predominantly from the use of HFCs as replacements for ozone-depleting substances (ODS). Growth in the use of HFCs will allow rapid phase-out of chlorofluorocarbons (CFCs), hydrochlorofluorocarbons (HCFCs), and halons in a number of important applications where other alternatives are not available.

HFCs are expected to be selected for applications where they provide superior technical (reliability) or safety

---

TABLE 5-6 U.S. CO_2 Emissions from All Sectors: 2000–2020 (Tg CO_2)

The growing relative share of natural gas emissions resulting from the increased use of natural gas is an important cause of the declining ratio of greenhouse gas—particularly CO_2—emissions to U.S. economic output.

Primary Fuel /Source	2000	2005	2010	2015	2020
Primary Fuel CO_2	5,725	6,210	6,728	7,206	7,655
Petroleum	2,411	2,584	2,829	3,063	3,266
Natural Gas	1,214	1,372	1,513	1,687	1,817
Coal	2,100	2,253	2,385	2,456	2,571
Non-energy CO_2	132	138	145	153	161
Total CO_2	5,857	6,348	6,873	7,359	7,816

Note: Totals may not sum due to independent rounding.

TABLE 5-7 Non-CO_2 Emissions: 2000-2020 (Tg CO_2 Eq.)

Emissions other than CO_2 include methane emissions from natural gas production and transmission, coal mine operation, landfills, and livestock operations; nitrous oxide emissions from agriculture and, to a lesser degree, transportation; and HFC, PFC, and SF_6 gases from industrial activities.

Non-CO_2 GHG/Source	2000	2005	2010	2015	2020
Methane Emissions	623	634	630	625	611
Natural Gas	116	115	115	117	119
Coal Mines	70	73	72	71	66
Landfills	214	219	213	202	186
Livestock Operations	163	167	171	175	178
Other	61	60	59	61	62
High GWP Substances	124	170	208	290	410
ODS Substitutes (HFCs)	58	119	171	266	392
Aluminum (PFCs)	8	7	6	6	5
HCFC-22 (HFC-23)	30	11	6	3	0
Stewardship Programs (Semiconductors, Magnesium, Electric Power Systems, New Programs; HFCs, PFCs, SF_6)	28	33	24	15	13
Nitrous Oxide	433	447	464	483	504
Agriculture	317	326	336	343	350
Mobile Combustion	62	62	66	74	83
Other	54	59	62	66	71
Total	1,180	1,250	1,302	1,398	1,686

Note: Totals may not sum due to independent rounding.

(low toxicity and flammability) performance. In many cases, HFCs provide equal or better energy efficiency compared to other available alternatives, and their acceptance in the market will reduce long-term environmental impacts. HFCs are expected to replace a significant portion of past and current demand for CFCs and HCFCs in insulating foams, refrigeration and air-conditioning, propellants used in metered dose inhalers, and other applications. Emissions of HFCs, PFCs, and SF_6 from all other industrial sources are expected to be reduced significantly below 1990 levels, despite high growth rates of manufacturing in some sectors.

Nitrous Oxide

Nitrous oxide emissions are expected to rise from 433 Tg CO_2 Eq. in 2000 to 504 Tg CO_2 Eq. in 2020—an increase of 16.3 percent over the 20-year projection period. Although the largest single source of these emissions is agricultural soils, emissions from this source are projected to grow at only 9.8 percent. The fastest-growing sources of nitrous oxide emissions are the transportation sector and adipic and nitric acid production. Emissions from each of these sources are projected to grow by about 33 percent over the 20-year period (U.S. EPA 2001b).

Carbon Sequestration

Improved management practices on forest and agricultural lands and the regeneration of previously cleared forests resulted in annual net uptake (i.e., sequestration) of carbon during the 1990s (Table 5-8). Land-use decisions influence net carbon uptake long after their application.

A trend toward managed growth on private land since the early 1950s has resulted in a near doubling of the biomass density in eastern U.S. forests. More recently, the 1970s and 1980s saw a resurgence of federally sponsored forest management and soil conservation

programs, which have focused on planting trees, improving timber management activities, combating soil erosion, and converting marginal croplands to forests. These efforts were maintained throughout the 1990s, and are expected to continue through the projection period. In addition, because most of the timber that is harvested from U.S. forests is used in wood products, and much of the discarded wood products are disposed of in landfills rather than by incineration, significant quantities of this harvested carbon are being transferred to long-term storage pools, rather than being released to the atmosphere.

Adjustments to Greenhouse Gas Emissions

Adjustments to the emissions reported in this chapter include adding the emissions—predominantly fuel-related—occurring in U.S. territories, and subtracting the international use of bunker fuels, both military and civilian (Table 5-9). Emissions from fuel use in U.S. territories are projected to grow from 51 Tg CO_2 Eq. in 2000 to 92 Tg

CO_2 Eq. in 2020.[2] Bunker fuels in excludable uses are estimated to produce emissions of 110 Tg CO_2 Eq. in 2000 and 143 Tg CO_2 Eq. in 2020.[3]

Future of the President's February 2002 Climate Change Initiative

On February 14, 2002, the President committed the United States to reduce its greenhouse gas intensity by 18 percent over the next decade and announced a series of voluntary programs to achieve that goal. This includes proposed enhancements to the existing emissions registry under section 1605(b) of the 1992 Energy Policy Act that would both protect entities that register reductions from penalty under a future climate policy, and create transferable credits for companies that show real emission reductions. It also included expanding sectoral challenges and renewed support for renewable energy and energy efficiency tax credits contained in the *National Energy Policy*. The President indicated that progress would be evaluated in 2012 and that additional policies, including a broad,

---

TABLE 5-8 Projections of Carbon Sequestration (Tg CO_2)

Improved management practices on forest and agricultural lands and the regeneration of previously cleared forests resulted in annual net uptake (i.e., sequestration) of carbon during the 1990s. These practices are expected to continue throughout the projection period.

	2000	2005	2010	2015	2020
Carbon Sequestration (-)	1,205	1,175	1,144	1,049	1,053

Note: The above land-use sequestration estimates and projections are based on the U.S. government's August 1, 2000, submission to the UNFCCC on methodological issues related to the treatment of carbon sinks (U.S. DOS 2000). The projections are not directly comparable to the estimates provided in Chapter 3 of this report for two reasons: (1) the values provided in Chapter 3 use updated inventory information, and these projections have not been revised to reflect this new information; and (2) these projections are for a slightly different set of forest areas and activities than are accounted for in the national greenhouse gas inventory. A new set of projections that will be consistent with updated inventory estimates will be available from the USDA's Forest Service in early 2002. The trends provided in these projections serve to illustrate the impact of forces that are likely to influence carbon sequestration rates over the next decades.

TABLE 5-9 Adjustments to U.S. Greenhouse Gas Emissions (Tg CO_2 Eq.)

Adjustments to the emissions reported in this chapter include adding the emissions—predominantly fuel-related—occurring in U.S. territories, and subtracting the international use of bunker fuels, both military and civilian.

Type of Adjustment	2000	2005	2010	2015	2020
Emissions in U.S. Territories	+ 51	+ 59	+ 69	+ 79	+ 92
International Bunker Fuels	-110	-117	-128	- 136	-143
Net Adjustments	- 59	- 58	- 59	- 57	- 51

---

[2] The projected annual growth rate is 3 percent (U.S. DOE).
[3] The projected annual growth rate is 1.3 percent (U.S. DOE).

market-based program, would be considered in light of the adequacy of these voluntary programs and developments in our understanding of the science surrounding climate change. The consequences of this announcement have not yet been incorporated in current emission forecasts.

KEY UNCERTAINTIES AFFECTING PROJECTIONS

Any projection of future emissions is subject to considerable uncertainty. In the short term (less than 5 years), the key factors that can increase or decrease estimated net emissions include unexpected changes in retail energy prices, shifts in the price relationship between natural gas and coal used for electricity generation, changes in the economic growth path, abnormal winter or summer temperatures, and imperfect forecasting methods. Additional factors may influence emission rates over the longer term, such as technology developments, shifts in the composition of economic activity, and changes in government policies.

Technology Development (+ or -)

Forecasts of net U.S. emissions of greenhouse gases take into consideration likely improvements in technology over time. For example, technology-based energy efficiency gains, which have contributed to reductions in U.S. energy intensity for more than 30 years, are expected to continue. However, while long-term trends in technology are often predictable, the specific areas in which significant technology improvements will occur and the specific new technologies that will become dominant in commercial markets are impossible to forecast accurately, especially over the long term.

Unexpected scientific breakthroughs can cause technology changes and shifts in economic activity that have sometimes had dramatic effects on patterns of energy production and use. Such breakthroughs could enable the United States to dramatically reduce future greenhouse gas emissions. While government and private support of research and develop-

Sample National Energy Policy Initiatives

The May 2001 *National Energy Policy* (NEP) is a long-term, comprehensive strategy to increase energy supplies; advance the development of new, environmentally friendly, energy-conservation technologies; and encourage cleaner, more efficient energy use (NEPD Group 2001). The NEP identified the major energy challenges facing the United States and developed 105 recommendations for addressing these challenges. When fully implemented, many of these recommendations will reduce domestic and international greenhouse gas emissions. Following is a snapshot of the NEP's proposed initiatives.

Reduce U.S. Energy Consumption

- Expand the ENERGY STAR® program to additional buildings, equipment, and services.
- Improve energy efficiency for appliances, and expand the scope of the appliance standards program.
- Encourage the use of combined heat-and-power operations and other clean-energy forms.
- Mitigate transportation congestion by both roadway improvements and information technology.
- Promote the purchase of fuel-efficient vehicles, including fuel-cell power plants for personal and heavy vehicles.
- Increase energy conservation in government facilities.

Increase U.S. Energy Supplies

- Enhance the reliability of U.S. energy supplies, and reduce U.S. reliance on energy imports.
- Increase domestic production of oil, natural gas, and coal.
- Expand support for advanced clean-coal technology research.
- Support the expansion of safe nuclear power technologies.
- Increase funding for research and development of renewable and alternative energy resources.
- Optimize the use of hydroelectric generation.
- Undertake long-term education and research into hydrogen fuels, advanced fuel cells, and fusion power.
- Extend tax credits for the production of electricity from biomass and wind resources.
- Create federal tax incentives to encourage landfill methane recovery.

Strengthen Global Alliances

- Expand international cooperation for energy research and development.
- Promote continued research on the science of global climate change.
- Cooperate with allies to develop cutting-edge technologies, market-based incentives, and other innovative approaches to address climate change.

ment efforts can accelerate the rate of technology change, the effect of such support on specific technology developments is difficult to predict.

The Administration has established a National Climate Change Technology Initiative (NCCTI) to strengthen basic research and develop advanced mitigation technologies for reducing green-

house gas emissions. Success under the NCCTI could dramatically expand low-cost emission-reduction opportunities for the United States and the rest of the world.

In a modest high-technology case examined as part of the projections, energy use in 2020 under the high-technology regime is 5.6 percent lower than in the reference case. By 2020, carbon emissions from energy use are 507 Tg CO_2 lower than in the reference case.

Regulatory or Statutory Changes (+ or -)

The current forecast of U.S. greenhouse gas emissions does not include the effects of any legislative or regulatory action that was not finalized before July 1, 2001. Consequently, the forecast does not include any increase in the stringency of equipment efficiency standards, even though existing law requires DOE to periodically strengthen its existing standards and issue new standards for other products. Similarly, the forecast does not assume any future increase in new building or auto fuel economy standards, even though such increases are required by law or are under consideration. Electric utility regulation is another area where further federal and state regulatory policy changes are anticipated, but are not reflected in the emissions forecast. Finally, the U.S. Congress is considering a broad range of legislative proposals, including many contained in the *National Energy Policy*, that will affect U.S. greenhouse gas emissions. Until specific legislative mandates are enacted, the forecast of emissions will not reflect their likely effects.

Energy Prices (+ or -)

The relationship between energy prices and emissions is complex. Lower energy prices generally reduce the incentive for energy conservation and tend to encourage increased energy use and related emissions. However, reduction in the price of natural gas relative to other fuels also encourages fuel switching that can reduce carbon emissions.

The AEO 2002 projections do not assume any dramatic changes in the energy price trends or the inter-fuel

prices ratio that existed during most of the 1990s (U.S. DOE/EIA 2001a). Nor do they assume that the dramatic increases in energy prices that occurred from mid-2000 through the beginning of 2001 will persist. This view is supported by the precipitous decline in oil prices that occurred during the second half of 2001.

While some analysts project that further decreases in delivered energy prices will result from increased competition in the electric utility sector and improved technology, others project that large energy price increases may result from the faster-than-expected depletion of oil and gas resources, or from political or other disruptions in oil-producing countries.

Economic Growth (+ or -)

Faster economic growth increases the future demand for energy services, such as vehicle miles traveled, amount of lighted and ventilated space, and process heat used in industrial production. However, faster growth also stimulates capital investment and reduces the average age of the capital stock, increasing its average energy efficiency. The energy-service demand and energy-efficiency effects of higher growth work in offsetting directions. The effect on service demand is the stronger of the two, so that levels of primary energy use are positively correlated with the size of the economy.

In addition to the reference case used in developing the updated baseline, the AEO 2002 provides high and low economic growth cases, which vary the annual GDP growth rate from the reference case. The high-growth case raises the GDP growth rate by 0.4 percent. The low-growth case reduces the GDP growth rate by 0.6 percent.

- In the high-growth case 2020 energy use is 5.6 percent higher than in the reference case. By 2020, the high-

growth economy is 8.8 percent larger than the reference economy, and carbon emissions from energy use are 462 Tg CO_2 Eq. greater than in the reference case.

- In the low-growth case 2020 energy use is 5.4 percent lower than in the reference case. By 2020, the low-growth economy is 9.7 percent smaller than the reference economy, and carbon emissions from energy use are 395 Tg CO_2 Eq. lower than in the reference case.

Faster-than-expected growth during the late 1990s was the major cause of higher-than-expected U.S. greenhouse gas emissions during this period. The U.S. economic slowdown in 2001 and post-September 11 fallout may well result in lower-than-expected greenhouse gas emissions during 2002 and the immediately following years. However, the long-run economic growth path remains unchanged.

Weather (+ or -)

Energy use for heating and cooling is directly responsive to weather variation. The forecast of emissions assumes 30-year average values for population-weighted heating and cooling degree-days. Unlike other sources of uncertainty, for which deviations between assumed and actual trends may follow a persistent course over time, the effect of weather on energy use and emissions in any particular year is largely independent year to year. For the United States, a swing in either direction of the magnitude experienced in individual years during the 1990s could raise or lower annual emissions by 70 Tg CO_2 Eq. relative to a year with average weather that generates typical heating and cooling demands. While small relative to total emissions, a change of this magnitude is significant relative to the year-to-year growth of total emissions.

Chapter 6
Impacts and Adaptation

Uncertainties in Estimates of the Timing, Magnitude, and Distribution of Future Warming

While current analyses are unable to predict with confidence the timing, magnitude, or regional distribution of climate change, the best scientific information indicates that if greenhouse gas concentrations continue to increase, changes are likely to occur. The U.S. National Research Council has cautioned, however, that "because there is considerable uncertainty in current understanding of how the climate system varies naturally and reacts to emissions of greenhouse gases and aerosols, current estimates of the magnitude of future warmings should be regarded as tentative and subject to future adjustments (either upward or downward)" (NRC 2001a). Moreover, there is perhaps even greater uncertainty regarding the social, environmental, and economic consequences of changes in climate. (See Chapter 1, page 4, "The Science" box.)

Uncertainties in Regional and Local Projections of Climate Change

One of the weakest links in our knowledge is the connection between global and regional predictions of climate change. The National Research Council's response to the President's request for a review of climate change policy specifically noted that fundamental scientific questions remain regarding the specifics of regional and local projections (NRC 2001a). Predicting the potential impacts of climate change is compounded by a lack of understanding of the sensitivity of many environmental systems and resources—both managed and unmanaged—to climate change. (See Chapter 1, page 6.)

In its June 2001 report, the Committee on the Science of Climate Change, which was convened by the National Research Council (NRC) of the National Academy of Sciences, concluded that "[h]uman-induced warming and associated sea level rises are expected to continue through the 21st century." The Committee recognized that there remains considerable uncertainty in current understanding of how climate varies naturally and will respond to projected, but uncertain, changes in the emissions of greenhouse gases and aerosols. It also noted that the "impacts of these changes will be critically dependent on the magnitude of the warming and the rate with which it occurs" (NRC 2001a).

SUMMARY OF THE NATIONAL ASSESSMENT

To develop an initial understanding of the potential impacts of climate change for the United States during the

21st century, the U.S. Global Change Research Program has sponsored a wide-ranging set of assessment activities since the submission of the Second National Communication in 1997. These activities examined regional, sectoral, and national components of the potential consequences for the environment and key societal activities in the event of changes in climate consistent with projections drawn from the Intergovernmental Panel on Climate Change (IPCC). Regional studies ranged from Alaska to the Southeast and from the Northeast to the Pacific Islands. Sectoral studies considered the potential influences of climate change on land cover, agriculture, forests, human health, water resources, and coastal areas and marine resources. A national overview drew together the findings to provide an integrated and comprehensive perspective.

These assessment studies recognized that definitive prediction of potential outcomes is not yet feasible as a result of the wide range of possible future levels of greenhouse gas and aerosol emissions, the range of possible climatic responses to changes in atmospheric concentration, and the range of possible environmental and societal responses. These assessments, therefore, evaluated the narrower question concerning the vulnerability of the United States to a specified range of climate warming, focusing primarily on the potential consequences of climate scenarios that project global average warming of about 2.5 to almost 4°C (about 4.5–7°F). While narrower than the IPCC's full 1.4–5.8°C (2.5–10.4°F) range of estimates of future warming, the selection of the climate scenarios that were considered recognized that it is important to treat a range of conditions about the mid-range of projected warming, which was given by the NRC as 3°C (5.4°F). Similarly, assumption of a mid-range value of sea level rise of about 48 cm (19 inches) was near the middle of the IPCC's range of 9–88 cm (about 4–35 inches) (2001d).

Because of these ranges and their uncertainties, and because of uncertain-

ties in projecting potential impacts, it is important to note that this chapter cannot present absolute probabilities of what is likely to occur. Instead, it can only present judgments about the relative plausibility of outcomes in the event that the projected changes in climate that are being considered do occur. To the extent that actual emissions of greenhouse gases turn out to be lower than projected, or that climate change is at the lower end of the projected ranges and climate variability about the mean varies little from the past, the projected impacts of climate change are likely to be reduced or delayed, and continued adaptation and technological development are likely to reduce the projected impacts and costs of climate change within the United States. Even in this event, however, the long lifetimes of greenhouse gases already in the atmosphere and the momentum of the climate system are projected to cause climate to continue to change for more than a century. Conversely, if the changes in climate are toward the upper end of the projected ranges and occur rapidly or lead to unprecedented conditions, the level of disruption is likely to be increased. Because of the momentum in the climate system and natural climate variability, adapting to a changing climate is inevitable. The question is whether we adapt poorly or well. With either weak or strong warming, however, the U.S. economy should continue to grow, with impacts being reduced if actions are taken to prepare for and adapt to future changes.

Although successful U.S. adaptation to the changing climate conditions during the 20th century provides some context for evaluating potential U.S. vulnerability to projected changes, the assessments indicate that the challenge of adaptation is likely to be greater during the 21st century than in the past. Natural ecosystems appear to be the most vulnerable to climate change because generally little can be done to help them adapt to the projected rate and amount of change. Sea level rise at mid-range rates is projected to cause

additional loss of coastal wetlands, particularly in areas where there are obstructions to landward migration, and put coastal communities at greater risk of storm surges, especially in the southeastern United States. Reduced snowpack is very likely to alter the timing and amount of water supplies, potentially exacerbating water shortages, particularly throughout the western United States, if current water management practices cannot be successfully altered or modified. Increases in the heat index (which combines temperature and humidity) and in the frequency of heat waves are very likely. At a minimum, these changes will increase discomfort, particularly in cities; however, their health impacts can be ameliorated through such measures as the increased availability of air conditioning.

At the same time, greater wealth and advances in technologies are likely to help facilitate adaptation, particularly for human systems. In addition, highly managed ecosystems, such as crops and timber plantations, appear more robust than natural and lightly managed ecosystems, such as grasslands and deserts.

Some potential benefits were also identified in the assessments. For example, due to increased carbon dioxide (CO_2) in the atmosphere and an extended growing season, crop and forest productivities are likely to increase where water and nutrients are sufficient, at least for the next few decades. As a result, the potential exists for an increase in exports of some U.S. food products, depending on impacts in other food-growing regions around the world. Increases in crop production in fertile areas could cause prices to fall, benefiting consumers. Other potential benefits could include extended seasons for construction and warm-weather recreation, and reduced heating requirements and cold-weather mortality.

While most studies conducted to date have primarily had an internal focus, the United States also recognizes that its well-being is connected to the world through the global economy, the common global environment, shared resources, historic roots and continuing family relations, travel and tourism, migrating species, and more. As a result, in addition to internal impacts, the United States is likely to be affected, both directly and indirectly and both positively and detrimentally, by the potential consequences of climate change on the rest of the world. To better understand those potential consequences and the potential for adaptation worldwide, we are conducting and participating in research and assessments both within the United States and internationally (see Chapter 8). To alleviate vulnerability to adverse consequences, we are undertaking a wide range of activities that will help nationally and internationally, from developing medicines for dealing with infectious disease to promoting worldwide development through trade and assistance. As described in Chapter 7, the United States is also offering many types of assistance to the world community, believing that information about and preparation for climate change can help reduce adverse impacts.

INTRODUCTION

This chapter provides an overview of the potential impacts of climate change affecting the United States. The chapter also summarizes current measures and future adaptation and response options that are designed to increase resilience to climate variations and reduce vulnerability to climate change. The chapter is not intended to serve as a separate assessment in and of itself, but rather is drawn largely from analyses prepared for the U.S. National and IPCC Assessments, where more detailed consideration and specific references to the literature can be found (see NAST 2000, 2001 and IPCC 2001d, including the review of these results presented in NRC 2001a and IPCC 2001a).

As indicated by the findings presented here, considerable scientific progress has been made in gaining an understanding of the potential consequences of climate change. At the same time, considerable uncertainties remain because the actual impacts will depend on how emissions change, how the cli-

mate responds at global to regional scales, how societies and supporting technologies evolve, how the environment and society are affected, and how much ingenuity and commitment societies show in responding to the potential impacts. While the range of possible outcomes is very broad, all projections prepared by the IPCC (2001d) indicate that the anthropogenic contribution to global climate change will be greater during the 21st century than during the 20th century. Although the extents of climate change and its impacts nationally and regionally remain uncertain, it is generally possible to undertake "if this, then that" types of analyses. Such analyses can then be used to identify plausible impacts resulting from projected changes in climate and, in some cases, to evaluate the relative plausibility of various outcomes.

Clear and careful presentation of uncertainties is also important. Because the information is being provided to policymakers and because the limited scientific understanding of the processes involved generally precludes a fully quantitative analysis, extensive consideration led both the IPCC and the National Assessment experts to express their findings in terms of the relative likelihood of an outcome's occurring. To integrate the wide variety of information and to differentiate more likely from less likely outcomes, a common lexicon was developed to express the considered judgment of the National Assessment experts about the relative likelihood of the results. An advantage of this approach is that it moves beyond the vagueness of ill-defined terms, such as *may* or *might*, which allow an interpretation of the likelihood of an outcome's occurring to range from, for example, 1 to 99 percent, and so provide little basis for differentiating the most plausible from the least plausible outcomes.

In this chapter, which uses a lexicon similar to that developed for the National Assessment, the term *possible* is intended to indicate there is a finite likelihood of occurrence of a potential

consequence, the term *likely* is used to indicate that the suggested impact is more plausible than other outcomes, and the term *very likely* is used to indicate that an outcome is much more plausible than other outcomes. Although the degree of scientific understanding regarding most types of outcomes is not complete, the judgments included here have been based on an evaluation of the consistency and extent of available scientific studies (e.g., field experiments, model simulations), historical trends, physical and biological relationships, and the expert judgment of highly qualified scientists actively engaged in relevant research (see NAST 2000, 2001). Because such judgments necessarily have a subjective component, the indications of relative likelihood may change as additional information is developed or as new approaches to adaptation are recognized.

Because this chapter is an overview, it generally focuses on types of outcomes that are at least considered *likely*, leaving discussion of the consequences of potential outcomes with lower likelihood to the more extensive scientific and assessment literature. However, it is important to recognize that there are likely to be unanticipated impacts of climate change that occur. Such "surprises," positive or negative, may stem from either (1) unforeseen changes in the physical climate system, such as major alterations in ocean circulation, cloud distribution, or storms; or (2) unpredicted biological consequences of these physical climate changes, such as pest outbreaks. For this reason, the set of suggested consequences presented here should not be considered comprehensive. In addition, unexpected social or economic changes, including major changes in wealth, technology, or political priorities, could affect society's ability to respond to climate change.

This chapter first describes the weather and climate context for the analysis of impacts, and then provides a summary of the types of consequences that are considered plausible across a range of sectors and regions. The chapter then concludes with a brief summary of actions being taken at the national level to learn more about the potential consequences of climate change and to encourage steps to reduce vulnerability and increase resilience to its impacts. Although the federal government can support research that expands understanding and the available set of options and that provides information about the potential consequences of climate change and viable response strategies, many of the adaptation measures are likely to be implemented at state and local levels and by the private sector. For these reasons and because of identified uncertainties, the results presented should not be viewed as definitive. Nonetheless, the more plausible types of consequences and impacts resulting from climate change and the types of steps that might be taken to reduce vulnerability and increase adaptation to climate variations and change are identified.

WEATHER AND CLIMATE CONTEXT

The United States experiences a wide variety of climate conditions. Moving across from west to east, the climates range from the semi-arid and arid climates of the Southwest to the continental climates of the Great Plains and the moister conditions of the eastern United States. North to south, the climates range from the Arctic climate of northern Alaska to the extensive forests of the Pacific Northwest to the tropical climates in Hawaii, the Pacific Islands, and the Caribbean. Although U.S. society and industry have largely adapted to the mean and variable climate conditions of their region, this has not been without some effort and cost. In addition, various extreme events each year still cause significant impacts across the nation. Weather events causing the most death, injury, and damage include hurricanes (or more generally tropical cyclones) and associated storm surges, lightning, tornadoes and other windstorms, hailstorms, severe winter storms, deep snow and avalanches, and extreme summer temperatures. Heat waves, floods, landslides, droughts, fires, land

subsidence, coastal inundation and erosion, and even dam failures also can result when extremes persist over time.

To provide an objective and quantitative basis for an assessment of the potential consequences of climate change, the U.S. National Assessment was organized around the use of climate model scenarios that specified changes in the climate that might be experienced across the United States (NAST 2001). Rather than simply considering the potential influences of arbitrary changes in temperature, precipitation, and other variables, the use of climate model scenarios ensured that the set of climate conditions considered was internally consistent and physically plausible. For the National Assessment, the climate scenarios were primarily drawn from results available from the climate models developed and used by the United Kingdom's Hadley Centre and the Canadian Centre for Climate Modeling and Analysis. In addition, some analyses also drew on results from model simulations carried out at U.S. centers, including the National Center for Atmospheric Research, the National Oceanic and Atmospheric Administration's (NOAA's) Geophysical Fluid Dynamics Laboratory, and the National Aeronautics and Space Administration's (NASA's) Goddard Institute for Space Studies.

Use of these model results is not meant to imply that they provide accurate *predictions* of the specific changes in climate that will occur over the next 100 years. Rather, the models are considered to provide plausible *projections* of potential changes for the 21st century.[1] For some aspects of climate, all models, as well as other lines of evidence, are in agreement on the types of changes to be expected. For example, compared to changes during the 20th century, all climate model results suggest that warming during the 21st century across the country is very likely to be greater, that

sea level and the heat index are going to rise more, and that precipitation is more likely to come in the heavier categories experienced in each region. Also, although there is not yet close agreement about how regional changes in climate can be expected to differ from larger-scale changes, the model simulations indicate some agreement in projections of the general seasonal and subcontinental patterns of the changes (IPCC 2001d).

This consistency has lent some confidence to these results. For some aspects of climate, however, the model results differ. For example, some models, including the Canadian model, project more extensive and frequent drought in the United States, while others, including the Hadley model, do not. As a result, the Canadian model suggests a hotter and drier Southeast during the 21st century, while the Hadley model suggests warmer and wetter conditions. Where such differences arise, the primary model scenarios provide two plausible, but different alternatives. Such differences proved helpful in exploring the particular sensitivities of various activities to uncertainties in the model results.

Projected Changes in the Mean Climate

The model scenarios used in the National Assessment project that the continuing growth in greenhouse gas emissions is likely to lead to annual-average warming over the United States that could be as much as several degrees Celsius (roughly 3–9°F) during the 21st century. In addition, both precipitation and evaporation are projected to increase, and occurrences of unusual warmth and extreme wet and dry conditions are expected to become more frequent. For areas experiencing these changes, they would feel similar to an overall northern shift in weather

---

[1] For the purposes of this chapter, *prediction* is meant to indicate forecasting of an outcome that will occur as a result of the prevailing situation and recent trends (e.g., tomorrow's weather or next winter's El Niño event), whereas *projection* is used to refer to potential outcomes that would be expected *if* some scenario of future conditions were to come about (e.g., concerning greenhouse gas emissions). In addition to uncertainties in how the climate is likely to respond to a changing atmospheric concentration, projections of climate change necessarily encompass a wide range because of uncertainties in projections of future emissions of greenhouse gases and aerosols and because of the potential effects of possible future agreements that might limit such emissions.

systems and climate condition. For example, the central tier of states would experience climate conditions roughly equivalent to those now experienced in the southern tier, and the northern tier would experience conditions much like the central tier. Figure 6-1 illustrates how the summer climate of Illinois might change under the two scenarios. While the two models roughly agree on the amount of warming, the differences between them arise because of differences in projections of changing summertime precipitation.

Recent analyses indicate that, as a result of an uncertain combination of natural and human-induced factors, changes of the type that are projected for the 21st century were occurring to some degree during the 20th century. For example, over the last 100 years most areas in the contiguous United States warmed, although there was cooling in the Southeast. Also, during the 20th century, many areas experienced more periods of very wet or very dry conditions, and most areas experienced more intense rainfall events. While warming over the 48 contiguous states amounted to about 0.6°C (about 1°F), warming in interior Alaska was as much as 1.6°C (about 3°F), causing changes ranging from the thawing of permafrost to enhanced coastal erosion resulting from melting of sea ice.

Model simulations project that minimum temperatures are likely to continue to rise more rapidly than maximum temperatures, extending the trend that started during the 20th century. Although winter temperatures are projected to increase somewhat more rapidly than summer temperatures, the summertime heat index is projected to rise quite sharply because the rising absolute humidity will make summer conditions feel much more uncomfortable, particularly across the southern and eastern United States.

Although a 0.6°C (1°F) warming may not seem large compared to daily variations in temperature, it caused a decline of about two days per year in the number of days that minimum temperatures fell below freezing. Across the United States, this change was most apparent in winter and spring, with little change in autumn. The timing of the last spring frost changed similarly, with earlier cessation of spring frosts contributing to a lengthening of the frost-free season over the country. Even these seemingly small temperature-related changes have had some effects on the natural environment, including shorter duration of lake ice, a northward shift in the distributions of some species of butterflies, changes in the timing of bird migrations, and a longer growing season.

With respect to changes in precipitation, observations for the 20th century indicate that total annual precipitation has been increasing, both worldwide and over the country. For the contiguous United States, total annual precipitation increased by an estimated 5–10 percent over the past 100 years. With the exception of localized decreases in parts of the upper Great Plains, the Rocky Mountains, and Alaska, most regions experienced greater precipitation (Figure 6-2). This increased precipitation is evident in daily precipitation rates and in the number of rain days. It has caused widespread increases in stream flow for all levels of flow conditions, particularly during times of low to moderate flow conditions—changes that have generally improved water resource conditions and have reduced situations of hydrologic drought.

For the 21st century, models project a continuing increase in global precipitation, with much of the increase occurring in middle and high latitudes. The models also suggest that the increases are likely to be evident in rainfall events that, based on conditions in each region, would be considered heavy (Figure 6-3). However, estimates of the regional pattern of changes vary significantly. While there are some indications that wintertime precipitation in the southwestern United States is likely to increase due to

---

FIGURE 6-1 Potential Effects of 21st-Century Warming on the Summer Climate of Illinois

This schematic illustrates how the summer climate of Illinois would shift under two plausible climate scenarios. Under the Canadian Climate Centre model's hot-dry climate scenario, the changes in summertime temperature and precipitation in Illinois would resemble the current climatic conditions in southern Missouri by the 2030s and in Oklahoma by the 2090s. For the warm-moist climate scenario projected by U.K.'s Hadley Centre model, summer in Illinois would become more like current summer conditions in the central Appalachians by the 2030s and North Carolina by the 2090s. Both shifts indicate warming of several degrees, but the scenarios differ in terms of projected changes in precipitation.

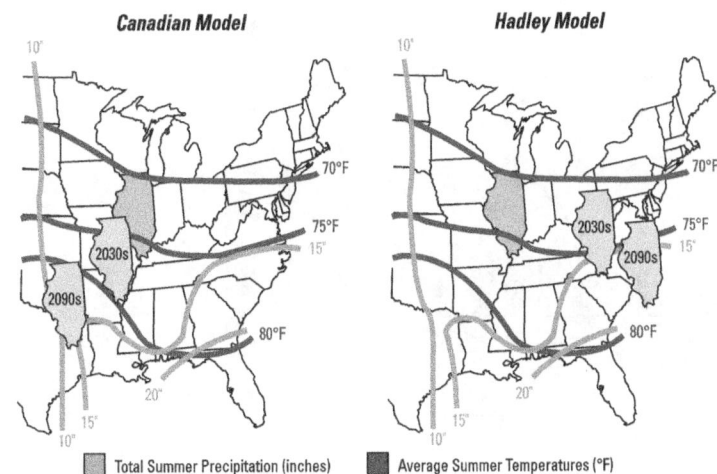

Note: The baseline climatic values are for the period 1961–90.
Source: D.J. Wuebbles, University of Illinois Urbana-Champaign, as included in NAST 2000.

warming of the Pacific Ocean, changes across key U.S. forest and agricultural regions remain uncertain.

Soil moisture is critical for agriculture, vegetation, and water resources. Projections of changes in soil moisture depend on precipitation and runoff; changes in the timing and form of the precipitation (i.e., rain or snow); and changes in water loss by evaporation, which in turn depends on temperature change, vegetation, and the effects of changes in CO_2 concentration on evapotranspiration. As a result of the many interrelationships, projections remain somewhat uncertain of how changes in precipitation are likely to affect soil moisture and runoff, although the rising summertime temperature is likely to create additional stress by significantly increasing evaporation.

FIGURE 6-2 Observed Changes in Precipitation: 1901–1998

The geographical pattern of observed changes in U.S. annual precipitation during the 20th century indicates that, although local variations are occurring, precipitation has been increasing in most regions. The results are based on data from 1,221 Historical Climatology Network stations. These data are being used to derive estimates of a 100-year trend for each U.S. climate division.

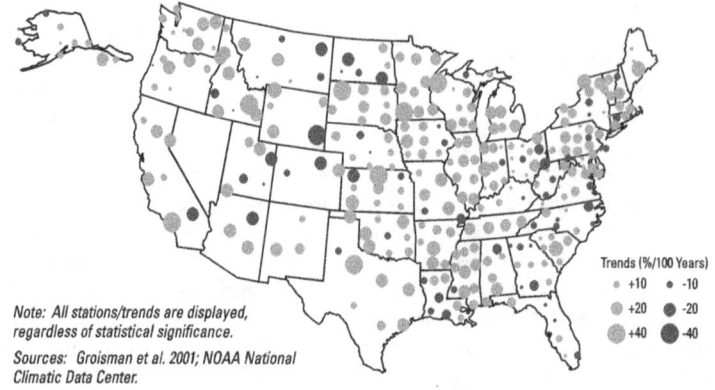

Note: All stations/trends are displayed, regardless of statistical significance.

Sources: Groisman et al. 2001; NOAA National Climatic Data Center.

FIGURE 6-3 Projected Changes in the Intensity of U.S. Precipitation for the 21st Century

The projected changes in precipitation over the United States as calculated by two models indicate that most of the increase is likely to occur in the locally heaviest categories of precipitation. Each bar represents the percentage change of precipitation in a different category of storm intensity. For example, the two bars on the far right indicate that the Canadian Centre model projects an increase of over 20 percent in the 5 percent most intense rainfall events in each region, whereas the Hadley Centre model projects an increase of over 55 percent in such events. Because both historic trends and future projections from many global climate models indicate an increase in the fraction of precipitation occurring during the heaviest categories of precipitation events in each region, a continuation of this trend is considered likely. Although this does not necessarily translate into an increase in flooding, higher river flows are likely to be a consequence.

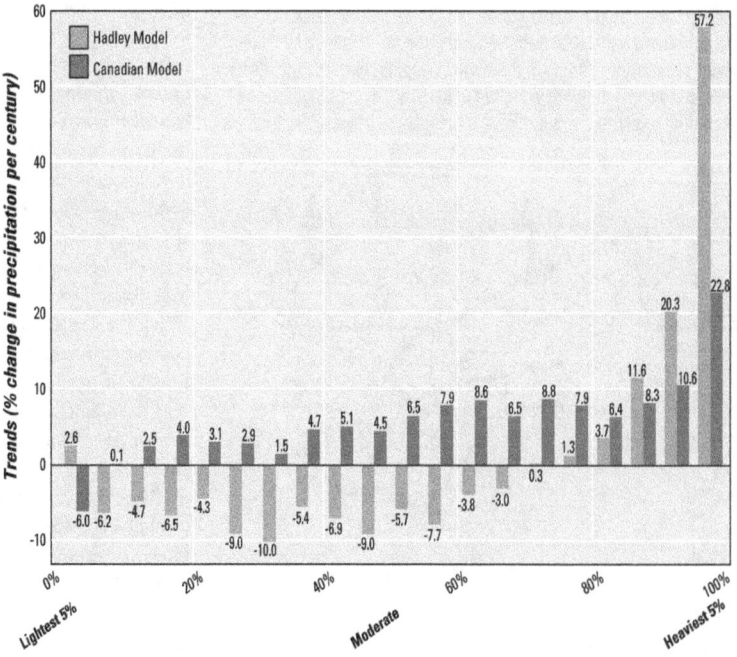

Source: Byron Gleason, NOAA National Climatic Data Center (updated from NAST 2000).

Projected Changes in Climate Variability

As in other highly developed nations, U.S. communities and industries have made substantial efforts to reduce their vulnerability to normal weather and climate fluctuations. However, adaptation to potential changes in weather extremes and climate variability is likely to be more difficult and costly. Unfortunately, projections of such changes remain quite uncertain, especially because variations in climate differentially affect different regions of the country. Perhaps the best-known example of a natural variation of the climate is caused by the El Niño–Southern Oscillation (ENSO), which is currently occurring every several years. ENSO has reasonably well-established effects on seasonal climate conditions across the country. For example, in the El Niño phase, unusually high sea-surface temperatures (SSTs) in the eastern and central equatorial Pacific act to suppress the occurrence of Atlantic hurricanes (Figure 6-4) and result in higher-than-average wintertime precipitation in the southwestern and southeastern United States, and in above-average temperatures in the Midwest (Figure 6-5). During a strong El Niño, effects can extend into the northern Great Plains.

During the La Niña phase, which is characterized by unusually low SSTs off the west coast of South America, higher-than-average wintertime temperatures prevail across the southern half of the United States, more hurricanes occur in the tropical Atlantic, and more tornadoes occur in the Ohio and Tennessee valleys (Figures 6-4 and 6-5). During the summer, La Niña conditions can contribute to the occurrence of drought in the eastern half of the United States.

Other factors that affect the inter-annual variability of the U.S. climate include the Pacific Decadal Oscillation (PDO) and the North Atlantic Oscillation (NAO).

The PDO is a phenomenon similar to ENSO, but is most apparent in the SSTs of the North Pacific Ocean. The PDO has a periodicity that is on the order of decades and, like ENSO, has

---

FIGURE 6-4 Likelihood of Hurricanes to Strike the United States Based on El Niño and La Niña Occurrence

The frequency at which various numbers of hurricanes struck the United States during the 20th century has been found to depend on whether El Niño or La Niña events were occurring. Because of this observed relationship, changes in the frequency and intensity of these events are expected to affect the potential for damaging hurricanes striking the United States.

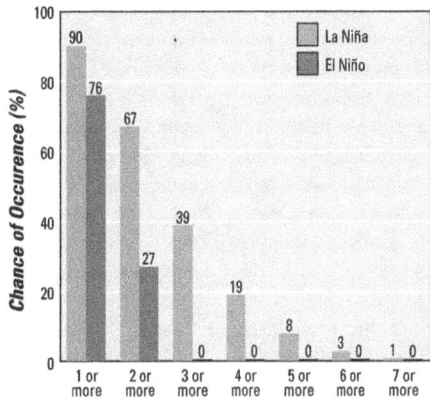

Source: Bove et al. 1998.   ***Number of Hurricanes per Year***

FIGURE 6-5 Climatic Tendencies across North America during El Niño and La Niña Events

Temperature and precipitation across North America have tended to vary from normal wintertime conditions as a result of El Niño (warmer-than-normal) and La Niña (colder-than-normal) events in the equatorial eastern Pacific Ocean. For many regions, the state of ocean temperatures in the equatorial Pacific Ocean has been found to be the most important determinant of whether winter conditions are relatively wet or dry, or relatively warm or cold. For example, winters in the Southeast tend to be generally cool and wet during El Niño (warm) events, and warm and dry during La Niña (cold) events.

Cold-Event Winter (La Niña)   *Warm-Event Winter (El Niño)*

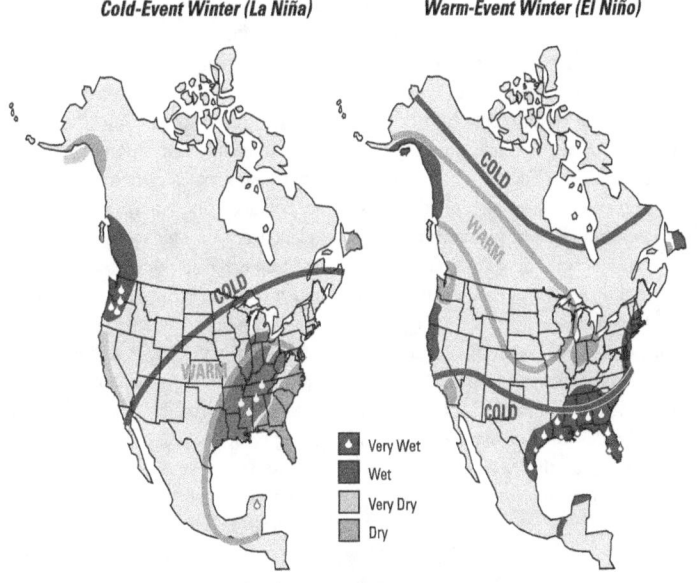

Source: Florida State University, Center for Ocean–Atmospheric Prediction Studies. <http://www.coaps.fsu.edu>

two distinct phases—a warm phase and a cool phase. In the warm phase, oceanic conditions lead to an intensification of the storm-generating Aleutian Low, higher-than-average winter temperatures in the Pacific Northwest, and relatively high SSTs along the Pacific Coast. The PDO also leads to dry winters in the Pacific Northwest, but wetter conditions both north and south of there. Essentially, the opposite conditions occur during the cool phase.

The NAO is a phenomenon that displays a seesaw in temperatures and atmospheric pressure between Greenland and northern Europe. However, the NAO also includes effects in the United States. For example, when Greenland is warmer than normal, the eastern United States is usually colder, particularly in winter, and vice-versa.

Given these important and diverse interactions, research is being intensified to improve model simulations of natural climate variations, especially to improve projections of how such variations are likely to change. Although projections remain uncertain, the climate model of the Max Planck Institute in Germany, which is currently considered to provide the most realistic simulation of the ENSO cycle, calculates stronger and wider swings between El Niño and La Niña conditions as the global climate warms (Timmermann et al. 1999), while other models simply project more El Niño-like conditions over the eastern tropical Pacific Ocean (IPCC 2001d). Either type of result would be likely to cause important climate fluctuations across the United States.

Using the selected model scenarios as guides, but also examining the potential consequences of a continuation of past climate trends and of the possibility of exceeding particular threshold conditions, the National Assessment focused its analyses on evaluating the potential environmental and societal consequences of the climate changes projected for the 21st century, as described in the next section.

POTENTIAL CONSEQUENCES OF AND ADAPTATION TO CLIMATE CHANGE

Since the late 1980s, an increasing number of studies have been undertaken to investigate the potential impacts of climate change on U.S. society and the environment (e.g., U.S. EPA 1989, U.S. Congress 1993) and as components of international assessments (e.g., IPCC 1996a, 1998). While these studies have generally indicated that many aspects of the U.S. environment and society are likely to be sensitive to changes in climate, they were unable to provide in-depth perspectives of how various types of impacts might evolve and interact. In 1997, the interagency U.S. Global Change Research Program (USGCRP) initiated a National Assessment process to evaluate and synthesize available information about the potential impacts of climate change for the United States, to identify options for adapting to climate change, and to summarize research needs for improving knowledge about vulnerability, impacts, and adaptation (see Chapter 8). The findings were also undertaken to provide a more in-depth analysis of the potential time-varying consequences of climate change for consideration in scheduled international assessments (IPCC 2001a) and to contribute to fulfilling obligations under sections 4.1(b) and (e) of the United Nations Framework Convention on Climate Change.

The U.S. National Assessment was carried out recognizing that climate change is only one among many potential stresses that society and the environment face, and that, in many cases, adaptation to climate change can be accomplished in concert with efforts to adapt to other stresses. For example, climate variability and change will interact with such issues as air and water pollution, habitat fragmentation, wetland loss, coastal erosion, and reductions in fisheries in ways that are likely to compound these stresses. In addition, an aging national populace and rapidly growing populations in cities, coastal areas, and across the South and West are social factors that interact with and in some ways can increase the sensitivity of society to climate variability and change. In both evaluating potential impacts and developing effective responses, it is therefore important to consider interactions among the various stresses.

In considering the potential impacts of climate change, however, it is also important to recognize that U.S. climate conditions vary from the cold of an Alaskan winter to the heat of a Texas summer, and from the year-round near-constancy of temperatures in Hawaii to the strong variations in North Dakota. Across this very wide range of climate conditions and seasonal variation, American ingenuity and resources have enabled communities and businesses to develop, although particular economic sectors in particular regions can experience losses and disruptions from extreme conditions of various types. For example, the amount of property damage from hurricanes has been rising, although this seems to be mainly due to increasing development and population in vulnerable coastal areas. On the other hand, the number of deaths each year from weather extremes and from climatically dependent infectious diseases has been reduced sharply compared to a century ago, and total deaths relating to the environment are currently very small in the context of total deaths in the United States, even though the U.S. population has been rising. In addition, in spite of climate change, the productivity of the agriculture and forest sectors has never been higher and continues to rise, with excess production helping to meet global demand.

This adaptation to environmental variations and extremes has been accomplished because the public and private sectors have applied technological change and knowledge about fluctuating climate to implement a broad series of steps that have enhanced resilience and reduced vulnerability. For example, these steps have ranged from better design and construction of buildings and communities to greater availability of heating in

Key National Findings Adapted from the U.S. National Assessment

Increased warming is projected for the 21st century—Assuming continued growth in world greenhouse gas emissions, the primary climate models drawn upon for the analyses carried out in the U.S. National Assessment projected that temperatures in the contiguous United States will rise 3–5°C (5–9°F) on average during the 21st century. A wider range of outcomes, including a smaller warming, is also possible.

Impacts will differ across regions—Climate change and its potential impacts are likely to vary widely across the country. Temperature increases are likely to vary somewhat among regions. Heavy precipitation events are projected to become more frequent, yet some regions are likely to become drier.

Ecosystems are especially vulnerable—Many ecosystems are highly sensitive to the projected rate and magnitude of climate change, although more efficient water use will help some ecosystems. A few ecosystems, such as alpine meadows in the Rocky Mountains and some barrier islands, are likely to disappear entirely in some areas. Other ecosystems, such as southeastern forests, are likely to experience major species shifts or break up into a mosaic of grasslands, woodlands, and forests. Some of the goods and services lost through the disappearance or fragmentation of natural ecosystems are likely to be costly or impossible to replace.

Widespread water concerns arise—Water is an issue in every region, but the nature of the vulnerabilities varies. Drought is an important concern virtually everywhere. Floods and water quality are concerns in many regions. Snowpack changes are likely to be especially important in the West, Pacific Northwest, and Alaska.

Food supply is secure—At the national level, the agriculture sector is likely to be able to adapt to climate change. Mainly because of the beneficial effects of the rising carbon dioxide levels on crops, overall U.S. crop productivity, relative to what is projected in the absence of climate change, is very likely to increase over the next few decades. However, the gains are not likely to be uniform across the nation. Falling prices are likely to cause difficulty for some farmers, while benefiting consumers.

Near-term forest growth increases—Forest productivity is likely to increase over the next several decades in some areas as trees respond to higher carbon dioxide levels by increasing water-use efficiency. Such changes could result in ecological benefits and additional storage of carbon. Over the longer term, changes in larger-scale processes, such as fire, insects, droughts, and disease, could decrease forest productivity. In addition, climate change is likely to cause long-term shifts in forest species, such as sugar maples moving north out of the country.

Increased damage occurs in coastal and permafrost areas—Climate change and the resulting rise in sea level are likely to exacerbate threats to buildings, roads, power lines, and other infrastructure in climate-sensitive areas. For example, infrastructure damage is expected to result from permafrost melting in Alaska and from sea level rise and storm surges in low-lying coastal areas.

Adaptation determines health outcomes—A range of negative health impacts is possible from climate change. However, as in the past, adaptation is likely to help protect much of the U.S. population. Maintaining our nation's public health and community infrastructure, from water treatment systems to emergency shelters, will be important for minimizing the impacts of water-borne diseases, heat stress, air pollution, extreme weather events, and diseases transmitted by insects, ticks, and rodents.

Other stresses are magnified by climate change—Climate change is very likely to modify the cumulative impacts of other stresses. While it may magnify the impacts of some stresses, such as air and water pollution and conversion of habitat due to human development patterns, it may increase agricultural and forest productivity in some areas. For coral reefs, the combined effects of increased CO_2 concentration, climate change, and other stresses are very likely to exceed a critical threshold, causing large, possibly irreversible impacts.

Uncertainties remain and surprises are expected—Significant uncertainties remain in the science underlying regional changes in climate and their impacts. Further research would improve understanding and capabilities for projecting societal and ecosystem impacts. Increased knowledge would also provide the public with additional useful information about options for adaptation. However, it is likely that some aspects and impacts of climate change, both positive and negative, will be totally unanticipated as complex systems respond to ongoing climate change in unforeseeable ways.

Sources: NAST 2000, 2001.

winter and cooling in summer, and from better warnings about extreme events to advances in public health care. Because of this increasing resilience to climate variations and relative success in adapting to the modest changes in climate that were observed during the 20th century, information about likely future climate changes and continuing efforts to plan for and adapt to these changes are likely to prove useful in minimizing future impacts and preparing to take advantage of the changing conditions.

With these objectives in mind, the U.S. National Assessment process, which is described more completely in Chapter 8, initiated a set of regional, sectoral, and national activities. This page presents an overview of key national findings, and the following subsections elaborate on these findings, covering both potential consequences and the types of adaptive steps that are underway or could be pursued to moderate or deal with adverse outcomes. The subsections summarize the types of impacts that are projected, covering initially the potential impacts on land cover; then the potential impacts on agriculture, forest, and water resources, which are key natural resource sectors on which society depends; then potential impacts associated with coastal regions and human health that define the environment in which people live; and finally summarization of the primary issues that are specific to particular U.S. regions. A full list of regional, sectoral, and national reports prepared under the auspices of the U.S. National Assessment process and additional materials relating to research and assessment activities can be found at http://www.usgcrp.gov.

Potential Interactions with Land Cover

The natural vegetative cover of the United States is largely determined by the prevailing climate and soil. Where not altered by societal activities, climate conditions largely determine where individual species of plants and animals can live, grow, and reproduce. Thus, the collections of species that we are familiar with—e.g., the southeastern mixed deciduous forest, the desert ecosystems of the arid Southwest, the productive grasslands of the Great Plains—are primarily a consequence of present climate conditions. Past changes in ecosystems indicate that some species are so strongly influenced by the climate to which they are adapted that they are vulnerable even to modest changes in climate. For example, alpine meadows at high elevations in the West exist where they do entirely because the plants that comprise them are adapted to cold conditions that are too harsh for other species in the region. The desert vegetation of the Southwest is adapted to the region's high summer temperatures and aridity. Similarly, the forests in the East tend to have adapted to relatively high rainfall and soil moisture; if drought conditions were to persist, grasses and shrubs could begin to out-compete tree seedlings, leading to completely different ecosystems.

To provide a common base of information about potential changes in vegetation across the nation for use in the National Assessment (NAST 2000), specialized ecosystem models were used to evaluate the potential consequences of climate change and an increasing CO_2 concentration for the dominant vegetation types. Biogeography models were used to simulate potential shifts in the geographic distribution of major plant species and communities (ecosystem structure). And biogeochemistry models were used to simulate changes in basic ecosystem processes, such as the cycling of carbon, nutrients, and water (ecosystem function). Each type of model was used in considering the potential consequences of the two primary model-based climate scenarios. These scenarios represented conditions across much of the United States that were generally either warmer and moister, or hotter and drier. The results from both types of models indicated that changes in ecosystems would be likely to be significant.

Climate changes that affect the land surface and terrestrial vegetation will also have implications for fresh-water and coastal marine ecosystems that depend on the temperature of runoff water, on the amount of erosion, and on other factors dependent on the land cover. For example, in aquatic ecosystems, many fish can breed only in water that falls within a narrow range of temperatures. As a result, species of fish that are adapted to cool waters can quickly become unable to breed successfully if water temperatures rise. As another example, because washed-off soil and nutrients can benefit wetland species (within limits) and harm estuarine ecosystems, changes in the frequency or intensity of runoff events caused by changes in land cover can be important. Such impacts are described in the subsections dealing with climate change interactions with water resources and the coastal environment, while issues affecting terrestrial land cover are covered in the following subsection.

Redistribution of Land Cover

The responses of ecosystems to projected changes in climate and CO_2 are made up of the individual responses of their constituent species and how they interact with each other. Species in current ecosystems can differ substantially in their tolerances to changes in temperature and precipitation, and in their responses to changes in the CO_2 concentration. As a result, the ranges of individual species are likely to shift at different rates, and different species are likely to have different degrees of success in establishing themselves in new locations and environments. While changes in climate projected for the coming hundred years are very likely to alter current ecosystems, projecting these kinds of biological and ecological responses and the structure and functioning of the new plant communities is very difficult.

Analyses of present ecosystem distributions and of past shifts indicate that natural ecosystems are sensitive to changes in surface temperature, precipitation patterns, and other climate parameters and changes in the atmospheric CO_2 concentration. For example, changes in temperature and precipitation of the magnitude being projected are likely to cause shifts in the areas occupied by dominant vegetation types relative to their current distribution. Some ecosystems that are already constrained by climate, such as alpine meadows in the Rocky Mountains, are likely to face extreme stress and disappear entirely in some places. Other more widespread ecosystems are also likely to be sensitive to climate change. For example, both climate model scenarios suggest that the southwestern United States will become moister, allowing more vegetation to grow (Figure 6-6). Such a change is likely to transform desert landscapes into grasslands or shrublands, altering both their potential use and the likelihood of fire. In the northeastern United States, both climate scenarios suggest changes mainly in the species composition of the forests, including the northward displacement of sugar maples, which could lead to loss in some areas. However, the studies also indicate that conditions in this region will remain conducive to maintaining a forested landscape, mainly oak and hickory. In the southeastern United States, however, there was less agreement among the models: the hot-dry climate scenario was projected to lead to conditions that would be conducive to the potential breakup of the forest landscape into a mosaic of forests, savannas, and grasslands; in contrast, the warm-moist scenario was projected to lead to a northward expansion of the southeastern mixed forest cover. (See additional discussion in the Forest subsection.)

Basically, changes in land cover were projected to occur, at least to some degree, in all locations, and these changes cannot generally be prevented if the climate changes and vegetation responds as much as projected.

Effects on the Supply of Vital Ecosystem Goods and Services

In addition to the value of natural ecosystems in their own right, ecosystems of all types, from the most natural to the most extensively managed, provide a variety of goods and services that benefit society. Some products of ecosystems enter the market and

FIGURE 6-6 Potential Effects of Projected Climate Change on Ecosystem Distribution

Both the Hadley and the Canadian models project increasing wintertime precipitation in the U.S. Southwest toward the end of the 21st century and a conversion of desert ecosystems to shrub and grassland ecosystems.

Current Ecosystems

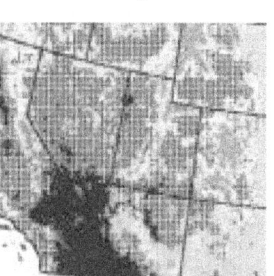

Source: R.P. Neilson, USDA Forest Service, Corvallis, Oregon, as presented in NAST 2000.

Canadian Model

Alpine   Shrubland
Forest   Grassland
Savanna   Arid Land

Hadley Model

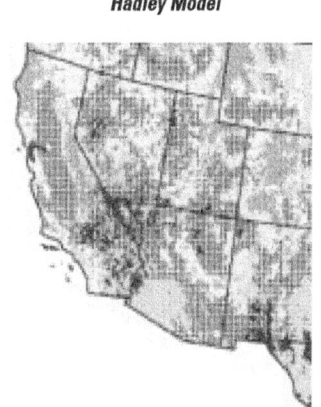

contribute directly to the economy. For example, forests serve as sources of timber and pulpwood, and agro-ecosystems serve as sources of food. Ecosystems also provide a set of unpriced services that are valuable but that typically are not traded in the marketplace. Although there is no current market, for example, for the services that forests and wetlands provide for improving water quality, regulating stream flow, providing some measure of protection from floods, and sequestering carbon, some of these services are very valuable to society. Ecosystems are also valued for recreational, aesthetic, and ethical reasons. For example, the bird life of the coastal marshes of the Southeast and the brilliant autumn colors of the New England forests are treasured components of the nation's regional heritages and important elements of our quality of life.

Based on the studies carried out, changes in land cover induced by climate change, along with an increased level of disturbances, could have varied impacts on ecosystem services, including the abilities of ecosystems to cleanse the air and water, stabilize landscapes against erosion, and store carbon. Even in such regions as the Southwest, where vegetation is expected to increase as a result of

increased rainfall and enhanced plant growth due to the rising CO_2 concentration, an important potential consequence is likely to be a heightened frequency and intensity of fires during the prolonged summer season. Increased fire frequency would likely be a threat not only to the natural land cover, but also to the many residential structures being built in vulnerable suburban and rural areas, and later would increase vulnerability to mudslides as a result of denuded hills. Considering the full range of available results, it is plausible that climate change-induced alterations to natural ecosystems could affect the availability of some ecosystem goods and services.

Effects of an Increased CO_2 Concentration on Plants

The ecosystem models used in the National Assessment considered the potential effects of increases in the atmospheric CO_2 concentration because the CO_2 concentration affects plant species via a direct physiological effect on photosynthesis (the process by which plants use CO_2 to create new biological material). Higher CO_2 concentrations generally enhance plant growth if the plants also have sufficient water and nutrients (such as nitrogen) that are needed to sustain this enhanced growth.

For example, the CO_2 level in commercial greenhouses is sometimes boosted to stimulate plant growth. In addition to enhancing plant growth, higher CO_2 levels can raise the efficiency with which plants use water and reduce their susceptibility to damage by air pollutants.

As a result of these various influences, different types of plants respond at different rates to increases in the atmospheric CO_2 concentration, resulting in a divergence of growth rates. Most species grow faster and increase biomass; however, the nutritional value of some of these plants could be altered. Both because of biochemical processing and because warming temperatures increase plant respiration, the beneficial effects of increased CO_2 on plants are also projected to flatten at some higher level of CO_2 concentration, beyond which continuing increases in the CO_2 concentration would not enhance plant growth.

While there is still much to be learned about the CO_2 "fertilization" effect, including its limits and its direct and indirect implications, many ecosystems are projected to benefit from a higher CO_2 concentration, and plants will use water more efficiently.

Effects on Storage of Carbon

In response to changes in climate and the CO_2 concentration, the biogeochemistry models used in the National Assessment generally simulated increases in the amount of carbon stored in vegetation and soils for the continental United States. The calculated increases were relatively small, however, and not uniform across the country. For example, one of the biogeochemistry models, when simulating the effects of hotter and drier conditions, projected that the southeastern forests would lose more carbon by respiration than they would gain by increased photosynthesis, causing an overall carbon loss of up to 20 percent by 2030. Such a loss would indicate that the forests were in a state of decline. The same biogeochemistry model, however, when calculating the potential effects of the warmer and moister climate scenario, projected that forests in the same part of the Southeast would likely gain between 5 and 10 percent in carbon over the next 30 years, suggesting a more vigorous forest.

Susceptibility of Ecosystems to Disturbances

Prolonged stress due to insufficient soil moisture can make trees more susceptible to insect attack, lead to plant death, and increase the probability of fire as dead plant material adds to an ecosystem's "fuel load." The biogeography models used in this analysis simulated at least part of this sequence of climate-triggered events in ecosystems as a prelude to calculating shifts in the geographic distribution of major plant species.

For example, one of the biogeography models projected that a hot, dry climate in the Southeast would be likely to result in the replacement of the current mixed evergreen and deciduous forests by savanna/woodlands and grasslands, with much of the change effected by an increased incidence of fire. Yet the same biogeography model projected a slight northward expansion of the mixed evergreen and deciduous forests of the Southeast in response to the warm, moist climate scenario, with no significant contraction along the southern boundary. Thus, in this region, changes in the frequency and

intensity of such disturbances as fire are likely to be major determinants of the type and rapidity of the conversion of the land cover to a new state.

As explained more fully in the sections on the interactions of climate change with coastal and water resources, aquatic ecosystems are also likely to be affected by both climate change and unusual disturbances, such as storms and storm surges.

Potential Adaptation Options to Preserve Prevailing Land Cover

The National Assessment concluded that the potential vulnerability of natural ecosystems is likely to be more important than other types of potential impacts affecting the U.S. environment and society. This importance arises because in many cases little can be done to help these ecosystems adapt to the projected rate and amount of climate change. While adjustments in how some systems are managed can perhaps reduce the potential impacts, the complex, interdependent webs that have been naturally generated over very long periods are not readily shifted from one place to another or easily recreated in new locations, even to regions of similar temperature and moisture. Although many regions have experienced changes in ecosystems as a result of human-induced changes in land cover, and people have generally become adapted to—and have even become defenders of—the altered conditions (e.g., reforestation of New England), the climate-induced changes during the 21st century are likely to affect virtually every region of the country—both the ecosystems where people live, as well as those in the protected areas that have been created as refuges against change.

Potential Interactions with Agriculture

U.S. croplands, grassland pasture, and range occupy about 420 million hectares (about 1,030 million acres), or nearly 55 percent of the U.S. land area, excluding Alaska and Hawaii (USDA/ERS 2000). Throughout the

20th century, agricultural production shifted toward the West and Southwest. This trend allowed regrowth of some forests and grasslands, generally enhancing wildlife habitats, especially in the Northeast, and contributing to sequestration of carbon in these regions.

U.S. food production and distribution comprise about 10 percent of the U.S. economy. The value of U.S. agricultural commodities (food and fiber) exceeds $165 billion at the farm level and over $500 billion after processing and marketing. Because of the productivity of U.S. agriculture, the United States is a major supplier of food and fiber for the world, accounting for more than 25 percent of total global trade in wheat, corn, soybeans, and cotton.

Changes in Agricultural Productivity

U.S. agricultural productivity has improved by over 1 percent a year since 1950, resulting in a decline in both production costs and commodity prices, limiting the net conversion of natural habitat to cropland, and freeing up land for the Conservation Reserve Program. Although the increased production and the two-thirds drop in real commodity prices have been particularly beneficial to consumers inside and outside the United States and have helped to reduce hunger and malnourishment around the world, the lower prices have become a major concern for producers and have contributed to the continuing decline in the number of small farmers across the country. Continuation of these trends is expected, regardless of whether climate changes, with continuing pressures on individual producers to further increase productivity and reduce production costs.

On the other hand, producers consider anything that might increase their costs relative to other producers or that might limit their markets as a threat to their economic well-being. Issues of concern include regulatory actions, such as efforts to control the off-site consequences of soil erosion, agricultural chemicals, and livestock wastes; extreme weather or climate events; new pests; and the development of pest resistance to existing pest control strategies.

Future changes in climate are expected to interact with all of these issues. In particular, although some factors may tend to limit growth in yields, rising CO_2 concentrations and continuing climate change are projected, on average, to contribute to extending the persistent upward trend in crop yields that has been evident during the second half of the 20th century. In addition, if all else remains equal, these changes could change supplies of and requirements for irrigation water, increase the need for fertilizers to sustain the gain in carbon production, lead to changes in surface-water quality, necessitate increased use of pesticides or other means to limit damage from pests, and alter the variability of the climate to which the prevailing agricultural sector has become accustomed. However, agricultural technology is currently undergoing rapid change, and future production technologies and practices seem likely to be able to contain or reduce these impacts.

Assuming that technological advances continue at historical rates, that there are no dramatic changes in federal policies or in international markets, that adequate supplies of nutrients are available and can be applied without exacerbating pollution problems, and that no prolonged droughts occur in major agricultural regions, U.S. analyses indicate that it is unlikely that climate change will imperil the ability of the United States to feed its population and to export substantial amounts of foodstuffs (NAAG 2002). These studies indicate that, at the national level, overall agricultural productivity is likely to increase as a result of changes in the CO_2 concentration and in climate projected for at least the next several decades. The crop models used in these studies assume that the CO_2 fertilization effect will be strongly beneficial and will also allow for a limited set of *on-farm* adaptation options, including changing planting dates and varieties, in res-ponse to the changing condi-

tions. These adaptation measures contribute small additional gains in yields of dry-land crops and greater gains in yields of irrigated crops. However, analyses performed to date have neither considered all of the consequences of possible changes in pests, diseases, insects, and extreme events that may result, nor been able to consider the full range of potential adaptation options (e.g., genetic modification of crops to enhance resistance to pests, insects, and diseases).

Recognizing these limitations, available evaluations of the effects of anticipated changes in the CO_2 concentration and climate on crop production and yield and the adaptive actions by farmers generally show positive results for cotton, corn for grain and silage, soybeans, sorghum, barley, sugar beets, and citrus fruits (Figure 6-7). The productivity of pastures may also increase as a result of these changes. For other crops, including wheat, rice, oats, hay, sugar cane, potatoes, and tomatoes, yields are projected to increase under some conditions and decrease under others, as explained more fully in the agriculture assessment (NAAG 2002).

The studies also indicate that not all U.S. agricultural regions are likely to be affected to the same degree by the projected changes in climate that have been investigated. In general, northern areas, such as the Midwest, West, and Pacific Northwest, are projected to show large gains in yields, while influences on crop yields in other regions vary more widely, depending on the climate scenario and time period. For example, projected wheat yields in the southern Great Plains could decline if the warming is not accompanied by sufficient precipitation.

These analyses used market-scale economic models to evaluate the overall economic implications for various crops. These models allow for a wide range of adaptations in response to changing productivity, prices, and resource use, including changes in irrigation, use of fertilizer and pesticides, crops grown and the location of cropping, and a variety of other farm management options. Based on studies to date, unless there is

inadequate or poorly distributed precipitation, the net effects of climate change on the agricultural segment of the U.S. economy over the 21st century are generally projected to be positive. These studies indicate that, economically, consumers are likely to benefit more from lower prices than producers suffer from the decline in profits. Complicating the analyses, however, the studies indicate that producer versus consumer effects will depend on how climate change affects production of these crops elsewhere in the world. For example, for crops grown in the United States, economic losses to farmers due to lower commodity prices are offset under some conditions by an increased advantage of U.S. farmers over foreign competitors, leading to an increased volume of exports.

Because U.S. food variety and supplies depend not only on foodstuffs produced nationally, the net effect of climate change on foods available for U.S. consumers will also depend on the effects of climate change on global production of these foodstuffs. These effects will in turn depend not only on international markets, but also on how farmers around the world are able to adapt to climate change and other factors they will face. While there are likely to be many regional variations, experience indicates that research sponsored by the United States and other nations has played an important role in promoting the ongoing, long-term increase in global agricultural productivity. Further research, covering opportunities ranging from genetic design to improving the salt tolerance of key crops, is expected to continue to enhance overall global production of foodstuffs.

Changes in Water Demands by Agriculture

Within the United States, a key determinant of agricultural productivity will be the ongoing availability of sufficient water where and when it is needed. The variability of the U.S. climate has provided many opportunities for learning to deal with a wide range of climate conditions, and the U.S. regions

FIGURE 6-7 Effects of Potential Changes in Climate on U.S Crop Yields

Results for 16 crops, given as the percentage differences between future yields for two periods (2030s and 2090s) and current yields indicate that warmer climate conditions are likely to lead to increased yields for most crops. The results consider the physiological responses of the crops to future climate conditions under either dry-land or irrigated cultivation, assuming a limited set of reasonable adaptive response by producers. Climate scenarios are drawn from two different climate models that are likely to span the range of changes of future conditions, ranging from the warm-moist changes projected by the U.K.'s Hadley Centre model (version 2) to the hot-dry changes projected by the Canadian Climate Centre model. The most positive responses resulted when conditions were warmer and wetter in key growing regions (e.g., cotton), when frost occurrence was reduced (e.g., grapefruit), and when northern areas warmed (e.g., silage from pasture improvement).

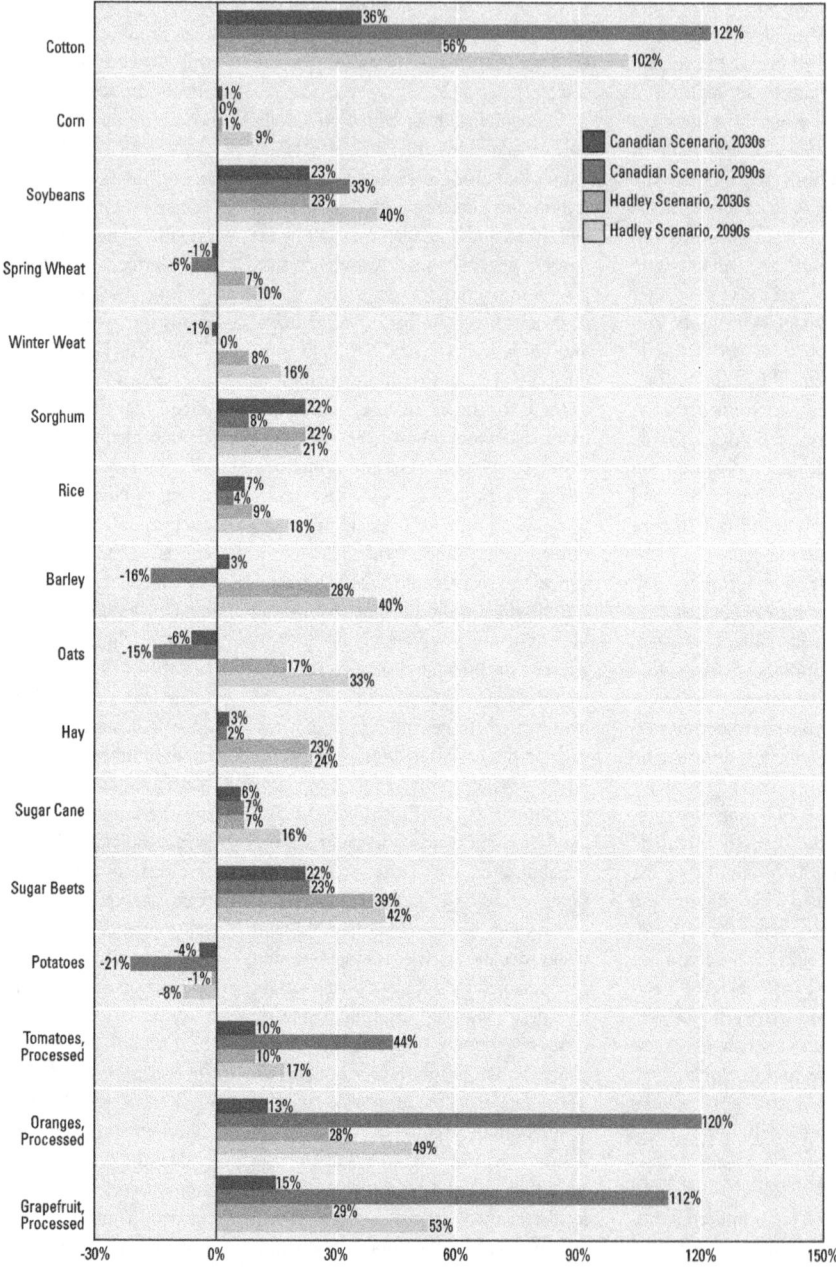

Source: NAAG 2002.

where many crops are grown have changed over time without disrupting production. In addition, steps to build up the amount of carbon in soils—which is likely to be one component of any carbon mitigation program—will enhance the water-holding capacity of soils and decrease erosion and vulnerability to drought, thereby helping to improve overall agricultural productivity. For areas that are insufficiently moist, irrigation has been used to enhance crop productivity. In addition, about 27 percent of U.S. cultivated land is currently under reduced tillage. Several projects, such as the Iowa Soil Carbon Sequestration Project, that are underway to promote conservation tillage practices as a means to mitigate climate change will have the ancillary benefits of reducing soil erosion and runoff while increasing soil water and nitrogen retention.

Analyses conducted for the National Assessment project that climate change will lead to changes in the demand for irrigation water and, if water resources are insufficient, to changes in the crops being grown. Although regional differences will likely be substantial, model projections indicate that, on average for the nation, agriculture's need for irrigation water is likely to slowly decline. At least two factors are responsible for this projected reduction: (1) precipitation will increase in some agricultural areas, and (2) faster development of crops due to higher temperatures and an increased CO_2 concentration is likely to result in a shorter growing period and consequently a reduced demand for irrigation water. Moreover, a higher CO_2 concentration generally enhance a plant's water-use efficiency. These factors can combine to compensate for the increased transpiration and soil water loss due to higher air temperatures. However, a decreased period of crop growth also leads to decreased yields, although it may be possible to overcome this disadvantage through crop breeding.

Changes in Surface-Water Quality due to Agriculture

Potential changes in surface-water quality as a result of climate change is an issue that has only started to be investigated. For example, in recent decades, soil erosion and excess nutrient runoff from crop and livestock production have severely degraded Chesapeake Bay, a highly valuable natural resource. In simulations for the National Assessment, loading of excess nitrogen from corn production into Chesapeake Bay is projected to increase due to both the change in average climate conditions and the effects of projected changes in extreme weather events, such as floods or heavy downpours that wash large amounts of fertilizers and animal manure into surface waters. Across the country, changes in future farm practice (such as no-till or reduced-till agriculture) that enhance buildup and retention of soil moisture, and better matching of the timing of a crop's need for fertilizer with the timing of application are examples of approaches that could reduce projected adverse impacts on water quality. In addition, the potential for reducing adverse impacts of fertilizer application and soil erosion by using genetically modified crops has not yet been considered.

Changes in Pesticide Use by Agriculture

Climate change is projected to cause farmers in most regions to increase their use of pesticides to sustain the productivity of current crop strains. While this increase is expected to result in slightly poorer overall economic performance, this effect is minimal because pesticide expenditures are a relatively small share of production costs. Neither the potential changes in environmental impacts as a result of increased pesticide use nor the potential for genetic modification to enhance pest resistance have yet been evaluated.

Effects of Changes in Climate Variability on Agriculture

Based on experience, agriculture is also likely to be affected if the extent and occurrence of climate fluctuations and extreme events change. The vulnerability of agricultural systems to climate and weather extremes varies with location because of differences in soils, production systems, and other factors. Changes in the form (rain, snow, or hail), timing, frequency, and intensity of precipitation, and changes in wind-driven events (e.g., wind storms, hurricanes, and tornadoes) are likely to have significant consequences in particular regions. For example, in the absence of adaptive measures, an increase in heavy precipitation events seems likely in some areas to aggravate erosion, water-logging of soils, and leaching of animal wastes, pesticides, fertilizers, and other chemicals into surface and ground water. Conversely, lower precipitation in other areas may reduce some types of impacts.

A major source of U.S. climate variability is the El Niño–Southern Oscillation (ENSO). The effects of ENSO events vary widely across the country, creating wet conditions in some areas and dry conditions in others that can have significant impacts on agricultural production. For example, over the past several decades, average corn yield has been reduced by about 15–30 percent in years with widespread floods or drought. Better prediction of such variations is a major focus of U.S. and international research activities (e.g., through the International Research Institute for Climate Prediction) because, in part, such information could increase the range of adaptive responses available to farmers. For example, given sufficient warning of climate anomalies (e.g., of conditions being warm and dry, cool and moist, etc.), crop species and crop planting dates could be optimized for the predicted variation, helping to reduce the adverse impact on yields and overall production. Because long-term projections suggest that ENSO variations may become even stronger as global average temperature increases, achieving even better predictive skill in the future will be especially important to efforts to maximize production in the face of climate fluctuations.

Potential Adaptation Strategies for Agriculture

To ameliorate the deleterious effects of climate change generally, such adaptation strategies as changing planting dates and varieties are likely to help to significantly offset economic losses and increase relative yields. Adaptive measures are likely to be particularly critical for the Southeast because of the large reductions in yields projected for some crops if summer precipitation declines. With the wide range of growing conditions across the United States, specific breeding for response to CO_2 is likely to be required to more fully benefit from the CO_2 fertilization effect detected in experimental crop studies. Breeding for tolerance to climatic stress has already been exploited, and varieties that do best under ideal conditions usually also out-perform other varieties under stress conditions.

Although many types of changes can likely be adapted to, some adaptations to climate change and its impacts may have negative secondary effects. For example, an analysis of the potential effects of climate change on water use from the Edward's aquifer region near San Antonio, Texas, found increased demand for ground-water resources. Increased water use from this aquifer would threaten endangered species dependent on flows from springs supported by the aquifer.

In addition, in the absence of genetic modification of available crop species to counter these influences, pesticide and herbicide use is likely to increase with warming. Greater chemical inputs would be expected to increase the potential for chemically contaminated runoff reaching prairie wetlands and ground water, which, if not controlled by on-site measures, could pollute rivers and lakes, drinking-water supplies, coastal waters, recreation areas, and waterfowl habitat.

As in the past, farmers will need to continue to adapt to the changing conditions affecting agriculture, and changing climate is likely to become an increasingly influential factor. Presuming adaptation to changing climate conditions is successful, the U.S. agricultural sector should remain strong—growing more on less land while continuing to lower prices for the consumer, exporting large amounts of food

to help feed the world, and storing carbon to enhance resilience to drought and contribute to the slowing of climate change.

Potential Interactions with Forests

Forests cover nearly one-third of the United States, providing wildlife habitat; clean air and water; carbon storage; and recreational opportunities, such as hiking, camping, and fishing. In addition, harvested products include timber, pulpwood, fuelwood, wild game, ferns, mushrooms, berries, and much more. This wealth of products and services depends on forest productivity and biodiversity, which are in turn strongly influenced by climate.

Across the country, native forests are adapted to the local climates in which they developed, such as the cold-tolerant boreal forests of Alaska, the summer drought-tolerant forests of the Pacific Northwest, and the drought-adapted piñon-juniper forests of the Southwest. Given the overall importance of the nation's forests, the potential impacts from climate change are receiving close attention, although it is only one of several factors meriting consideration.

A range of human activities causes changes in forests. For example, significant areas of native forests have been converted to agricultural use, and expansion of urban areas has fragmented forests into smaller, less contiguous patches. In some parts of the country, intensive management and favorable climates have resulted in development of highly productive forests, such as southern pine plantations, in place of the natural land cover. Fire suppression, particularly in southeastern, midwestern, and western forests, has also led to changes in forest area and in species composition. Harvesting methods have also changed species composition, while planting trees for aesthetic and landscaping purposes in urban and rural areas has expanded the presence of some species. In addition, large areas, particularly in the Northeast, have become reforested

as forests have taken over abandoned agricultural lands, allowing reestablishment of the ranges of many wildlife species.

Changes in climate and in the CO_2 concentration are emerging as important human-induced influences that are affecting forests. These factors are interacting with factors already causing changes in forests to further affect the socioeconomic benefits and the goods and services forests provide, including the extent, composition, and productivity of forests; the frequency and intensity of such natural disturbances as fire; and the level of biodiversity (NFAG 2001). Based on model projections of moderate to large warming, Figure 6-8 gives an example of the general character of changes that could occur for forests in the eastern United States by the late 21st century.

Effects on Forest Productivity

A synthesis of laboratory and field studies and modeling indicates that the fertilizing effect of atmospheric CO_2 will increase forest productivity. However, increases are likely to be strongly tempered by local conditions, such as moisture stress and nutrient availability. Across a wide range of scenarios, modest warming is likely to result in increased carbon storage in most U.S. forests, although under some of the warmer model scenarios, forests in the Southeast and the Northwest could experience drought-induced losses of carbon, possibly exacerbated by increased fire disturbance. These potential gains and losses of carbon would be in addition to changes resulting from changes in land use, such as the conversion of forests to agricultural lands or development.

Other components of environmental change, such as nitrogen deposition and ground-level ozone concentrations, are also affecting forest processes. Models used in the forest sector assessment suggest a synergistic fertilization response between CO_2 and nitrogen enrichment, leading to further increases in productivity (NFAG 2001). However, ozone acts in the opposite direction. Current ozone levels, for example, have important effects on many herbaceous species and are estimated to decrease production in southern pine plantations by 5 percent, in northeastern forests by 10 percent, and in some western forests by even more. Interactions among these physical and chemical changes and other components of global change will be important in projecting the future state of U.S. forests. For example, a higher CO_2 concentration can tend to suppress the impacts of ozone on plants.

Effects on Natural Disturbances

Natural disturbances having the greatest effects on forests include insects, disease, non-native species, fires, droughts, hurricanes, landslides, wind storms, and ice storms. While some tree species are very susceptible to fire, others, such as lodgepole pine, are dependent on occasional fires for successful reproduction.

Over millennia, local, regional, and global-scale changes in temperature and precipitation have influenced the occurrence, frequency, and intensity of these natural disturbances. These changes in disturbance regimes are a natural part of all ecosystems. However, as a consequence of climate change, forests may soon be facing more rapid alterations in the nature of these disturbances. For example, unless there is a large increase in precipitation, the seasonal severity of fire hazard is projected to increase during the 21st century over much of the country, particularly in the Southeast and Alaska.

The consequences of drought depend on annual and seasonal climate changes and whether the current adaptations of forests to drought will offer resistance and resilience to new conditions. The ecological models used in the National Assessment indicated that increases in drought stresses are most likely to occur in the forests of the Southeast, southern Rocky Mountains, and parts of the Northwest. Hurricanes,

ice storms, wind storms, landslides, insect infestations, disease, and introduced species are also likely to be climate-modulated influences that affect forests. However, projection of changes in the frequencies, intensities, and locations of such factors and their influences are difficult to project. What is clear is that, as climate changes, alterations in these disturbances and in their effects on forests are possible.

Effects on Forest Biodiversity

In addition to the very large influences of changes in land cover, changes in the distribution and abundance of plant and animal species are a result of both (1) the birth, growth, death, and dispersal rates of individuals in a population and (2) the competition between individuals of the same species and other species. These can all be influenced in turn by weather, climate, contaminants, nutrients, and other abiotic factors. When aggregated, these processes can result in the local disappearance or

---

FIGURE 6-8 Potential Effects of Projected Climate Change on Dominant Forest Types

Both the warm-moist climate change scenario from the Hadley climate model and the hot-dry scenario from the Canadian climate model suggest a significant northward shift in prevailing forest types. For example, the maple-beech-birch forest type is projected to shift north into Canada and no longer be dominant in the late 21st century in the northeastern United States.

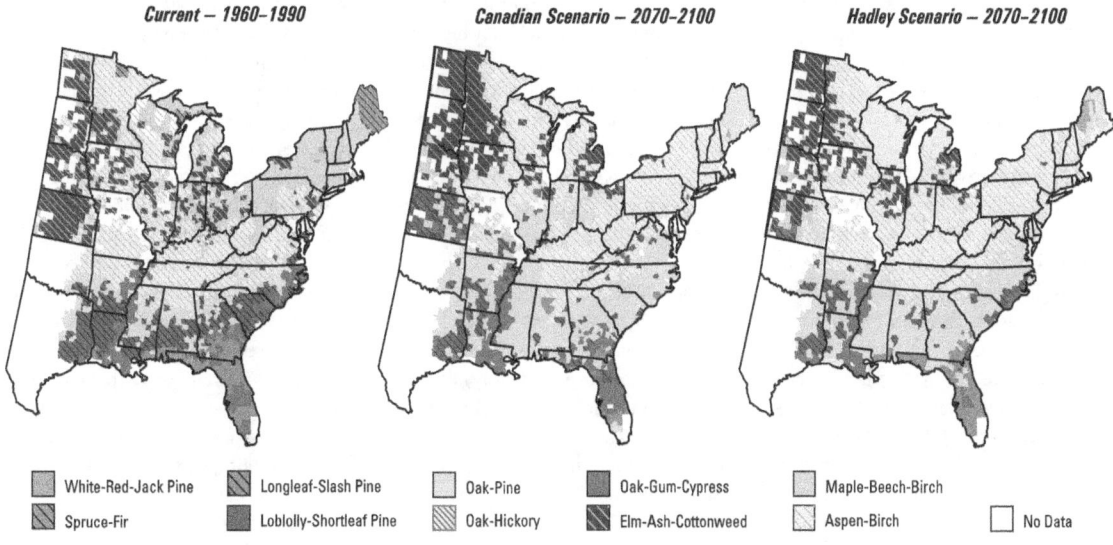

Current – 1960–1990 *Canadian Scenario – 2070–2100* *Hadley Scenario – 2070–2100*

White-Red-Jack Pine Longleaf-Slash Pine Oak-Pine Oak-Gum-Cypress Maple-Beech-Birch

Spruce-Fir Loblolly-Shortleaf Pine Oak-Hickory Elm-Ash-Cottonweed Aspen-Birch No Data

Note: All cases were calculated using the DISTRIB tree species distribution model, which calculates the most likely dominant types of vegetation for the given climatic conditions, assuming they have persisted for several decades.

Source: A.M. Prasad and L. R. Iverson, Northeastern Research Station, USDA Forest Service, Delaware, Ohio, as reported in NAST 2000.

introduction of a species, and ultimately determine the species' range and influence its population.

Although climate and soils exert strong controls on the establishment and growth of plant species, the response of plant and animal species to climate change will be the result of many interacting and interrelated processes operating over several temporal and spatial scales. Movement and migration rates, changes in disturbance regimes and abiotic environmental variables, and interactions within and between species will all affect the distributions and populations of plants and animals.

Analyses conducted using ecological models indicate that plausible climate scenarios are very likely to cause shifts in the location and area of the potential habitats for many tree species. For example, potential habitats for trees acclimated to cool environments are very likely to shift northward. Habitats of alpine and sub-alpine spruce-fir in the contiguous United States are likely to be reduced and, possibly in the long term, eliminated as their mountain habitats warm. The extents of aspen, eastern birch, and sugar maple are likely to contract dramatically in the United States and largely shift into Canada, with the shift in sugar maple causing loss of syrup production in northern New York and New England. In contrast, oak/hickory and oak/pine could expand in the East, and Ponderosa pine and arid woodland communities could expand in the West. How well these species track changes in their potential habitats will be strongly influenced by the viability of their mechanisms for dispersal to other locations and the disturbances to these alternative environments.

Because of the dominance of nonforest land uses along migration routes, the northward shift of some native species to new habitats is likely to be disrupted if the rate of climate change is too rapid. For example, conifer encroachment, grazing, invasive species, and urban expansion are currently displacing sagebrush and aspen communities. The effects of climate change on the rate and magnitude of disturbance (forest damage and destruction associated with fires, storms, droughts, and pest outbreaks) will be important factors in determining whether transitions from one forest type to another will be gradual or abrupt. If the rate and type of disturbances in New England do not increase, for example, a smooth transition from the present maple, beech, and birch tree species to oak and hickory may occur. Where the frequency or intensity of disturbances increases, however, transitions are very likely to occur more rapidly. As these changes occur, invasive (weedy) species that disperse rapidly are likely to find opportunities in newly forming ecological communities. As a result, the species composition of these communities will likely differ significantly in some areas from those occupying similar habitats today.

Changes in the composition of ecosystems may, in turn, have important effects on wildlife. For example, to the extent that climate change and a higher CO_2 concentration increase forest productivity, this might result in reduced overall land disturbance and improved water quality, tending to help wildlife, at least in some areas. However, changes in composition can also affect predator–prey relationships, pest types and populations, the potential for non-native species, links in the chain of migratory habitats, the health of keystone species, and other factors. Given these many possibilities, much remains to be examined in projecting influences of climate change on wildlife.

Socioeconomic Impacts

North America is the world's leading producer and consumer of wood products. U.S. forests provide for substantial exports of hardwood lumber, wood chips, logs, and some types of paper. Coming the other way, the United States imports, for example, about 35 percent of its softwood lumber and more than half of its newsprint from Canada.

The U.S. market for wood products will be highly dependent upon the future area in forests, the species composition of forests, future supplies of wood, technological changes in production and use, the availability of such substitutes as steel and vinyl, national and international demands for wood products, and competitiveness among major trading partners. Analyses indicate that, for a range of climate scenarios, forest productivity gains are very likely to increase timber inventories over the next 100 years (NFAG 2001). Under these scenarios, the increased wood supply leads to reductions in log prices, helping consumers, but decreasing producers' profits. The projected net effect on the economic welfare of participants in timber markets increases by about 1 percent above current values.

Analyses conducted for the forest sector assessment indicate that land use will likely shift between forestry and agriculture as these economic sectors adjust to climate-induced changes in production. U.S. hardwood and softwood production is projected to generally increase, although the projections indicate that softwood output will only increase under moderate warming. Timber output is also projected to increase more in the South than in the North, and saw-timber volume is projected to increase more than pulpwood volume.

Patterns and seasons of outdoor, forest-oriented recreation are likely to be modified by the projected changes in climate. For example, changes in forest-oriented recreation, as measured by aggregate days of activities and total economic value, are likely to be affected and are likely to vary by type of recreation and location. In some areas, higher temperatures are likely to shift typical summer recreation activities, such as hiking, northward or to higher elevations and into other seasons. In winter, downhill skiing opportunities are very likely to shift geographically because of fewer cold days and reduced snowpack in many existing ski areas. Therefore, costs to maintain skiing opportunities are likely to rise, especially for the more southern areas. Effects on fishing are also likely to vary. For example, warmer waters are likely to increase fish production and opportunities to fish for some warm-water species, but decrease habitat and opportunities to fish for cold-water species.

Possible Adaptation Strategies to Protect Forests

Even though forests are likely to be affected by the projected changes in climate, the motivation for adaptation strategies is likely to be most strongly influenced by the level of U.S. economic activity. This level, in turn, is intertwined with the rate of population growth, changes in taste, and general preferences, including society's perceptions about these changes. Market forces have proven to be powerful when it comes to decisions involving land use and forestry and, as such, will strongly influence adaptation on private lands. For forests valued for their current biodiversity, society and land managers will have to decide whether more intense management is necessary and appropriate for maintaining plant and animal species that may be affected by climate change and other factors.

If new technologies and markets are recognized in a timely manner, timber producers could adjust and adapt to climate change under plausible climate scenarios. One possible adaptation measure could be to salvage dead and dying timber and to replant species adapted to the changed climate conditions. The extent and pattern of U.S. timber harvesting and prices will also be influenced by the global changes in forest productivity and prices of overseas products.

Potential climate-induced changes in forests must also be put into the context of other human-induced pressures, which will undoubtedly change significantly over future decades. While the potential for rapid changes in natural disturbances could challenge current management strategies, these changes will occur simultaneously with human activities, such as agricultural and urban encroachment on forests, multiple uses of forests, and air pollu-tion. Given these many interacting factors, climate-induced changes should be manageable if planning is proactive.

Potential Interactions with Water Resources

Water is a central resource supporting human activities and ecosystems, and adaptive management of this resource has been an essential aspect of societal development. Increases in global temperatures during the 20th century have been accompanied by more precipitation in the middle and high latitudes in many regions of North America. For example, U.S. precipitation increased by 5–10 percent, predominantly from the spring through the autumn. Much of this increase resulted from a rise in locally heavy and very heavy precipitation events, which has led to the observed increases in low to moderate stream flow that have been characteristic of the warm season across most of the contiguous United States.

Local to global aspects of the hydrologic cycle, which determine the availability of water resources, are likely to be altered in important ways by climate change (NWAG 2000). Because higher concentrations of CO_2 and other greenhouse gases tend to warm the surface, all models project that the global totals of both evaporation and precipitation will continue to increase, with increases particularly likely in middle and high latitudes.

The regional patterns of the projected changes in precipitation remain uncertain, however, although there are some indications that changes in atmospheric circulation brought on by such factors as increasing Pacific Ocean temperatures may bring more precipitation to the Southwest and more winter precipitation to the West. Continuing trends first evident during the 20th century, model simulations project that increases in precipitation are likely to be most evident in the most intense rainfall categories typical of various regions. To the extent such increases occur during the warm season when stream flows are typically low to moderate, they could augment available water resources. If increases in precipitation occur during high stream flow or saturated soil conditions, the results suggest a greater potential for flooding in susceptible areas where additional control measures are not taken, especially because under these conditions the relative increase in runoff is generally observed to be greater than the relative increase in precipitation.

Effects on Available Water Supplies

Water is a critical national resource, providing services to society for refreshment, irrigation of crops, nourishment of ecosystems, creation of hydroelectric power, industrial processing, and more. Many U.S. rivers and streams do not have enough water to satisfy existing water rights and claims. Changing public values about preserving in-stream flows, protecting endangered species, and settling Indian water rights claims have made competition for water supplies increasingly intense. Depending on how water managers are able to take adaptive measures, the potential impacts of climate change could include increased competition for water supplies, stresses on water quality in areas where flows are diminished, adverse impacts on groundwater quantity and quality, an increased possibility of flooding in the winter and early spring, a reduced possibility of flooding later in the spring, and more water shortages in the summer. In some areas, however, an increase in precipitation could outweigh these factors and increase available supplies.

Significant changes in average temperature, precipitation, and soil moisture resulting from climate change are also likely to affect water demand in most sectors. For example, demand for water associated with electric power generation is projected to increase due to the increasing demand for air conditioning with higher summer temperatures. Climate change is also likely to reduce water levels in the Great Lakes and summertime river levels in the central United States, thereby adversely affecting navigation, general water supplies, and populations of aquatic species.

Effects on Water Quality

Increases in heavy precipitation events are likely to flush more contaminants and sediments into lakes and rivers, degrading water quality. Where uptake of agricultural chemicals and other nonpoint sources could be exacerbated, steps to limit water pollution are likely to be needed. In some regions, however, higher average flows will likely dilute pollutants and, thus, improve water quality. In coastal regions where river flows are reduced, increased salinity could also become more of a problem. Flooding can also cause overloading of storm-water and wastewater systems, and can damage water and sewage treatment facilities, mine tailing impoundments, and landfills, thereby increasing the risks of contamination and toxicity.

Because the warmer temperatures will lead to increased evaporation, soil moisture is likely to be reduced during the warm season. Although this effect is likely be alleviated somewhat by increased efficiency in water use and reduced demand by native plants for water, the drying is likely to create a greater susceptibility to fire and then loss of the vegetation that helps to control erosion and sediment flows. In agricultural areas, the CO_2-induced improvement of water-use efficiency by crops is likely to decrease demands for water, particularly for irrigation water. In addition, in some regions, increasing no-till or reduced-till agriculture is likely to improve the water-holding capacity of soils, regardless of whether climate changes, thereby reducing the susceptibility of agricultural lands to erosion from intensified heavy rains (NAAG 2002, NWAG 2000).

Effects on Snowpack

Rising temperatures are very likely to affect snowfall and increase snowmelt conditions in much of the western and northern portions of the country that depend on winter snowpack for runoff. This is particularly important because snowpack provides a natural reservoir for water storage in mountainous areas, gradually releasing its water in spring

and even summer under current climate conditions.

Model simulations project that snowpack in western mountain regions is likely to decrease as U.S. climate warms (Figure 6-9). These reductions are projected, despite an overall increase in precipitation, because (1) a larger fraction of precipitation will fall as rain, rather than snow; and (2) the snowpack is likely to develop later and melt earlier. The resulting changes in the amount and timing of runoff are very likely to have significant implications in some basins for water management, flood protection, power production, water quality, and the availability of water resources for irrigation, hydropower, communities, industry, and the sustainability of natural habitats and species.

Effects on Ground-Water Quantity and Quality

Several U.S. regions, including parts of California and the Great Plains, are dependent on dwindling ground-water supplies. Although ground-water supplies are less susceptible to short-term climate variability than surface-water supplies, they are more affected by long-term trends. Ground water serves as the base flow for many streams and rivers. Especially in areas where springtime snow cover is reduced and where higher summer temperatures increase evaporation and use of ground water for irrigation, ground-water levels are very likely to fall, thus reducing seasonal stream flows. River and stream temperatures fluctuate more rapidly with reduced volumes of water, affecting fresh-water and estuarine habitats. Small streams that are heavily influenced by ground water are more likely to have reduced flows and changes in seasonality of flows, which in turn is likely to damage existing wetland habitats.

Pumping ground water at a faster rate than it can be recharged is already a major concern, especially in parts of the country where other water resources are limited. In the Great Plains, for example, model projections indicate that drought is likely to be more frequent and intense,

which will create additional stresses because ground-water levels are already dropping in parts of important aquifers, such as the Ogallala.

The quality of ground water is being diminished by a variety of factors, including chemical contamination. Salt-water intrusion is another key ground-water quality concern, particularly in coastal areas where changes in fresh-water flows and increases in sea level will both occur. As ground-water pumping increases to serve municipal demand along the coast and less recharge occurs, coastal ground-water aquifers are increasingly being affected by sea-water intrusion. Because the ground-water resource has been compromised by many factors, managers are increasingly looking to surface-water supplies, which are more sensitive to climate change and variability.

Effects on Floods, Droughts, and Heavy Precipitation Events

Projected changes in the amount, timing, and distribution of rainfall and snowfall are likely to lead to changes in the amount and timing of high and low water flows—although the relationships of changes in precipitation rate to changes in flood frequency and intensity are uncertain, especially due to uncertainties in the timing and persistence of rainfall events and river levels and capacities. Because changes in climate extremes are more likely than changes in climate averages to affect the magnitude of damages and raise the need for adaptive measures at the regional level, changes in the timing of precipitation events, as well as increases in the intensity of precipitation events, are likely to become increasingly important considerations.

Climate change is likely to affect the frequency and amplitude of high stream flows, with major implications for infrastructure and emergency management in areas vulnerable to flooding. Although projections of the number of hurricanes that may develop remain uncertain, model simulations indicate that, in a warmer climate, hurricanes that do develop are likely to have

FIGURE 6-9 Projected Reductions in Western Snowpack Resulting from Potential Changes in Climate

Climate model scenarios for the 21st century project significant decreases from the 1961–1990 baseline in the average April 1 snowpack for four mountainous areas in the western United States. Scenarios from the Canadian model, which simulates warming toward the upper end of IPCC projections, and from the Hadley model, which simulates warming near the middle of IPCC projections, provide similar results. Such a steep reduction in the April 1 snowpack would significantly shift the time of peak runoff and reduce average river flows in spring and summer.

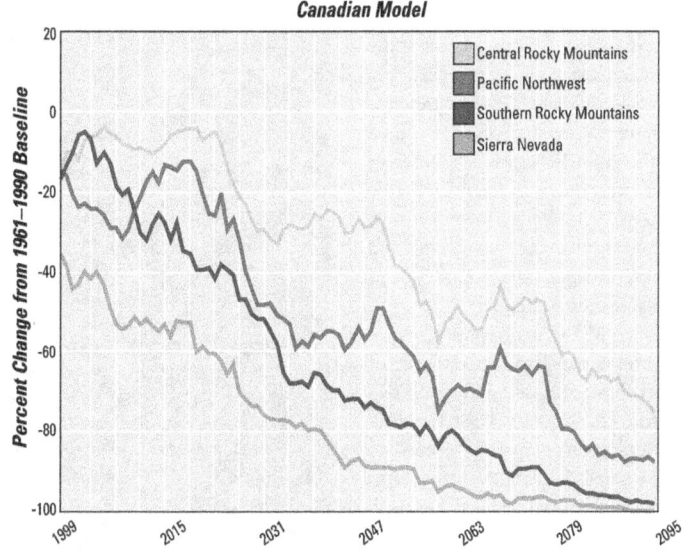

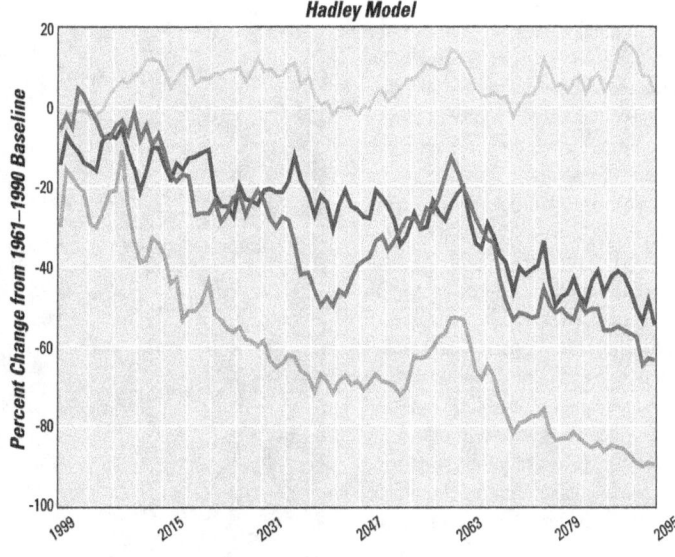

Source: Redrawn from McCabe and Wolock 1999, as presented in NAST 2000.

higher wind speeds and produce more rainfall. As a result, they are likely to cause more damage, unless more extensive (and therefore more costly) adaptive measures are taken, including reducing the increasing exposure of property to such extreme events. Historical records indicate that improved warning has been a major factor in reducing the annual number of deaths due to storms, and that the primary cause of the increasing property damage in recent decades has been the increase in at-risk structures, such as widespread construction of vacation homes on barrier islands.

Despite the overall increase in precipitation and past trends indicating an increase in low to moderate stream flow, model simulations suggest that increased air temperatures and more intense evaporation are likely to cause many interior portions of the country to experience more frequent and longer dry conditions. To the extent that the frequency and intensity of these conditions lead to an increase in droughts, some areas are likely to experience wide-ranging impacts on agriculture, water-based transportation, and ecosystems, although the effects on vegetation (including crops and forests) are likely to be mitigated under some conditions by increased efficiency in water use due to higher CO_2 levels.

Water-driven Effects on Ecosystems

Species live in the larger context of ecosystems and have differing environmental needs. In some ecosystems, existing stresses could be reduced if increases in soil moisture or the incidence of freezing conditions are reduced. Other ecosystems, including some for which extreme conditions are critical, are likely to be most affected by changes in the frequency and intensity of flood, drought, or fire events. For example, model projections indicate that changes in temperature, moisture availability, and the water demand from vegetation are likely to lead to significant changes in some ecosystems in the coming decades

(NAST 2000). As specific examples, the natural ecosystems of the Arctic, Great Lakes, Great Basin, and Southeast, and the prairie potholes of the Great Plains appear highly vulnerable to the projected changes in climate (see Figures 6-6 and 6-8).

The effects of changes in water temperatures are also important. For example, rising water temperatures are likely to force out some cold-water fish species (such as salmon and trout) that are already near the threshold of their viable habitat, while opening up additional areas for warm-water species. Increasing temperatures are also likely to decrease dissolved oxygen in water, degrade the health of ecosystems, reduce ice cover, and alter the mixing and stratification of water in lakes—all of which are key to maintaining optimal habitat and suitable nutrient levels. In addition, warmer lake waters combining with excess nutrients from agricultural fertilizers (washed into lakes by heavy rains) would be likely to create algal blooms on the lake surfaces, further depleting some lake ecosystems of life-sustaining oxygen.

Potential Adaptation Options to Ensure Adequate Water Resources

In contrast to the vulnerability of natural ecosystems, humans have exhibited a significant ability to adapt to the availability of different amounts of water. There are many types of water basins across the country, and many approaches are already in use to ensure careful management of water resources. For example, more than 80,000 dams and reservoirs and millions of miles of canals, pipes, and tunnels have been developed to store and transport water. Some types of approaches that studies have indicated might prove useful are highlighted on this page.

Strategies for adapting to climate change and other stresses include changing the operation of dams and reservoirs, re-evaluating basic engineering assumptions used in facility construction, and building new infrastructure (although for a variety of reasons, large dams are no longer generally viewed as a cost-

effective or environmentally acceptable solution to water supply problems). Other potentially available options include conserving water; changing water pricing; using reclaimed wastewater; using water transfers; and developing markets for water, which can lead to increased prices that discourage wasteful practices.

Existing or new infrastructure can also be used to dampen the impacts of climate-induced influences on flow regimes and aquatic ecosystems of many of our nation's rivers. While significant adaptation is possible, its cost could be reduced if the probable effects of climate change are factored in before making major long-term investments in repairing, maintaining, expanding, and operating existing water supply and management infrastructure.

Because of the uncertainties associated with the magnitude and direction of changes in precipitation and runoff due to climate change, more flexible institutional arrangements may be needed to ensure optimal availability of water as supplies and demand change. Although social, equity, and environmental considerations must be addressed, market solutions offer the potential for resolving supply problems in some parts of the country. However, because water rights systems vary from state to state and even locally, water managers will need to take the lead in selecting the most appropriate adaptive responses.

Because the United States shares water resources with Canada and Mexico, it participates in a number of institutions designed to address common water issues. These institutions, which include the U.S.–Canada Great Lakes Commission and joint commissions and agreements covering the Colorado and Rio Grande rivers, could provide the framework for designing adaptive measures for responding to the effects of climate change. For example, the U.S.–Canada Great Lakes Commission has already conducted studies to evaluate options for dealing with the potential for increased evaporation, shorter duration of lake ice, and other climate changes that are projected to affect the

Great Lakes–St. Lawrence River basin. Close coordination will be needed to efficiently manage the levels of these crucial water resources to ensure adequate water supplies for communities and irrigation, high water quality, needed hydroelectric power, high enough levels for recreation and

Potential Adaptation Options for Water Management

Following are some potential adaptation options for water management in response to climate change and other stresses:

- Improve capacity for moving water within and between water-use sectors (including agriculture to urban).
- Use pricing and market mechanisms proactively to decrease waste.
- Incorporate potential changes in demand and supply in long-term planning and infrastructure design.
- Create incentives to move people and structures away from flood plains.
- Identify ways to sustainably manage supplies, including ground water, surface water, and effluent.
- Restore and maintain watersheds to reduce sediment loads and nutrients in runoff, limit flooding, and lower water temperature.
- Encourage the development of institutions to confer property rights to water. This would be intended to encourage conservation, recycling, and reuse of water by all users, as well as to provide incentives for research and development of such conservation technologies.
- Reduce agricultural demand for water by focusing research on development of crops and farming practices for minimizing water use, for example, via precision agricultural techniques that closely monitor soil moisture.
- Reuse municipal wastewater, improve management of urban storm-water runoff, and promote collection of rain water for local use.
- Increase the use of forecasting tools for water management. Some weather patterns, such as those resulting from El Niño, can now be predicted, allowing for more efficient management of water resources.
- Enhance monitoring efforts to improve data collection for weather, climate, and hydrologic modeling to aid understanding of water-related impacts and management strategies.

Source: Adapted from NWAG 2000.

shipping, low enough levels to protect communities and shorelines from flooding and wave-induced erosion, and more.

Potential Interactions with Coastal Areas and Marine Resources

The United States has over 95,000 miles of coastline and over 3.4 million square miles of ocean within its territorial waters. These areas provide a wide range of goods and services to the U.S. economy. Approximately 53 percent of the U.S. population lives on the 17 percent of land in counties that are adjacent to or relatively near the coast. Over recent decades, populations in these coastal counties have been growing more rapidly than elsewhere in the country. As a result of this population growth and increased wealth, demands on coastal and marine resources for both leisure activities and economic benefits are rapidly intensifying, while at the same time exposure to coastal hazards is increasing.

Coastal and marine environments are intrinsically linked to the prevailing climate in many ways. Heat given off by the oceans warms the land during the winter, and ocean waters help to keep coastal regions cooler during the summer. Moisture evaporated from the oceans is the ultimate source of precipitation, and the runoff of precipitation carries nutrients, pollutants, and other materials from the land to the ocean. Sea level exerts a major influence on the coastal zone, shaping barrier islands and pushing salt water up estuaries and into aquifers. For example, cycles of beach and cliff erosion along the Pacific Coast have been linked to the natural sequence of El Niño events that alter storm tracks and temporarily raise average sea levels by several inches in this region (NCAG 2000). During the 1982–83 and 1997–98 El Niño events, erosion damage was widespread along the Pacific coastline.

Climate change will affect interactions among conditions on the land and sea and in the atmosphere. Warming is likely to alter coastal weather and could affect the intensity, frequency, and extent of severe storms. Melting of glaciers and ice sheets and thermal expansion of ocean waters will cause sea level to rise, which is likely to intensify erosion and endanger coastal structures. Rising sea level and higher temperatures are also likely to affect the ecology of estuaries and coastal wetlands. Higher temperatures coupled with increasing CO_2 concentrations are likely to severely stress coral reefs, and the changing temperature patterns are likely to cause fisheries to relocate and alter fish migration patterns. While quantifying these consequences is difficult, indications of the types of outcomes that are possible have emerged from U.S. assessments (NCAG 2000).

Effects on Sea Level

Global sea level rose by 10–20 cm (about 4–8 inches) during the 20th century, which was significantly more than the rate of rise that was typical over the last few thousand years. Even in the absence of a change in Atlantic storminess, the deeper inundation that has resulted from recent storms has exacerbated flooding and has led to damage to fixed coastal structures from storms that were previously inconsequential.

Looking to the future, climate models project that global warming will increase sea level by 9–88 cm (4–35 inches) during the 21st century, with mid-range values more likely than the very high or very low estimates (IPCC 2001d). Because of the long time constants involved in ocean warming and glacier and ice sheet melting, further sea level rise is likely for several centuries, even after achieving significant limitations in emissions of CO_2 and other greenhouse gases. However, these global changes are only one factor in what determines sea level change at any particular coastal location. For example, along the Mid-Atlantic coast, where land levels are subsiding, relative sea level rise will be somewhat greater; conversely, in New England, where land levels are rising, relative sea level rise will be somewhat less.

Not surprisingly, an increased rate of global sea level rise is likely to have the most dramatic impacts in regions where subsidence and erosion problems already exist. Estuaries, wetlands, and shorelines along the Atlantic and Gulf coasts are especially vulnerable. Impacts on fixed structures will intensify, even in the absence of an increase in storminess. However, because the slope of these areas is so gentle, even a small rise in sea level can produce a large inland shift of the shoreline. The rise will be particularly important if the frequency or intensity of storm surges or hurricanes increases.

Increases in the frequency or intensity of El Niño events would also likely exacerbate the impacts of long-term sea level rise. Coastal erosion increases the threats to coastal development, transportation infrastructure, tourism, freshwater aquifers, fisheries (many of which are already stressed by human activities), and coastal ecosystems. Coastal cities and towns, especially those in storm-prone regions, such as the Southeast, are particularly vulnerable. Intensive residential and commercial development in these regions is placing more and more lives and property at risk (Figure 6-10).

Effects on Estuaries

Climate change and sea level rise could present significant threats to valuable, productive coastal ecosystems. For example, estuaries filter and purify water and provide critical nursery and habitat functions for many commercially important fish and shellfish populations. Because the temperature increase is projected to be greater in the winter than in the summer, a narrowing of the annual water temperature range of many estuaries is likely. This, in turn, is likely to cause a shift in species' ranges and to increase the vulnerability of some estuaries to invasive species (NCAG 2000).

Changes in runoff are also likely to adversely affect estuaries. Unless new agricultural technologies allow reduced use of fertilizers, higher rates of runoff are likely to deliver greater amounts of nutrients such as nitrogen and

FIGURE 6-10 Projected Rates of Annual Erosion along U.S. Shorelines

The U.S. coastal areas that are most vulnerable to future increases in sea level are those with low relief and those that are already experiencing rapid erosion rates, such as the Southeast and Gulf Coast.

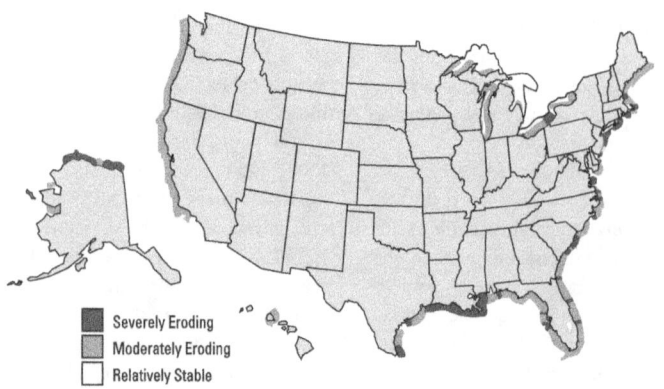

- ■ Severely Eroding
- ■ Moderately Eroding
- □ Relatively Stable

Source: U.S. Geological Survey Coastal Geology Program, as presented in NAST 2000.

phosphorus to estuaries, while simultaneously increasing the stratification between fresh-water runoff and marine waters. Such conditions would be likely to increase the potential for algal blooms that deplete the water of oxygen. These conditions would also increase stresses on sea grasses, fish, shellfish, and other organisms living in lakes, streams, and oceans (NCAG 2000, and regional assessment reports listed at http://www.usgcrp.gov). In addition, decreased runoff is likely to reduce flushing, decrease the size of estuarine nursery zones, and increase the range of estuarine habitat susceptible to predators and pathogens of shellfish.

Effects on Wetlands

Coastal wetlands (marshes and mangroves) are highly productive ecosystems, particularly because they are strongly linked to the productivity of fisheries. Dramatic losses of coastal wetlands have occurred along the Gulf Coast due to subsidence, alterations in flow and sediment load caused by dams and levees, dredge and fill activities, and sea level rise. Louisiana alone has been losing land at rates of about 68–104 square kilometers (24–40 square miles) per year for the last 40 years, accounting for as much as 80 percent of the total U.S. coastal wetland loss.

In general, coastal wetlands will survive if soil buildup equals the rate of relative sea level rise or if they are able to migrate inland (although this migration necessarily displaces other ecosystems or land uses). However, if soil accumulation does not keep pace with sea level rise, or if bluffs, coastal development, or shoreline protective structures (such as dikes, sea walls, and jetties) block wetland migration, wetlands may be excessively inundated and, thus, lost. The projected increase in the current rate of sea level rise is very likely to exacerbate the nationwide rate of loss of existing coastal wetlands, although the extent of impacts will vary among regions, and some impacts may be moderated by the inland formation of new wetlands.

Effects on Coral Reefs

The demise or continued deterioration of reefs could have profound implications for the United States. Coral reefs play a major role in the environment and economies of Florida and Hawaii as well as in most U.S. territories in the Caribbean and Pacific. They support fisheries, recreation, and tourism and protect coastal areas. In addition, coral reefs are one of the largest global storehouses of marine biodiversity, sheltering one-quarter of all marine life and containing extensive untapped genetic resources.

The last few years have seen unprecedented declines in the health of coral reefs. The 1998 El Niño was associated with record sea-surface temperatures and associated coral bleaching (which occurs when coral expel the algae that live within them and that are necessary to their survival). In some regions, as much as 70 percent of the coral may have died in a single season. There has also been an upsurge in the variety, incidence, and virulence of coral diseases in recent years, with major die-offs in Florida and much of the Caribbean region (NCAG 2000).

Other factors that are likely to be contributing to the decline of coral reefs include increased sediment deposition, sewage and agricultural runoff, excessive harvesting of fish, and damage from ships and tourists. In addition to the potential influences of further global warming, increasing atmospheric CO_2 concentrations are likely to decrease the calcification rates of the reef-building corals, resulting in weaker skeletons, reduced growth rates, and increased vulnerability to wave-induced damage. Model results suggest that these effects would likely be most severe at the current margins of coral reef distribution, meaning that it is unlikely coral reefs will be able to spread northward to reach cooler waters. While steps can be taken to reduce the impacts of some types of stress on coral reefs (e.g., by creating Marine Protected Areas, as called for in Executive Order 13158, and constructing artificial reefs to provide habitat for threatened species), damage to coral reefs from climate change and the increasing CO_2 concentration may be moderated to some extent only by significantly reducing other stresses.

Effects on Marine Fisheries

Based on studies summarized in the coastal sector assessment, recreational and commercial fishing has contributed approximately $40 billion a year to the U.S. economy, with total marine landings averaging about 4.5 million metric tons over the last decade. Climate change is very likely to substantially alter the distribution and abundance of major fish stocks, many of which are a shared international resource.

Along the Pacific Coast, impacts to fisheries related to the El Niño–Southern Oscillation illustrate how climate directly affects marine fisheries on short time scales. For example, elevated sea-surface temperatures associated with the 1997–98 El Niño had a tremendous impact on the distribution and abundance of market squid. Although California's largest fishery by volume, squid landings fell to less than 1,000 metric tons in the 1997–98 season, down from a record-breaking 110,000 metric tons in the 1996–97 season. Many other unusual events occurred during this same El Niño as a result of elevated sea-surface temperatures. Examples include widespread deaths of California sea lion pups, catches of warm-water marlin in the usually frigid waters off Washington State, and poor salmon returns in Bristol Bay, Alaska.

The changes in fish stocks resulting from climate change are also likely to have important implications for marine populations and ecosystems. Changes over the long term that will affect all nations are likely to include poleward shifts in distribution of marine populations, and changes in the timing, locations, and, perhaps, viability of migration paths and nesting and feeding areas for marine mammals and other species.

With changing ocean temperatures and conditions, shifts in the distribution of commercially important species are likely, affecting U.S. and international fisheries. For example, model projections suggest that several species of Pacific salmon are likely to have reduced distribution and productivity, while species that thrive in warmer waters, such as Pacific sardine and Atlantic menhaden, are likely to show an increased distribution. Presuming that the rate of climate change is gradual, the many efforts being made to bet-ter manage the world's fisheries might promote adaptation to climate change, along with helping to relieve the many other pressures on these resources.

Potential Adaptation Options for Coastal Regions

Because climate variability is currently a dominant factor in shaping coastal and marine systems, projecting the specific effects of climate change over the next few decades and evaluating the potential effectiveness of possible response options is particularly challenging. Effects will surely vary greatly among the diverse coastal regions of the nation. Human-induced disturbances also influence coastal and marine systems, often reducing the ability of systems to adapt, so that systems that might ordinarily be capable of responding to variability and change are less able to do so. In this context, climate change is likely to add to the cumulative impact of both natural and human-caused stresses on ecological systems and resources. As a result, strategies for adapting to the potential consequences of long-term climate change in the overall context of coastal development and management are only beginning to be considered (NCAG 2000).

However, as further plans are made for development of land in the coastal zone, it is especially urgent for governing bodies at all levels to begin to consider the potential changes in the coastal climate and sea level. For example, the U.S. Geological Survey is expanding its gathering and assembly of relevant coastal information, and the U.S. Environmental Protection Agency's Sea Level Rise project is dedicated to motivating adaptation to rising sea level. This project has assessed the probability and has identified and mapped vulnerable low-elevation coastal zones. In addition, cost-effective strategies and land-use planning approaches involving landward migration of wetlands, levee building, incorporation of sea level rise in beach conservation plans, engineered landward retreats, and sea walls have all been developed.

Several states have already included sea level rise in their planning, and some have already implemented adaptation activities. For example, in New Jersey, where relative sea level is rising approximately one inch (2.5 cm) every six years, $15 million is now set aside each year for shore protection, and the state discourages construction that would later require sea walls. In addition, Maine, Rhode Island, South Carolina, and Massachusetts have implemented various forms of "rolling easement" policies to ensure that wetlands and beaches can migrate inland as sea level rises, and that coastal landowners and conservation agencies can purchase the required easements. Other states have modified regulations on, for example, beach preservation, land reclamation, and inward migration of wetlands and beaches. Wider consideration of potential consequences is especially important, however, because some regulatory programs continue to permit structures that may block the inland shift of wetlands and beaches, and in some locations shoreline movement is precluded due to the high degree of coastal development.

To safeguard people and better manage resources along the coast, NOAA provides weather forecasts and remotely sensed environmental data to federal, state, and local governments, coastal resource managers and scientists, and the public. As part of its mandate and responsibilities to administer the National Flood Insurance Program, the Federal Emergency Management Agency (FEMA) prepares Flood Insurance Rate Maps that identify and delineate areas subject to severe (1 percent annual chance) floods. FEMA also maps coastal flood hazard areas as a separate flood hazard category in recognition of the additional risk associated with wave action. In addition, FEMA is working with many coastal cities to encourage steps to reduce their vulnerability to storms and floods, including purchasing vulnerable properties.

University and state programs are also underway across the country. This is particularly important because most coastal planning in the United States is the responsibility of state and local governments, with the federal government interacting with these efforts through the development of coastal zone management plans.

Potential Interactions with Human Health

Although the overall susceptibility of Americans to environmental health concerns dropped dramatically during the 20th century, certain health outcomes are still recognized to be associated with the prevailing environmental conditions. These adverse outcomes include illnesses and deaths associated with temperature extremes; storms and other heavy precipitation events; air pollution; water contamination; and diseases carried by mosquitoes, ticks, and rodents. As a result of the potential consequences of these stresses acting individually or in combination, it is possible that projected climate change will have measurable beneficial and adverse impacts on health (see NHAG 2000, 2001).

Adaptation offers the potential to reduce the vulnerability of the U.S. population to adverse health outcomes—including possible outcomes of projected climate change—primarily by ensuring strong public health systems, improving their responsiveness to changing weather and climate conditions, and expanding attention given to vulnerable subpopulations. Although the costs, benefits, and availability of resources for such adaptation must be found, and further research into key knowledge gaps on the relationships between climate/weather and health is needed, to the extent that the U.S. population can keep from putting itself at greater risk by where it lives and what it does, the potential impacts of climate change on human health can likely be addressed as a component of efforts to address current vulnerabilities.

Projections of the extent and direction of potential impacts of climate variability and change on health are extremely difficult to make with confidence because of the many confound-

ing and poorly understood factors associated with potential health outcomes. These factors include the sensitivity of human health to aspects of weather and climate, differing vulnerability of various demographic and geographic segments of the population, the international movement of disease vectors, and how effectively prospective problems can be dealt with. For example, uncertainties remain about how climate and associated environmental conditions may change. Even in the absence of improving medical care and treatment, while some positive health outcomes—notably, reduced cold-weather mortality—are possible, the balance between increased risk of heat-related illnesses and death and changes in winter illnesses and death cannot yet be confidently assessed. In addition to uncertainties about health outcomes, it is very difficult to anticipate what future adaptive measures (e.g., vaccines, improved use of weather forecasting to further reduce exposure to severe conditions) might be taken to reduce the risks of adverse health outcomes.

Effects on Temperature-Related Illnesses and Deaths

Episodes of extreme heat cause more deaths in the United States than any other category of deaths associated with extreme weather. In one of the most severe examples of such an event, the number of deaths rose by 85 percent during a five-day heat wave in 1995 in which maximum temperatures in Chicago, Illinois, ranged from 34 to 40°C (93 to 104°F) and minimum temperatures were nearly as high. At least 700 excess deaths (deaths in that population beyond those expected for that period) were recorded, most of which were directly attributable to heat.

For particular years, studies in certain urban areas show a strong association between increases in mortality and increases in heat, measured by maximum or minimum daily temperature and by heat index (a measure of temperature and humidity). Over longer periods, determination of trends is often difficult due to the episodic

nature of such events and the presence of complicating health conditions, as well as because many areas are taking steps to reduce exposure to extreme heat. Recognizing these complications, no nationwide trend in deaths directly attributed to extreme heat is evident over the past two decades, even though some warming has occurred.

Based on available studies, heat stroke and other health effects associated with exposure to extreme and prolonged heat appear to be related to environmental temperatures above those to which the population is accustomed. Thus, the regions expected to be most sensitive to projected increases in severity and frequency of heat waves are likely to be those in which extremely high temperatures occur only irregularly. Within heat-sensitive regions, experience indicates that populations in urban areas are most vulnerable to adverse heat-related health outcomes. Daily average heat indices and heat-related mortality rates are higher in these urban core areas than in surrounding areas, because urban areas remain warmer throughout the night compared to outlying suburban and rural areas. The absence of nighttime relief from heat for urban residents has been identified as a factor in excessive heat-related deaths. The elderly, young children, the poor, and people who are bedridden, who are on certain medications, or who have certain underlying medical conditions are at particular risk.

Plausible climate scenarios project significant increases in average summer temperatures, leading to new record highs. Model results also indicate that the frequency and severity of heat waves would be very likely to increase along with the increase in average temperatures. The size of U.S. cities and the proportion of U.S. residents living in them are also projected to increase through the 21st century. Because cities tend to retain daytime heat and so are warmer than surrounding areas, climate change is very likely to lead to an increase in the population potentially at risk from heat events. While the potential risk may increase, heat-related illnesses and deaths are largely preventable through behavioral adaptations, including use of air conditioning, increased fluid intake, and community warning and support systems. The degree to which these adaptations can be even more broadly made available and adopted than in the 20th century, especially for sensitive populations, will determine if the long-term trend toward fewer deaths from extreme heat can be maintained.

Death rates not only vary with summertime temperature, but also show a seasonal dependence, with more deaths in winter than in summer. This relationship suggests that the relatively large increases in average winter temperature could reduce deaths in winter months. However, the relationship between winter weather and mortality is not as clear as for summertime extremes. While there should be fewer deaths from shoveling snow and slipping on ice, many winter deaths are due to respiratory infections, such as influenza, and it is not clear how influenza transmission would be affected by higher winter temperatures. As a result, the net effect on winter mortality from milder winters remains uncertain.

Influences on Health Effects Related to Extreme Weather Events

Injury and death also result from natural disasters, such as floods and hurricanes. Such outcomes can result both from direct bodily harm and from secondary influences, such as those mediated by changes in ecological systems (such as bacterial and fungal proliferation) and in public health infrastructures (such as reduced availability of safe drinking water).

Projections of climate change for the 21st century suggest a continuation of the 20th-century trend toward increasing intensity of heavy precipitation events, including precipitation during hurricanes. Such events, in addition to the potential consequences listed above, pose an increased risk of floods and associated health impacts. However, much can be done to prepare for powerful storms and heavy precipitation events, both through community

design and through warning systems. As a result of such efforts, the loss of life and the relative amounts of damage have been decreasing. For the future, therefore, the net health impacts of extreme weather events hinge on continuing efforts to reduce societal vulnerabilities. For example, FEMA's Safe Communities program is promoting implementation of stronger building codes and improved warning systems, as well as enhancing the recovery capacities of the natural environment and the local population, which are also being addressed through disaster assistance programs.

Influences on Health Effects Related to Air Pollution

Current exposures to air pollution exceed health-based standards in many parts of the country. Health assessments indicate that ground-level ozone can exacerbate respiratory diseases and cause short-term reductions in lung function. Such studies also indicate that exposure to particulate matter can aggravate existing respiratory and cardiovascular diseases, alter the body's defense systems against foreign materials, damage lung tissue, lead to premature death, and possibly contribute to cancer. Health effects of exposure to carbon monoxide, sulfur dioxide, and nitrogen dioxide have also been related to reduced work capacity, aggravation of existing cardiovascular diseases, effects on breathing, respiratory illnesses, lung irritation, and alterations in the lung's defense systems.

Projected changes in climate would be likely to affect air quality in several ways, some of which are likely to be dealt with by ongoing changes in technology, and some of which can be dealt with, if necessary, through changes in regulations. For example, changes in the weather that affect regional pollution emissions and concentrations can be dealt with by controlling sources of emissions. However, adaptation will be needed in response to changes in natural sources of air pollution that result from changes in weather. Analyses show that hotter, sunnier days tend to

increase the formation of ground-level ozone, other conditions being the same. This creates a risk of higher concentrations of ground-level ozone in the future, especially because higher temperatures are frequently accompanied by stagnating circulation patterns. However, more specific projections of exposure to air pollutants cannot be made with confidence without more accurate projections of changes in local and regional weather and projections of the amounts and locations of future emissions, which will in turn be affected by the implementation and success of air pollution control policies designed to ensure air quality. Also, more extensive health-warning systems could help to reduce exposures, decreasing any potential adverse consequences.

In addition to affecting exposure to air pollutants, there is some chance that climate change will play a role in exposure to airborne allergens. For example, it is possible that climate change will alter pollen production in some plants and change the geographic distribution of plant species. Consequently, there is some chance that climate change will affect the timing or duration of seasonal allergies. The impact of pollen and of pollen changes on the occurrence and severity of asthma, the most common chronic disease among children, is currently very uncertain.

Effects on Water- and Food-borne Diseases

In the United States, the incidence of and deaths due to waterborne diseases declined dramatically during the 20th century. While much less frequent or lethal nowadays, exposure to waterborne disease can still result from drinking contaminated water, eating seafood from contaminated water, eating fresh produce irrigated or processed with contaminated water, and participating in such activities as fishing or swimming in contaminated water. Water-borne pathogens of current concern include viruses, bacteria (such as *Vibrio vulnificus*, a naturally occurring estuarine bacterium responsible for a high percentage of the deaths associated with shellfish consumption), and protozoa (such as *Cryptosporidium*, associated with gastrointestinal illnesses).

Because changes in precipitation, temperature, humidity, salinity, and wind have a measurable effect on water quality, future changes in climate have the potential to increase exposure to water-borne pathogens. In 1993, for example, *Cryptosporidium* contaminated the Milwaukee, Wisconsin, drinking-water supply. As a result, 400,000 people became ill. Of the 54 individuals who died, most had compromised immune systems because of HIV infection or other illness. A contributing factor in the contamination, in addition to treatment system malfunctions, was heavy rainfall and runoff that resulted in a decline in the quality of raw surface water arriving at the Milwaukee drinking-water plants.

In another example, during the strong El Niño winter of 1997–98, heavy precipitation and runoff greatly elevated the counts of fecal bacteria and infectious viruses in Florida's coastal waters. In addition, toxic red tides proliferate as sea-water temperatures increase. Reports of marine-related illnesses have risen over the past two and a half decades along the East Coast, in correlation with El Niño events. Therefore, climate changes projected to occur in the next several decades—in particular, the likely increase in heavy precipitation events—raise the risk of contamination events.

Effects on Insect-, Tick-, and Rodent-borne Diseases

Malaria, yellow fever, dengue fever, and other diseases transmitted between humans by blood-feeding insects, ticks, and mites were once common in the United States. The incidence of many of these diseases has been significantly reduced, mainly because of changes in land use, agricultural methods, residential patterns, human behavior, vector control, and public health systems. However, diseases that may be transmitted to humans from wild animals

continue to circulate in nature in many parts of the country. Humans may become infected with the pathogens that cause these diseases through transmission by insects or ticks (such as Lyme disease, which is tick-borne) or by direct contact with the host animals or their body fluids (such as hantaviruses, which are carried by numerous rodent species and transmitted to humans through contact with rodent urine, droppings, and saliva). The organisms that directly transmit these diseases are known as vectors.

The ecology and transmission dynamics of vector-borne infections are complex, and the factors that influence transmission are unique for each pathogen. Most vector-borne diseases exhibit a distinct seasonal pattern, which clearly suggests that they are weather-sensitive. Rainfall, temperature, and other weather variables affect both vectors and the pathogens they transmit in many ways. For example, epidemics of malaria are associated with rainy periods in some parts of the world, but with drought in others. Higher temperatures may increase or reduce vector survival rate, depending on each specific vector, its behavior, ecology, and many other factors. In some cases, specific weather patterns over several seasons appear to be associated with increased transmission rates. For example, in the Midwest, outbreaks of St. Louis encephalitis (a viral infection of birds that can also infect and cause disease in humans) appear to be associated with the sequence of warm, wet winters, cold springs, and hot, dry summers. Although the potential for such diseases seems likely to increase, both the U.S. National Assessment (NHAG 2000, 2001) and a special report prepared by the National Research Council (NRC 2001b) agree that significant outbreaks of these diseases as a result of climate change are unlikely because of U.S. health and community standards and systems. However, even with actions to limit breeding habitats of mosquitoes and other disease vectors and to carefully monitor for infectious diseases, the

continued occurrence of local, isolated incidences of such diseases probably cannot be fully eliminated.

Although the United States has been able to reduce the incidence of such climatically related diseases as dengue and malaria, these diseases continue to extract a heavy toll elsewhere (Figure 6-11). Accordingly, the U.S. government and other governmental and nongovernmental organizations are actively supporting efforts to reduce the incidence and impacts of such diseases. For instance, U.S. agencies and philanthropies are in the forefront of malaria research, including the search for vaccines and genome sequencing of the anopheles mosquitoes and the malaria parasite *Plasmodium falciparum*. Efforts such as these should help to reduce global vulnerability to malaria and other vector-borne diseases, and need to be considered in global adaptation strategies.

The results from this work will serve the world in the event that human-induced climate change, through whatever mechanism, increases the potential for malaria. This work will also be

---

FIGURE 6-11 Reported Cases of Dengue Fever: 1980–1999

In 1922, there were an estimated 500,000 cases of dengue fever in Texas. The mosquitoes that transmit this viral disease remain abundant. The striking contrast in incidence in Texas over the last two decades, and in three Mexican states that border Texas, illustrates the importance of factors other than climate in the incidence of vector-borne diseases.

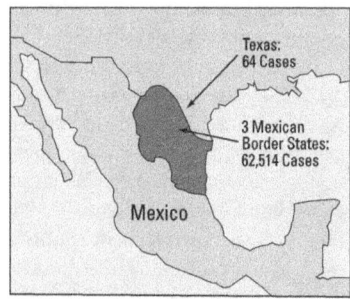

Texas: 64 Cases

3 Mexican Border States: 62,514 Cases

Mexico

Sources: National Institute of Health, Mexico, Texas Department of Health, U.S. Public Health Service, and unpublished data analyzed by the National Health Assessment Group and presented in NAST 2001.

beneficial for U.S. residents because our nation cannot be isolated from diseases occurring elsewhere in the world. Of significant importance, the potential for disease vectors to spread into the United States via travel and trade is likely to increase just as the natural, cold-winter conditions that have helped to protect U.S. residents are moderating.

Potential Adaptation Options to Ensure Public Health

The future vulnerability of the U.S. population to the health impacts of climate change will largely depend on maintaining—if not enhancing—the nation's capacity to adapt to potential adverse changes through legislative, administrative, institutional, technological, educational, and research-related measures. Examples include basic research into climate-sensitive diseases, building codes and zoning to prevent storm or flood damage, severe weather warning systems to allow evacuation, improved disease surveillance and prevention programs, improved sanitation systems, education of health professionals and the public, and research addressing key knowledge gaps in climate–health relationships.

Many of these adaptive responses are desirable from a public health perspective, irrespective of climate change. For example, reducing air pollution obviously has both short- and long-term health benefits. Improving warning systems for extreme weather events and eliminating existing combined sewer and storm-water drainage systems are other measures that can ameliorate some of the potential adverse impacts of current climate extremes and of the possible impacts of climate change. Improved disease surveillance, prevention systems, and other public health infrastructure at the state and local levels are already needed. Because of this, we expect awareness of the potential health consequences of climate change to allow adaptation to proceed in the normal course of social and economic development.

Potential Impacts in Various U.S. Regions

While some appreciation can be gained about the potential national consequences of climate change by looking at sectors such as the six considered above, the United States is a very large and diverse nation. There are both important commonalities and important differences in the climate-related issues and in the potential economic and environmental consequences faced by different regions across the country. Therefore, there are many different manifestations of a changing climate in terms of vulnerability and impacts, and the potential for adaptation. For example, while all coastal regions are at risk, the magnitude of the vulnerabilities and the types of adaptation necessary will depend on particular coastal conditions and development. Water is a key issue in virtually all regions, but the specific changes and impacts in the West, in the Great Lakes, and in the Southeast will differ.

With this variability in mind, 20 regional workshops that brought together researchers, stakeholders, and community, state, and national leaders were conducted to help identify key issues facing each region and to begin identifying potential adaptation strategies. These workshops were followed by the initiation of 16 regionally based assessment studies, some of which are already completed and others of which are nearing completion. Each of the regional studies has examined the potential consequences that would result from the climate model scenarios used in the national level analysis (the first finding in the Key National Findings on page 89), and from model simulations of how such climate changes would affect the types and distributions of ecosystems. The following page provides highlights of what has been learned about the regional mosaic of consequences from these studies. A much more comprehensive presentation of the results is included in the National Assessment regional reports (see http://www.usgcrp.gov).

In summarizing potential consequences for the United States, it is important to recognize that the U.S. government represents not only the 50 states, but also has trust responsibility for a number of Caribbean and Pacific islands and for the homelands of Native Americans. In particular, the U.S. government has responsibilities of various types for Puerto Rico, the American Virgin Islands, American Samoa, the Commonwealth of the Northern Mariana Islands, Guam, and more than 565 tribal and Alaska Native governments that are recognized as "domestic dependent nations."

For the island areas, the potential consequences are likely to be quite similar to those experienced by nearby U.S. states. With regard to Native Americans, treaties, executive orders, tribal legislation, acts of Congress, and decisions of the federal courts determine the relationships between the tribes and the federal government. These agreements cover a range of issues that will be important in facing the potential consequences of climate change, including use and maintenance of land and water resources. Although the diversity of land areas and tribal perspectives and situations makes generalizations difficult, a number of key issues have been identified for closer study concerning how climate variability and change will affect Native populations and their communities. These issues include tourism and community development; human health and extreme events; rights to and availability of water and other natural resources; subsistence economies and cultural resources; and cultural sites, wildlife, and natural resources. Closer examination of the potential consequences for tribes in the Southwest is the topic of one of the regional assessments now underway.

FEDERAL RESEARCH ACTIVITIES

The types and nature of impacts of climate change that are projected to affect the United States make clear that climate change is likely to become an

Key Regional Vulnerability and Consequence Issues

The following key vulnerability and consequence issues were identified across the set of regions considered in the U.S. National Assessment. Additional details may be found in the regional reports indexed at http://www.usgcrp.gov.

Northeast, Southeast, and Midwest—Rising temperatures are likely to increase the heat index dramatically in summer. Warmer winters are likely to reduce cold-related stresses. Both types of changes are likely to affect health and comfort.

Appalachians—Warmer and moister air is likely to lead to more intense rainfall events in mountainous areas, increasing the potential for flash floods.

Great Lakes—Lake levels are likely to decline due to increased warm-season evaporation, leading to reduced water supply and degraded water quality. Lower lake levels are also likely to increase shipping costs, although a longer shipping season is likely. Shoreline damage due to high water levels is likely to decrease, but reduced wintertime ice cover is likely to lead to higher waves and greater shoreline erosion.

Southeast—Under warmer, wetter scenarios, the range of southern tree species is likely to expand. Under hotter, drier scenarios, it is likely that grasslands and savannas will eventually displace southeastern forests in many areas, with the transformation likely accelerated by increased occurrence of large fires.

Southeast Atlantic Coast, Puerto Rico, and the Virgin Islands—Rising sea level and higher storm surges are likely to cause loss of many coastal ecosystems that now provide an important buffer for coastal development against the impacts of storms. Currently and newly exposed communities are more likely to suffer damage from the increasing intensity of storms.

Midwest/Great Plains—A rising CO_2 concentration is likely to offset the effects of rising temperatures on forests and agriculture for several decades, increasing productivity and thereby reducing commodity prices for the public. To the extent that overall production is not increased, higher crop and forest productivity is likely to lead to less land being farmed and logged, which may promote recovery of some natural environments.

Great Plains—Prairie potholes, which provide important habitat for ducks and other migratory waterfowl, are likely to become much drier in a warmer climate.

Southwest—With an increase in precipitation, the desert ecosystems native to this region are likely to be replaced in many areas by grasslands and shrublands, increasing both fire and agricultural potential.

Northern and Mountain Regions—It is very likely that warm-weather recreational opportunities like hiking will expand, while cold-weather activities like skiing will contract.

Mountain West—Higher winter temperatures are very likely to reduce late winter snowpack. This is likely to cause peak runoff to be lower, which is likely to reduce the potential for spring floods associated with snowmelt. As the peak flow shifts to earlier in the spring, summer runoff is likely to be reduced, which is likely to require modifications in water management to provide for flood control, power production, fish runs, cities, and irrigation.

Northwest—Increasing river and stream temperatures are very likely to further stress migrating fish, complicating current restoration efforts.

Alaska—Sharp winter and springtime temperature increases are very likely to cause continued melting of sea ice and thawing of permafrost, further disrupting ecosystems, infrastructure, and communities. A longer warm season could also increase opportunities for shipping, commerce, and tourism.

Hawaii and Pacific Trust Territories—More intense El Niño and La Niña events are possible and would be likely to create extreme fluctuations in water resources for island citizens and the tourists who sustain local economies.

---

increasingly important factor in the future management of our land and water resources. To better prepare for coming changes, it is important to enhance the basis of understanding through research and to start to consider the potential risks that may be created by these impacts in the making of short- and long-term decisions in such areas as planning for infrastructure, land use, and other natural resource management. To promote these steps, the U.S. government sponsors a wide range of related activities reaching across federal agencies and on to the states, communities, and the general public.

Interagency Research Subcommittees

At the federal level, climate change and, even more generally, global environmental change and sustainability are topics that have ties to many agencies across the U.S. government. To ensure coordination, the U.S. Congress passed the Global Change Research Act of 1990 (Public Law 101-606). This law provides for the interagency coordination of global change activities, including research on how the climate is likely to change and on the potential consequences for the environment and society. Responsibility is assigned to the Executive Office of the President and is implemented under the guidance of the Office of Science and Technology Policy (OSTP). To implement this coordination, OSTP has established several interagency subcommittees. The U.S. Global Change Research Program (USGCRP) provides a framework for coordination of research to reduce uncertainties about climate change and potential impacts on climate, ecosystems, natural resources, and society (see Chapter 8). A number of the activities of the other subcommittees are also related to the issues of vulnerability and adaptation to global climate change:

- *Natural Disaster Reduction*—This subcommittee promotes interagency efforts to assemble and analyze data and information about the occurrence and vulnerability of the United States to a wide range of weather- and climate-related events. Through its participating agencies, the subcommittee is also promoting efforts by communities, universities, and others to increase their preparation for, and resilience to, natural disasters. In that climate change may alter the intensity, frequency, duration and location of such disasters, enhancing resilience and flexibility will assist in coping with climate change.

- *Air Quality*—This subcommittee promotes interagency efforts to document and investigate the factors affecting air quality on scales from regional and subcontinental to intercontinental and global, focusing particularly on tropospheric ozone and particulate matter, both of which contribute to climate change as well as being affected by it.

- *Ecological Systems*—This subcommittee promotes interagency efforts to assemble information about ecological systems and services and their coupling to society and environmental change. It is sponsoring assessments that document the current state of the nation's ecosystems, and that provide scenarios of future conditions under various management and policy options, providing a baseline for the National Assessment studies concerning how ecosystems are likely to change over the long term.

Individual Agency Research Activities

In addition to their interagency activities, many of the USGCRP agencies have various responsibilities relating to the potential consequences of climate change and of consideration of responses and means for coping with and adapting to climate change.

U.S. Department of Agriculture

Research sponsored by the U.S. Department of Agriculture (USDA) focuses on understanding terrestrial systems and the effects of global change (including water balance, atmospheric deposition, vegetative quality, and UV-B radiation) on food, fiber, and forestry production in agricultural, forest, and range ecosystems. USDA research also addresses how resilient managed agricultural, rangeland, and forest ecosystems are to climate change and what adaptation strategies will be needed to adjust to a changing climate. Programs include long-term studies addressing the structure, function, and management of forest and grassland ecosystems; research in applied sciences, including soils, climate, food and fiber crops, pest management, forests and wildlife, and social sciences; implementation of ecosystem management on the national forests and grasslands; and human interaction with natural resources.

For example, U.S. Forest Service research has established a national plan of forest sustainability to continue to provide water, recreation, timber, and clean air in a changing environment. Two goals of this program are to improve strategies for sustaining forest health under multiple environmental stresses and to develop projections of future forest water quality and yield in light of potential changes in climate.

Similarly, research at the U.S. Agricultural Research Service (ARS) looks to determine the impacts of increased atmospheric CO_2 levels, rising temperatures, and water availability on crops and their interactions with other biological components of agricultural ecosystems. ARS also conducts research on characterizing and measuring changes in weather and the water cycles at local and regional scales, and determining how to manage agricultural production systems facing such changes.

National Oceanic and Atmospheric Administration

The National Oceanic and Atmospheric Administration (NOAA) supports *in situ* and remote sensing and monitoring, research, and assessment to improve the accuracy of forecasts of weather and intense storms, and projections of climate change; to improve the scientific basis for federal, state, and local management of the coastal and marine environment and its natural resources; and to ensure a safe and productive marine transportation system. In addition to direct responsibilities for managing National Marine Sanctuaries and for protecting threatened, endangered, and trust resources, NOAA works with states to implement their coastal zone management plans and with regional councils to ensure sustained productivity of marine fisheries. Climate change and variability influence all areas of NOAA's responsibilities, both through direct effects and through intensification of other stresses, such as pollution, invasive species, and land and resource use.

U.S. Department of Health and Human Services

Through the National Institutes of Health, the Department of Health and Human Services sponsors research on a wide variety of health-related issues ranging from research on treatments for existing and emerging diseases to studies of risks from exposures to environmental stresses. For example, the National Institute of Environmental Health Sciences (NIEHS) conducts research on the effects of exposure to environmental agents on human health. The core programs of the NIEHS provide data and understanding for risk assessments due to changes in human vulnerability and exposures. Climate change raises issues of susceptibility to disease and needs for ensuring public health services. Changes in crop production techniques can increase human exposures to toxic agents and to disease vectors.

U.S. Department of the Interior

The U.S. Department of the Interior (DOI) is the largest manager of land and the associated biological and other natural resources within the United States. Its land management agencies, which include the Bureau of Land Management, the U.S. Fish and Wildlife Service, and the National Park Service, cumulatively manage over 180 million hectares (445 million acres) or 20 percent of the nation's land area for a variety of purposes, including preservation, tourism and recreation, timber harvesting, migratory birds, fish, wildlife, and a multiplicity of other functions and uses.

DOI's Bureau of Reclamation is the largest supplier and manager of water in the 17 western states, delivering water to over 30 million people for agricultural, municipal, industrial, and domestic uses. The Bureau also generates over a billion dollars worth of hydroelectric power and is responsible for multipurpose projects encompassing flood control, recreation, irrigation, fish, and wildlife. Management of land, water, and other natural resources is of necessity an exercise in adaptive management (IPCC 1991).

Research related to climate change conducted by DOI's U.S. Geological Survey includes efforts to identify which parts of the natural and human-controlled landscapes, ecosystems, and coastlines are at the highest risk under potential changes in climate and climate variability, water availability, and different land and resource management practices.

U.S. Department of Transportation

The U.S. Department of Transportation has recognized that many of the nation's transportation facilities and operations, which are now generally exposed to weather extremes, are also likely to be affected as the climate changes. Among a long list of potential impacts, sea level rise is likely to affect many port facilities and coastal airports; higher peak stream flows are likely to affect bridges and roadways, whereas lower summertime levels of rivers and the Great Lakes are likely to inhibit barge and ship traffic; and higher peak temperatures and more intense storms are likely to adversely affect pavements and freight movement. An assessment of the potential significance of changes for the U.S. transportation system and of guidelines for improving resilience is being organized.

U.S. Environmental Protection Agency

The U.S. Environmental Protection Agency (EPA) works closely with other federal agencies, state and local governments, and Native American tribes to develop and enforce regulations under existing environmental laws, such as the Clean Air Act, the Clean Water Act, and the Safe Drinking Water Act. In line with EPA's mission to protect human health and safeguard the natural environment, EPA's Global Change Research Program is assessing the consequences of global change for human health, aquatic ecosystem health, air quality, and water quality. Recognizing the need for "place-based" information, these assessments will focus on impacts at appropriate geographic scales (e.g., regional, watershed). In addition, EPA is supporting three integrated regional assessments in the Mid-Atlantic, Great Lakes, and Gulf Coast regions. Finally, in support of these assessments, EPA laboratories and centers conduct research through intramural and extramural programs.

OTHER RESEARCH ACTIVITIES

In addition to federal activities, a number of local, state, and regional activities are underway. Many of these activities have developed from the various regional assessments sponsored by the USGCRP or with the encouragement of various federal agencies. In addition, the USGCRP and federal agencies have been expanding their education and outreach activities to the public and private sectors, as described in Chapter 9.

Recognizing our shared environment and the resources it provides, it is important that the nations of the world work together in planning and coordinating their steps to adapt to the changing climate projected for coming decades. As part of this effort, the United States has been co-chair of Working Group II of the Intergovernmental Panel on Climate Change, which is focused on impacts, adaptation, and vulnerability. For the IPCC's Fourth Assessment Report, the United States will co-chair IPCC Working Group I on Climate Science.

The United States is also a leader in organizing the Arctic Climate Impact Assessment (ACIA), which is being carried out under the auspices of the eight-nation Arctic Council to "evaluate and synthesize knowledge on climate variability, climate change, and increased ultraviolet radiation and their consequences.... The ACIA will examine possible future impacts on the environment and its living resources, on human health, and on buildings, roads and other infrastructure" (see http://www.acia.uaf.edu/). These and other assessments need to continue to be pursued in order to ensure the most accurate information possible for preparing for the changing climate.

Chapter 7
Financial Resources and Transfer of Technology

The United States is committed to working with developing countries and countries with economies in transition to address the challenge of global climate change. The U.S. government has participated actively in the Technology Transfer Consultative Process under the United Nations Framework Convention on Climate Change (UNFCCC), and has implemented international programs and activities to facilitate the transfer of environmentally sound technologies and practices that reduce growth in greenhouse gas emissions and address vulnerability to climate impacts.

Under Article 4.5 of the UNFCCC, Annex I Parties, such as the United States, committed to "take all practicable steps to promote, facilitate and finance, as appropriate, the transfer of, or access to, environmentally sound technologies and know-how to other Parties." The Parties defined *technology transfer* at the Second Meeting of the

Conference of the Parties to the FCCC (COP-2) in Geneva as follows:

The term "transfer of technology" encompasses practices and processes such as "soft" technologies, for example, capacity building, information networks, training and research, as well as "hard" technologies, for example, equipment to control, reduce, or prevent anthropogenic emissions of greenhouse gases in energy, transport, forestry, agriculture, and industry sectors, to enhance removal by sinks, and to facilitate adaptation.

This chapter summarizes efforts undertaken by the United States in support of its strong commitment to technology cooperation and transfer. It also reports financial flows from the United States to different international bodies, foreign governments, and institutions that support climate-friendly activities.

Between 1997 and 2000, the U.S. government appropriated $285.8 million to the Global Environment Facility (GEF). A significant portion of overall GEF financing has been dedicated to climate-related activities. It provided nearly $4.5 billion to multilateral institutions and programs, such as the United Nations and affiliated multilateral banks, to address climate change and related international development priorities.

In addition, during the years 1997–2000 U.S. direct, bilateral, and regional assistance in support of climate change mitigation, adaptation, and crosscutting activities totaled $4.1 billion. Commercial sales for technologies that supported emissions mitigation and reduced vulnerability amounted to $3.6 billion. Over this same period, the United States leveraged $954.3 million in indirect financing through U.S. government-based financial instruments.

Some important highlights of U.S. assistance described in this chapter include:

- The U.S. Initiative on Joint Implementation, accepting 52 pioneering projects in 26 countries, with substantial cooperation and support from U.S. and host-country governments, nongovernmental organizations (NGOs), and the private sector.

- The U.S. Country Studies Program, which has helped 56 countries meet their UNFCCC obligations to report climate trends.

- The U.S. Agency for International Development's Climate Change Initiative, a program to leverage $1 billion in development assistance to address climate change through activities supporting renewable- and clean-energy activities, energy efficiency, forest and biodiversity conservation, and reduced vulnerability to climate impacts.

- A variety of public–private partnership programs that provide access to funding and expertise from the private sector, government, and NGOs to facilitate cooperation and foster innovation in climate-friendly sustainable development.

- Targeted programs to assist developing countries that are particularly vulnerable to the adverse effects of climate change, through weather forecasting and warning systems, climate and vulnerability modeling, and disaster preparedness and response.

This chapter also provides success stories to illustrate programs that demonstrate significant achievement and innovation in climate change mitigation and adaptation activities under U.S. leadership.

TYPES AND SOURCES OF U.S. ASSISTANCE

The United States recognizes that effectively addressing global climate change requires assistance to developing countries and countries with economies in transition to limit their net greenhouse gas emissions and reduce their vulnerability to climate impacts. As such, U.S. government agencies, private foundations, NGOs, research institutions, and businesses channel significant financial and technical resources to these countries to promote technology transfer that helps address the challenges posed by global climate change. In addition to the transfer of "hard" technologies, the United States supports extensive "soft"

technology transfer, such as the sharing of technical experience and know-how for targeted capacity building and strengthening of in-country institutions.

U.S. financial flows to developing and transition economies that support the diffusion of climate-friendly technologies include official development assistance (ODA) and official aid (OA), government-based project financing, foundation grants, NGO resources, private-sector commercial sales, commercial lending, foreign direct investment, foreign private equity investment, and venture capital. Financial resources are also provided indirectly in the forms of U.S. government-supported credit enhancements (loan and risk guarantees) and investment insurance. U.S. ODA and OA provide grants for a variety of technology transfer programs, while U.S. government-supported project financing and credit enhancements, commercial sales, commercial lending, foreign private equity investment, and foreign direct investment typically involve investments in physical capital, such as plants and equipment.[1] Note that this chapter provides only a partial monetary accounting of the flow types mentioned above, and does not account for commercial lending, foreign private equity investment, or venture capital, except for some brief illustrative examples. Further detail on how these flows are accounted is provided in the section of this chapter entitled "U.S. Financial Flow Information: 1997–2000."

ODA and OA are important to help create the economic, legal, and regulatory environment that is necessary to attract potential foreign investors, and enable larger flows of private financial resources to be leveraged in recipient countries. Private-sector participation is critical to the successful transfer of much-needed technical know-how and technologies in most regions of the world because it finances, produces, and supplies most climate-friendly

---

[1] The financial flow types reported in this chapter reflect those described in chapter 2 of IPCC 2000.

technologies, and thus can provide much of the human and financial capital for their effective deployment. U.S. government agencies, foundations, NGOs, and businesses each play a different role in promoting climate technology transfer to developing and transition economies.

U.S. Government Assistance

The U.S. government has facilitated technology transfer initiatives in developing and transition economies by forming partnerships and creating incentives for investment in climate-friendly technologies. U.S. government climate change projects support core U.S. development assistance priorities and the essential elements needed to achieve sustainable development. These priorities include supporting economic growth and social development that protects the resources of the host country; supporting the design and implementation of policy and institutional frameworks for sustainable development; and strengthening in-country institutions that involve and empower citizenry.

Official Development Assistance and Official Aid

The U.S. government provides ODA and OA to foreign governments and provides financial support to U.S. and host-country NGOs that have expertise in climate change mitigation and adaptation measures. Through this kind of assistance, the U.S. government has facilitated technology transfer in developing and transition economies by advancing the market for climate-friendly technologies and by forming partnerships and creating incentives for investment in climate-friendly technologies. U.S. ODA and OA strive to build local capacity as well as the policy frameworks and regulatory reforms needed to ensure that developing and transition economies can grow economically while limiting their net greenhouse gas emissions. U.S. ODA and OA are especially important in sectors where private-sector flows are comparatively low.

U.S. Agency for International Development. To date, U.S. bilateral assistance has primarily been implemented through the U.S. Agency for International Development (USAID), the foreign assistance arm of the U.S. government. Since 1997, USAID has implemented many new programs in developing and transition economies to address climate change. Specifically, USAID launched a $1 billion Climate Change Initiative to expand the Agency's already extensive efforts to help developing and transition economies. The goals of this initiative have been to help USAID-assisted countries reduce their net greenhouse gas emissions and their vulnerability to the impacts of climate change, and increase their participation in the UNFCCC. Between 1998 (when the initiative began) and 2000, USAID had committed $478.6 million to support climate change objectives throughout its programs and $6.3 million in leveraged credit. (Additional information about USAID's Climate Change Initiative is provided in the following sections.) USAID also works closely with other U.S. government agencies to leverage additional resources and expertise in addressing a variety of climate-related issues.

U.S. Department of Energy. In addition to providing funding support for interagency activities such as the U.S. Initiative on Joint Implementation (USIJI), the U.S. Country Studies Program (CSP), and the Technology Cooperation Agreement Pilot Project (TCAPP), the U.S. Department of Energy (DOE) works directly with foreign governments and institutions to promote dissemination of energy-efficiency, renewable-energy, and clean-energy technologies and practices. DOE's International Clean Cities program, for example, works with foreign governments, industry, and NGOs to help them implement viable activities that address climate change, transportation needs, local air quality, and related health risks.

U.S. Environmental Protection Agency. The U.S. Environmental Protection Agency (EPA) supports bilateral climate change programs, as well as such international programs as USIJI, CSP, and TCAPP. EPA is instrumental in designing and implementing innovative programs on a variety of global environmental challenges, including efforts to reduce greenhouse gas emissions and local air pollution and efforts to protect marine resources.

U.S. Department of Agriculture. The U.S. Department of Agriculture (USDA) supports international efforts to promote forest conservation and sustainable forestry, agroforestry, and improved agricultural practices. Such activities have provided meaningful benefits in addressing both climate change mitigation, through improved carbon sequestration, and adaptation to climate impacts, often related to food supply and conservation of agricultural resources. USDA is also instrumental in establishing food security warning systems.

National Oceanic and Atmospheric Administration. The National Oceanic and Atmospheric Administration (NOAA) has played an important role as a world leader in the study and provision of meteorological and hydrological forecasting and modeling; satellite imaging and analysis; climate change assessment, analysis, and modeling; and hazardous weather prediction. Critical information gained from these activities is made available to developing and transition country partners to address areas of vulnerability to climate-related impacts.

National Aeronautics and Space Administration. Like NOAA, the National Aeronautics and Space Administration (NASA) provides important technical information; satellite imaging and other surveillance; analysis and research related to climate changes, predictions, and weather trends; as well as analysis of shifts in the conditions of forests, natural areas, and agricultural zones.

Trade and Development Financing

U.S. government agencies also provide trade and development financing to developing and transition economies. These agencies facilitate the transfer of climate-friendly technologies by providing OA, export credits, project financing, risk and loan guarantees, and investment insurance to U.S. companies as well as credit enhancements for host-country financial institutions. Trade and development financing leverages foreign direct investment, foreign private equity investment, or host-country and non-U.S. private capital by decreasing the risk involved in long-term, capital-intensive projects or projects in nontraditional sectors. Several agencies engage in this type of financing.

Overseas Private Investment Corporation. The Overseas Private Investment Corporation (OPIC) provides project financing, political risk insurance, and investment guarantees for U.S. company projects covering a range of investments, including clean-energy projects in developing countries. OPIC also supports a variety of funds that make direct equity and equity-related investments in new, expanding, and privatizing companies in emerging market economies.

Export-Import Bank. The Export–Import Bank (Ex–Im) provides loan guarantees to U.S. exporters, guarantees the repayment of loans, and makes loans to foreign purchasers of U.S. goods and services. It also provides credit insurance that protects U.S. exporters against the risks of nonpayment by foreign buyers for political or commercial reasons. Ex–Im has provided project loans and risk guarantees related to climate change mitigation for clean-energy and renewable-energy projects in developing and transition economies.

USAID Development Credit Authority. USAID's Development Credit Authority (DCA) provides partial loans and risk guarantees to host-country and international financial intermediaries to encourage project finance in nontradi-

tional sectors, such as energy efficiency. In addition to this immediate financial leverage benefit, DCA facilitates long-term relationships with the private sector that outlive USAID's project involvement, allowing USAID to contribute to the direction of investment of the ever-increasing global private capital flows.

U.S. Trade and Development Agency. The U.S. Trade and Development Agency (TDA) helps U.S. companies pursue overseas business opportunities through OA. By supporting feasibility studies, orientation visits, specialized training grants, business workshops, and technical assistance, TDA enables American businesses to compete for infrastructure and industrial projects in developing countries. TDA has promoted the transfer of climate-friendly technology in the energy, environment, and water resources sectors.

U.S. Department of Commerce. The U.S. Department of Commerce (DOC) recently established an International Clean Energy Initiative that links U.S. companies with foreign markets to facilitate dissemination of clean-energy technologies, products, and services. The initiative seeks to realize a vision for enhanced exports of clean-energy technology.

NGO Assistance

U.S. foundations and NGOs have played a pivotal role in helping countries undertake sustainable development projects that have increased their ability to mitigate and adapt to the effects of global climate change. These organizations help improve host-country capacity by implementing small-scale, targeted initiatives related to the mitigation of and adaptation to climate change impacts. Following are some examples of these organizations.

W. Alton Jones Foundation

The W. Alton Jones Foundation supports the development of climate-friendly energy in developing countries. The Foundation also seeks to

build the capacity of entrepreneurs in developing countries to bring renewable-energy technologies to market.

Rockefeller Brothers Fund

The Rockefeller Brothers Fund seeks to help developing countries define and pursue locally appropriate development strategies. In East Asia, the Fund provides grants for coastal zone management and integrated watershed planning efforts that will help these countries prepare to adapt to the effects of global climate change.

The Nature Conservancy

The Nature Conservancy (TNC), in partnership with the U.S. private sector,[2] is working to lower net CO_2 emissions in Belize (the Río Bravo Carbon Sequestration Pilot Project) and Bolivia (the Noel Kempff Mercado Climate Action Project) through the prevention of deforestation and sustainable forest management practices. These projects are also helping to conserve local biodiversity, improve local environmental quality, and meet sustainable development goals.

Conservation International

Through its innovative partnerships with donors, businesses, and foundations, Conservation International (CI) protects biodiversity and promotes cost-effective emission reductions with a special emphasis on conservation and restoration of critical forest ecosystems. CI implements programs through its conservation financing mechanism, the Conservation Enterprise Fund. It has also established the Center for Environmental Leadership in Business, a CI/Ford Motor Company joint venture that promotes collaborative business practices that reduce industry's ecological impacts, contribute to conservation efforts, and create economic value for the companies that adopt them.

---

[2] U.S. private-sector investors participating in these activities have included Cinergy, Detroit Edison, PacifiCorp, Suncor, Utilitree Carbon Company, Wisconsin Electric/Wisconsin Gas (formerly Wisconsin Electric Power Company), and American Electric Power.

Private-Sector Assistance

As part of their normal business practices, many U.S. private-sector entities seek opportunities to expand their markets outside of the United States. As a result, these companies are contributing to the transfer of climate-friendly technologies through foreign direct investment, commercial lending, private equity investment, venture capital investment, and commercial sales of "hard" technology in developing and transition economies. Consequently, many technologies have been transferred to the industrial, energy supply, transportation, agriculture, and water supply sectors.

Foreign direct investment and commercial lending together represent the primary means for long-term, private-sector technology transfer. U.S. companies like the Global Environment Fund are making investments in foreign private equity through such funds as the Global Environment Strategic Technology Partners, LP fund. This Fund seeks investments in U.S.-based companies whose technologies promote improvements in economic efficiency, the environment, health, and safety. It seeks new equity investment opportunities in the range of $1–$2 million.[3]

Among the member countries of the Organization for Economic Cooperation and Development (OECD), venture capital—normally reserved for high-risk, long-term investments—is most prominent in the United States. U.S. venture capital firms have begun to make innovative and high-risk investments in the environmental sector in developing countries. For example, the Corporación Financiera Ambiental, capitalized in part by U.S. investors, invests in small and medium-sized private enterprises that undertake environmental projects in Central America.[4] Investments range from $100,000 to $800,000 per project.

The United States is the largest producer of environmental technologies and services. In 2000, commercial sales of these technologies represented $18 billion of U.S. export flows (Business Roundtable 2001). Typical U.S. climate change mitigation and adaptation exports include wastewater treatment, water supply, renewable energy, and heat/energy savings and management equipment. For mitigation technologies in the commercial, industrial, residential use, energy supply, and transportation sectors, U.S. developing country market share in 2000 was estimated to be $5.3 billion, or 18 percent of the entire market for these technologies in developing and transition economies (USAID 2000b).

MAJOR U.S. GOVERNMENT INITIATIVES

Three major U.S. government initiatives are the U.S. Initiative on Joint Implementation, the U.S. Country Studies Program, and the Climate Change Initiative.

U.S. Initiative on Joint Implementation

Launched in 1993 as part of the U.S. Climate Change Action Plan, the U.S. Initiative on Joint Implementation (USIJI)[5] supports the development of voluntary projects that reduce, avoid, or sequester greenhouse gas emissions. These projects are implemented between partners located in the United States and their counterparts in other countries. USIJI is a flexible, nonregulatory pilot program that encourages U.S. businesses and NGOs to voluntarily use their resources and innovative technologies and practices to reduce greenhouse gas emissions and promote sustainable development. USIJI also promotes projects that test and evaluate methodologies for measuring and tracking greenhouse gas reductions and verifying the costs and benefits of projects.

USIJI is the largest and most developed worldwide program exploring the potential of project-based mechanisms. It is administered by an interagency secretariat co-chaired by DOE and EPA, with significant participation from USAID and the U.S. Departments of Agriculture, Commerce, Interior, State, and Treasury.[6]

Between 1994 and 2000, the USIJI project portfolio included 52 projects in the following 26 countries: Argentina (3), Belize (2), Bolivia (3), Chile (3), Columbia (1), Costa Rica (7), Czech Republic (1), Djibouti (1), Ecuador (1), El Salvador (1), Equatorial Guinea (1), Guatemala (3), Honduras (3), India (1), Indonesia (1), Mali (1), Mauritius (1), Mexico (4), Nicaragua (1), Panama (1), Peru (1), Philippines (1), the Russian Federation (6), South Africa (1), Sri Lanka (1) and Uganda (2). On-site implementation has begun for 24 of these projects. In addition, eight new projects are currently under development (USIJI 2000).[7] To support USIJI, the U.S. government provided more than $15.9 million in funding. Seven projects leveraged a total of $8.5 million in financing from private sources.[8]

USIJI projects involve a range of participants and are funded through several different mechanisms. Projects include participants and technical experts from U.S. and host-government agencies, private-sector companies, industry associations, NGOs, state and local governments, universities, research

---

[3] http://www.globalenvironmentfund.com/funds.htm.

[4] http://www.cfa-fund.com.

[5] The concept of "Joint Implementation" (JI) was introduced early in the negotiations leading up to the 1992 Earth Summit in Rio de Janeiro, and was formally adopted into the text of the UNFCCC. The United States joined more than 150 countries in signing the UNFCCC, which explicitly provides through Article 4(2)(a) for signatories to meet their obligation to reduce greenhouse gas emissions "jointly with other Parties." The term has been used subsequently to describe a wide range of possible arrangements between entities in two or more countries, leading to the implementation of cooperative development projects that seek to reduce, avoid, or sequester greenhouse gas emissions (http://www.gcrio.org/usiji/about/whatisji.html).

[6] http://www.gcrio.org/usiji/about/whatisji.html.

[7] This designation could mean, for example, that although project implementation activities (e.g., construction and planning) have begun, greenhouse gas benefits have not yet necessarily begun to accrue. The remaining projects have not yet initiated on-site activities, and are classified as "mutually agreed."

[8] Because information about private-sector investment in such projects is proprietary, the full breadth of leveraged funding under USIJI cannot be ascertained.

institutes, national laboratories, and financing organizations. Project funding is typically based on the anticipated sale of carbon offsets; revenues generated directly by project activities (e.g., the sale of timber, other biomass resources, and energy); investment capital from private-sector companies; loans provided by commercial banks and multilateral organizations, such as the International Finance Corporation; government incentives; endowments; and grants.[9] Past technical assistance under USIJI generally has consisted of workshops, guidance documents, issue papers, hotline assistance, and meetings.[10]

USIJI projects span the land-use change and forestry, energy, waste, and agricultural sectors, and involve a range of activities that achieve greenhouse gas benefits. As of 2000, the aggregate USIJI projects were anticipated to generate greenhouse gas reductions totaling at least 259.8 teragrams of CO_2 over a period of approximately 60 years, including 5.7 teragrams of CH_4, and 4.6 gigagrams of N_2O. These benefits are equivalent to 350.5 teragrams of CO_2, which are expected to accrue over project lifetimes that vary from 10 to 60 years if fully funded and implemented (U.S. IJI 2000). For example, the Noel Kempff Mercado Climate Action Project, which conserves forest area in the Bolivian Amazon covering over 600,000 hectares (nearly 15 million acres), is expected to have a net carbon benefit of 15 teragrams of carbon over the next 30 years.

U.S. Country Studies Program

The UNFCCC requires all signatory countries to provide to the Secretariat of the Convention a national inventory of greenhouse gas emissions by sources and removals by sinks, and to describe the steps they are taking to implement the Convention, including mitigation and adaptation measures. The U.S. Country Studies Program (CSP) provided assistance to developing and transition economies to help meet this commitment, and to fulfill U.S. obligations under the UNFCCC to provide additional financial and technical

resources to developing countries. The first round of two-year studies began in October 1993 after the United Nations Conference on Environment and Development (UNCED, the Earth Summit) in Rio de Janeiro in 1992.

The CSP has helped 56 countries build the human and institutional capacities necessary to assess their vulnerability to climate change and opportunities to mitigate it. Under the CSP, the United States has helped countries develop inventories of their anthropogenic greenhouse gas emissions, evaluate their response options for mitigating and adapting to climate change, assess their vulnerability to climate change, perform technology assessments,[11] develop National Communications, and disseminate analytical information to further national and international discussions on global strategies for reducing the threat of climate change.[12] Technical assistance was delivered through workshops, research, major country reports, guidance documents, technical papers, consultations with technical experts, analytic tools, data, equipment, and grants to support and facilitate climate change studies around the world.[13]

In all, the CSP has helped other countries and international institutions produce over 160 major country reports, 10 guidance documents, 60 workshop and conference proceedings, and 16 special journal editions. In 1997, the CSP completed a report entitled *Global Climate Change Mitigation Assessment Results for Fourteen Transition and Developing Countries* (U.S. CSP 1997), and in 1998 produced *Climate Change Assessments by Developing and Transition Countries* (U.S. CSP 1998). These and numerous other reports continue to make important contributions to the work of the GEF, the Intergovernmental Panel on Cli-

mate Change (IPCC), and the Subsidiary Bodies to the Convention.

In response to requests from developing and transition economies, the U.S. government supplemented the CSP activity by helping countries develop their national climate change action plans. Building on the experience of the CSP, the Support for National Action Plans (SNAP) program provided financial and technical assistance to help countries use the results of their climate change country studies to develop action plans and technology assessments for implementing a portfolio of mitigation and adaptation measures. An objective of the SNAP phase is to promote diffusion of mitigation and adaptation technologies by assisting countries with assessments of opportunities for technology exchange and diffusion. Countries can use these studies, action plans, and technology assessments as a basis for developing their national communications, and to meet their obligations under the UNFCCC. Eighteen countries participated in the SNAP phase of the CSP.[14] The CSP activity has been completed, and the information gained from the program is being converted to an electronic database available for future use.

Oversight for the program was provided by the U.S. Country Studies Management Team, which was composed of technical experts from EPA, DOE, USAID, USDA, NOAA, the National Science Foundation, and the Departments of State, Interior, and Health and Human Services. Between 1997 and 2000, these agencies jointly provided a total of $9.4 million in funding for the CSP.

Climate Change Initiative

In 1998 USAID launched the Climate Change Initiative (CCI), a

---

[9] http://www.gcrio.org/usiji/about/whatisji.html.
[10] http://www.gcrio.org/usiji/about/whatben.html.
[11] http://www.gcrio.org/CSP/ap.html.
[12] http://www.epa.gov/globalwarming/actions/international/countrystudies/index.html.
[13] http://www.gcrio.org/CSP/ap.html. See also http://www.epa.gov/globalwarming/actions/international/countrystudies/index.html.
[14] http://www.gcrio.org/CSP/ap.html. These countries include Bolivia, Bulgaria, China, Czech Republic, Egypt, Hungary, Indonesia, Kazakhstan, Mauritius, Mexico, Micronesia, Philippines, Russian Federation, Tanzania, Thailand, Ukraine, Uruguay, and Venezuela. See also http://www.epa.gov/globalwarming/actions/international/countrystudies/index.html.

$1 billion, five-year program to collaborate with developing nations and countries with economies in transition to reduce the threat of climate change. This multi-agency initiative supports activities that address climate change in more than 40 countries and regions around the world. Its overarching objective is to promote sustainable development that minimizes the associated growth in greenhouse gas emissions and to reduce vulnerability to climate change.

Through the CCI, USAID has helped countries to mitigate greenhouse gas emissions from the energy sector, industries, and urban areas; protect forests and farmland that can sequester CO_2 from the atmosphere; participate more effectively in the UNFCCC; and reduce their vulnerability to the impacts of climate change. An important aspect of the CCI is continued support for technology transfer and public–private partnerships that work to achieve the UNFCCC's goals. The initiative has strengthened the U.S. government's ability to measure the impact of its global assistance work to address climate change, and has helped fulfill U.S. obligations to assist and collaborate with developing countries under the UNFCCC.

From 1998 to 2000, USAID committed $478.6 million under the CCI to support climate change objectives. In addition, USAID leveraged approximately $2.9 billion to support climate change activities in developing and transition economies. This funding was directly leveraged from other bilateral and multilateral donors, the private sector, foundations, NGOs, and host-country governments. USAID also indirectly leveraged $5.3 billion in further investments from outside sources that built on projects it originally initiated.

In addition to the funding leveraged under the CCI, USAID used credit instruments available through the Agency's Development Credit Authority (DCA) to leverage funding for "climate-friendly" investment in developing and transition economies. DCA is a credit enhancement mechanism that provides greater flexibility in choosing the appropriate financing tool, such as loans, guarantees, grants, or a combination of these, for climate change and other sustainable development projects. Since its inception in 1999, DCA credit enhancements have leveraged $6.3 million in climate-friendly private-sector financed activities.

PUBLIC–PRIVATE PARTNERSHIP ACTIVITIES

An important U.S. objective is to leverage the private sector's financial and technical capabilities to promote sustainable development and help address climate change in developing and transition economies. The U.S. government and its partners do this through programs designed to facilitate dialogue, build partnerships, and support direct investment in climate-friendly and other sustainable development projects. Examples of such projects include the Technology Cooperation Agreement Pilot Project, the U.S.–Asia Environment Partnership, EcoLinks, and several energy and forest conservation partnerships.

The U.S. government also makes significant efforts to engage the private sector directly in many of its ongoing development assistance programs, both as key implementation partners and as a source of supplemental funding for climate-related activities. For example, USAID leveraged over $3 million from outside sources to support its Maya Biosphere Reserve project in Guatemala, and used a two-to-one matching-fund program with several organizations to collect $1.8 million in additional funding. USAID also helped the Mgahinga and Bwindi Impenetrable Forest Conservation Trust in Uganda grow to approximately $6 million, and leveraged an additional $1 million from the Government of Denmark to support USAID's community conservation in 25 parishes adjacent to the Bwindi and Mgahinga National Parks. In Ukraine, USAID also leveraged $18 million from the World Bank to support energy efficiency in government buildings in Kyiv, and helped private sugar mills in India obtain $66 million in loans to construct new bagasse cogeneration units.

Technology Cooperation Agreement Pilot Project

The Technology Cooperation Agreement Pilot Project (TCAPP) was a bilateral program initiated in 1997 as a collaborative effort of USAID, EPA, and DOE.[15] TCAPP's primary goal was to assist developing country partners in defining clean-technology priorities. To encourage the transfer of clean technologies, it focused on helping countries remove market barriers and promote direct private investment.[16] The pilot project was successful in building support for a country-driven, market-oriented, technology transfer approach under the UNFCCC. Building on lessons learned from TCAPP, which ended in 2001, these agencies continue to support efforts to accelerate adoption of clean-energy technologies and practices in partner countries.

Between 1997 and 2000, the U.S. government provided $2.9 million to TCAPP to support technology transfer activities in Brazil, China, Egypt, Kazakhstan, Mexico, Philippines, and South Korea. Through TCAPP, the U.S. government has facilitated the development of more than 20 clean-energy business investment projects in participating countries. Overall, TCAPP has engaged more than 400 U.S. and international business representatives to collaborate in developing new investment projects and to assist with implementation of actions to remove market barriers. Examples of TCAPP successes include renewable-energy policy reforms in the Philippines, development of an industrial energy services company (ESCO) pilot program in Mexico, financial support for sugar mill co-generation projects in Brazil, training for conducting energy audits in

---

[15] http://www.epa.gov/globalwarming/actions/international/techcoop/tcapp.html and http://www.nrel.gov/tcapp.
[16] http://www.epa.gov/globalwarming/actions/international/techcoop/tcapp.html and http://www.nrel.gov/tcapp.

Korea, training to verify the performance of wind turbines manufactured in China, and development of refinery energy efficiency pilot projects in Egypt.

Climate Technology Initiative

The Climate Technology Initiative (CTI), a voluntary, multilateral cooperative program, supports implementation of the UNFCCC by fostering international cooperation for accelerated development and diffusion of climate-friendly technologies and practices.[17] The United States, the European Commission, and 22 other OECD nations established the CTI at the First Meeting of the Conference of Parties to the UNFCCC (COP-1) in Berlin in 1995.[18] They agreed to work collaboratively to "accelerate development, application and diffusion of climate-friendly technologies in all relevant sectors."[19]

The CTI has become an international model of multilateral support for technology transfer and has built developing country support for a market-relevant approach to technology transfer implementation. An important component of the CTI is the reduction of market barriers and other obstacles to the transfer of climate-friendly technologies consistent with UNFCCC objectives.[20] Committed to focusing on areas where it can make a significant difference, the CTI works in voluntary partnership with stakeholders, including the private sector, NGOs, and other international organizations. While the CTI was designed to address all greenhouse gases from a variety of sources, its primary focus to date has been on efficient and renewable-energy technologies.

Within the U.S. government, support for the CTI is provided jointly by DOE, EPA, and USAID. Since 1998, these agencies have committed over $2 million to capacity-building activities, such as providing regional technology training courses, conducting technology needs assessments, and developing in-country technology implementation plans. These plans define opportunities for accelerating

implementation of such technologies as energy-efficient and photovoltaic lighting, efficient motors and boilers, energy-efficient housing, solar energy, biomass electricity generation, and natural gas. They also propose actions to improve technical capacity, increase access to funding, or reduce policy barriers to investment. More recently, the CTI has been working with the Southern Africa Development Community (SADC) to promote investment in climate-friendly technologies through public–private partnerships. This extensive effort under the CTI's Cooperative Technology Implementation Plan program was initiated in response to a request by SADC energy and environment ministers participating in a March 1999 CTI/Joint Industry seminar in Zimbabwe. Since then, the United States has provided approximately $320,000 in support of this effort.

U.S.–Asia Environmental Partnership

The United States–Asia Environmental Partnership (US–AEP) promotes environmentally sustainable development in Asia by building public–private partnerships, developing technical capacity, and promoting policy reforms that lead to environmentally sound investments, including climate-friendly technologies. US–AEP is jointly implemented by several U.S. government agencies, under the leadership of USAID.[21] Overall, US–AEP has supported climate change activities in Bangladesh, Hong Kong, India, Indonesia, South Korea, Malaysia, Nepal, Philippines, Singapore, Sri Lanka, Taiwan, Thailand, and Vietnam.[22]

US–AEP was created with the recog-

nition of Asia's growing commitment to sustainable development and growing U.S. interest in sharing its experience, technology, and management practices. With the participation of governments, NGOs, academia, and the private sector, US–AEP has become a flexible, responsive vehicle for delivering timely answers to environmental questions. US–AEP's mission has been to promote a "clean revolution" in Asia, transforming how Asia industrializes and protects its environment through the continuing development and adoption of less polluting and more resource-efficient products, processes, and services.[23]

A significant number of US–AEP activities address climate change by targeting the efficient use of energy resources, and the conversion of waste to energy. Other activities include waste minimization, power-sector reform, efficient electricity generation and transmission, and renewable energy. In 1999, for example, US–AEP activities led to $6.6 million in confirmed sales of energy-efficiency and related climate-friendly technologies and services. Additionally, US–AEP contributed $1.5 million to the USAID mission in Bangladesh to launch a major energy program there. Among its technology transfer activities, US–AEP also directly engaged small- to medium-sized U.S. private-sector firms to provide training and demonstrations of climate-related technologies and practices in 11 Asian countries, most of which involved converting waste to either energy or products, and recycling, recovering, and reusing materials. Also, 29 climate-related professional exchanges and study tours were conducted through US–AEP's

[17] http://www.climatetech.org/home.shtml.

[18] http://www.climatetech.org/about/index.shtmla.

[19] http://www.epa.gov/globalwarming/actions/international/techcoop/cti.html. See http://www.climatetech.org/home.shtml.

[20] http://www.epa.gov/globalwarming/actions/international/techcoop/cti.html. The CTI is intended to implement and support a number of objectives of the UNFCCC, including, for example, the requirement under Article 4.1.c, which calls for Parties to "Promote and cooperate in the development, application and diffusion, including transfer, of technologies, practices and processes." Similarly, the CTI furthers the goals of Article 4.5, which states that Annex I Parties "shall take all practicable steps to promote, facilitate and finance, as appropriate, the transfer of, or access to environmentally sound technologies and know-how." http://www.epa.gov/globalwarming/actions/international/techcoop/cti.html.

[21] http://www.usaep.org/about.htm.

[22] US–AEP Secretariat.

[23] http://www.usaep.org/about.htm.

Environmental Exchange Program. The majority of these activities addressed the conversion of waste to energy and products, and enhancing the efficient use of energy and resources.

EcoLinks

Launched in 1998, Eurasian–American Partnerships for Environmentally Sustainable Economies (EcoLinks) is a USAID initiative to help solve urban and industrial environmental problems through improved access to financial resources, trade and investment, and information technology. The program promotes sustainable, market-based partnerships among businesses, local governments, and associations in Central and Eastern Europe and in Eurasia with U.S. businesses to identify environmental problems and adopt best management practices and technologies. As these partnerships mature, trade and investment in environmental goods and services are expected to increase.[24] EcoLinks provides support through technology transfer and investment activities, partnership grants, and an information technology initiative. Countries participating in EcoLinks include Bulgaria, Croatia, the Czech Republic, Hungary, Kazakhstan, Macedonia, Poland, Romania, Russia (Far East), and Ukraine (USAID 2000a and 2001a).

While EcoLinks does not specifically target climate change, a large percentage of its technology transfer activities provide climate benefits. For example, EcoLinks addresses inadequate wastewater treatment capacity, inefficient and highly polluting industries and public utilities, poor waste management practices, and weak environmental management and regulatory systems. Some examples of EcoLinks' trade and investment support and grants activities include:

- *In a Bulgarian municipality*—Developing environmental management systems for mitigating greenhouse gas emissions.

- *In Bulgaria*—Developing landfill gas extraction systems.
- *In Romania*—Introducing a comprehensive energy audit methodology.
- *In Croatia*—Assessing water turbines in water delivery systems.
- *In all participating countries*—Facilitating technology demonstrations in energy efficiency and alternative energy.
- *In the Czech Republic*—Promoting landfill gas utilization technology.
- *In Kazakhstan*—Promoting cleaner production in the oil and gas industry.
- *In Hungary*—Facilitating a $1.2 million loan to a joint U.S.–Hungarian company promoting a new wastewater treatment technology (USAID 2000a and 2001a).

Funding and implementation for EcoLinks are jointly provided by USAID, the U.S. Department of Commerce, the Environmental Export Council, the Global Environment and Technology Foundation, the Institute for International Education, and the Regional Environment Center for Central and Eastern Europe. Since EcoLinks began, four grant cycles have been completed, 135 grants have been awarded, and currently more than 100 active projects are funded (USAID 2001a). In 2000 alone, EcoLinks awarded 41 Challenge Grants to participating country institutions totaling nearly $2 million. EcoLinks also provided over $536,000 in Quick Response Awards in 2000 throughout the region (USAID 2001c).

Energy Partnership Program

Funded by USAID and implemented by the United States Energy Association (USEA), the Energy Partnership Program is an important public–private partnership activity with climate benefits. This program establishes practitioner-to-practitioner, multi-year partnerships between U.S. and developing country utilities and regulatory agencies in Asia, Africa, Latin America, Central and Eastern Europe, and the former Soviet Union. Its main objective is to provide a mechanism for the U.S. energy industry (utilities, regulators,

and policymakers) to transfer its experience in market-based energy production, transmission, and distribution to its international counterparts, while providing U.S. participants with the opportunity to learn about the energy industry in another country. Regional program activities encompass such topics as regulation, the environment, system reliability and efficiency, renewable energy, customer service, and financial management, with an emphasis on mitigating greenhouse gas emissions.

Working with USAID, USEA identifies and matches utilities or regulatory agencies in the United States and overseas according to the compatibility of their needs and capabilities, the similarity of their energy systems, potential common business interest, and other criteria. The benefits to the foreign partners include the opportunity for senior executives of foreign utilities and regulatory agencies to observe how their U.S. counterparts are structured, financed, managed, and regulated under free-market conditions. The program also offers U.S. energy executives the opportunity to understand the dynamics of non-U.S. energy markets and to forge strategic international alliances. Once selected, the participating organizations execute partnership agreements and commit to cooperate for a two-year period, during which the partners focus their exchange activities on several key issues. Following are some examples of these efforts.

- *In India*—Corporate restructuring, increased energy efficiency through reduction of distribution losses, improved plant operations, development of India's National Institute for Power Systems and Distribution Management, and joint-venture and pilot projects with U.S. partners.
- *In Indonesia*—Managing a distribution company in a privatized environment, utility decision making from the private company perspective, regulation and trading mechanisms, and privatization of the gas industry.
- *In the Philippines*—Management and corporate restructuring, quality of service, and customer service.

---

[24] http://www.ecolinks.org/about.html.

- *In Senegal*—Generating capacity through independent power production, improved efficiency, and improved system reliability through enhanced water, fuel, and materials analysis.
- *In Brazil*—Delegation of regulatory powers to Brazil's states, staff development and training, and generation resource portfolio planning.

Forest Conservation Partnerships

Among the leading U.S. innovative programs to address climate change through forest conservation activities are those being implemented through NGOs, such as The Nature Conservancy (TNC) and Conservation International (CI), often in partnership with the U.S. government and the private sector.

International Partnership Program

TNC's International Partnership Program (IPP) aims to strengthen the capacity of local organizations through collaborative efforts to preserve biological diversity and forest resources—efforts with valuable climate benefits. Through the IPP, TNC now works with more than 70 partner organizations in 26 countries throughout the Asia Pacific, Caribbean, and Latin America regions. The program specifically emphasizes the opportunities for promoting local leadership in biodiversity conservation, and improving access to technical information and expertise. As a result of the program, TNC and its partners have protected more than 32,375 hectares (over 80 million acres) of land in these locations that include climate projects to preserve forests, protect carbon sinks, and provide jobs; ecotourism training that enables fishermen to thrive by protecting rivers and coastal areas; and community-led marine conservation that empowers villagers to manage the fisheries that support their livelihoods.[25]

EcoEnterprise Fund

TNC's relatively new EcoEnterprise Fund is a joint initiative with the Inter-American Development Bank that seeks to use venture capital to protect natural areas in Latin America and the Caribbean. The Fund includes two components: (1) an investment fund that provides venture capital to profitable businesses involved in sustainable agriculture, sustainable forestry, ecotourism, and other environmentally compatible businesses; and (2) limited technical assistance funds to provide business advisory services to prospective projects. Participating companies are required to collaborate with a nonprofit conservation or community partner, by paying fees for monitoring services, by sharing profits, or by other financial arrangements. The Fund invests in ventures at all stages of development with prospective sales revenues up to $3 million. It gives preference to businesses that are unable to secure financing from conventional sources due to their small size, the innovative nature of their business, and/or the financial risks involved.[26]

Conservation Enterprise Fund

Similar to TNC's EcoEnterprise Fund, CI's Conservation Enterprise Fund (CEF) was created in 1999 with a $1 million loan from the International Finance Corporation's Small and Medium Enterprise Global Environmental Facility program. The CEF is a development tool that enables conservation enterprises to expand their operations through financial leveraging. CI acts as the financial intermediary to provide $25,000–$250,000 in debt and equity financing to small and medium-sized enterprises (possessing $5 million or less in assets) that are strategically important to conservation. For instance, a CEF loan helped coffee farmers in Chiapas, Mexico, finance post-harvest expenses in 1999. CEF funds are also directed to businesses engaged in agroforestry, ecotourism, and wild-harvest products.[27]

U.S. GOVERNMENT ASSISTANCE ADDRESSING VULNERABILITY AND ADAPTATION

Assisting countries that are particularly vulnerable to the adverse effects of climate change is a high priority for the United States. The U.S. government has provided extensive financial and technical support to such countries for many years, primarily through a number of programs designed to address disaster preparedness and relief, food security and sustainable agricultural production, biodiversity conservation, water resources management, and climate research and weather prediction programs. These activities involve numerous government agencies, such as USAID, NOAA, USDA, DOE, and EPA.

For example, under the U.S. Country Studies Program, the U.S. government has provided support to developing countries to conduct assessments of climate change vulnerability and adaptation options. Under the UNFCCC and pursuant to guidance from the GEF, donor nations are obligated to help developing nations participate in research and systematic observation of climate change, assess their vulnerability, prepare adaptation strategies, and implement adaptation measures. The results of these assessments and studies have been highly successful at promoting more meaningful participation by developing countries in the UNFCCC process, and at more accurately gauging potential risks and adaptation measures to address long- and short-term climate impacts. More detail on these activities is provided later in this chapter. More specific financial information about U.S. adaptation activities appears in Appendix C and in the section of this chapter concerning financial flows.

---

[25] http://nature.org/international/specialinitiatives/.
[26] http://nature.org/international/specialinitiatives/ecofund/.
[27] http://www.conservation.org/WEB/FIELDACT/C-C_PROG/ECON/fund.htm.

U.S. FINANCIAL FLOW INFORMATION, 1997–2000

This chapter presents financial resource information for the years 1997–2000. This information is also presented in Tables 7-1, 7-2, and 7-3. For the table on financial flows to specific countries and regions (Table 7-3 in Appendix C), this chapter goes beyond the minimum guidance requirement of presenting each flow by year, country, and sector. To provide a more complete description of these financial flows, further detail has been included to show both the type of flow and its source. To provide a framework for analysis, the chapter follows the approach of *Methodological and Technological Issues in Technology Trends* (IPCC 2000).

Financial information provided in this chapter is derived from the U.S. government, foundations, and other sources of financing to institutions supporting climate change mitigation, adaptation, and technology transfer activities in developing and transition economies. To a limited extent, this report also includes information about financial flows from the U.S. private sector, which if fully accounted for would be expected to far outweigh all other financial flows. Because private-sector financial and investment information is mostly proprietary and not available to the public, only two of these flows to climate change mitigation and adaptation activities are even partly accounted for in the tables that follow.[28]

Recipients of U.S. financial resources include the GEF (reported in Table 7-1), multilateral institutions (reported in Table 7-2), as well as NGOs, universities, research institutions, and foreign governments. While some of this funding is provided to U.S.-based institutions, only those activities providing assistance directly to developing countries and countries with economies in transition are reported here.

Due to the difficulty in identifying exact expenditures under most U.S. government programs, financial information provided in this report refers only to those activities for which funding was obligated in the given year, from 1997 to 2000, and in some cases 2001. In most cases, U.S. government information referred to the fiscal year for which funding was obligated—i.e., beginning October 1 in the year prior to and ending September 30 in the calendar year in question. For example, Fiscal Year 1997 began October 1, 1996, and ended September 30, 1997. In most other cases, including funding from U.S. foundations and other public and private institutions, information relates to the calendar year in which funding was awarded.

Financial Contributions to the Global Environment Facility

The Global Environment Facility (GEF) was established in 1991 to forge international cooperation and finance actions for addressing critical threats to the global environment resulting from the loss of biological diversity, climate change, degradation of international waters, and ozone depletion. It also provides funding to address the pervasive problem of land degradation. The GEF is now the interim financial mechanism for the Protocol on Persistent Organic Pollutants and acts as the financial mechanism of both the Convention on Biological Diversity and the UNFCCC. The GEF leverages its resources through co-financing and cooperation with other donor groups and the private sector. In 1998, 36 nations pledged a total of $2.75 billion in funding to protect the global environment and promote sustainable development. The United States has been a member country and supporter of the GEF since 1994. As of December 2000, 167 countries were participating members of the GEF.[29]

Aggregated U.S. Government Funding

Between 1997 and 2000, the U.S. government has provided $285.8 million to the GEF. Recently, President Bush announced his Administration's intention to fully fund payment for arrears incurred during the previous Administration. The President's budget request for fiscal year 2003 includes $70 million for the first installment of this payment.

U.S. government funding to the GEF, as all donors' funding, is provided in aggregate and not differentiated by type of activity. However, a significant portion of GEF activities addresses climate change, both directly through the climate change focal area and indirectly through other focal areas. For instance, programs that address biological diversity and coastal zone management also help address vulnerability and adaptation of numerous species to changing climatic conditions. Currently approximately 38 percent of GEF grants support activities specifically related to climate change. This is only surpassed by GEF support for biodiversity activities, which comprise 42 percent of the overall portfolio. Table 7-1 provides annual U.S. contributions to the GEF for the years 1997 through 2000.

Financial Contributions to Multilateral Institutions and Programs

The U.S. government provides direct financial support to multilateral institutions, such as the United Nations and development banks, in recognition of their important role in meeting the goals of sustainable economic development, poverty alleviation, and protection of the global environment (Table 7-2).

---

[28] The information reported here was collected and analyzed from primary sources, including surveys of various U.S. government agencies, foundations, NGOs, private-sector companies, and queries of official U.S. government databases. In the case of commercial sales flows, the United States queried the U.S. International Trade Commission's database for U.S. export values for the energy (renewables and process efficiency) and water supply/wastewater sectors based on internationally agreed-upon harmonized tariff system codes (HTS). The United States chose the appropriate codes (HTS6 and HTS10) at the most detailed level possible to best select and account for only climate-friendly exports. The United States referenced both its own and OECD's analyses on environmental export values in creating this query (US–AEP 2000, OECD 2000).

[29] http://www.gefweb.org/.

Aggregated U.S. Government Funding for Multilateral Institutions

Between 1997 and 2000, the U.S. government provided funds to numerous multilateral banks and institutions through block grants. The funding is not specifically disaggregated by type of activity because donors meet their commitments by providing annual contributions that do not include earmarks for specific activities. Therefore, those activities that supported greenhouse gas emissions mitigation or addressed vulnerability and adaptation to climate impacts in developing and transition economies represent a portion of the total funding shown.

Between 1997 and 2000, the U.S. government also provided $3.9 million to the supplementary UNFCCC trust fund to support general participation in the Convention. These activities included support for the development of National Communications by non-Annex I (developing) countries, as well as information systems and databases of national greenhouse gas emission inventories.

Other Funding for Multilateral Scientific, Technological, and Training Programs

In 2000, the U.S. government provided grant funding to the World Mete-orological Organization in support of climate forecasting at the Drought Monitoring Center in Nairobi, Kenya (DMC-N). In collaboration with Columbia University's International Institute for Climate Prediction, this activity seeks to improve the capabilities of the DMC-N to provide reliable forecasts and early warning of extreme climate events, such as drought and floods.

Bilateral and Regional Financial Contributions

This section provides information on bilateral and regional financial contributions by U.S. foundations, NGOs, universities, the private sector, and the U.S. government related to climate change mitigation and adaptation activities. U.S. financial flows by year, country, and type of activity are presented in Table 7-3 in Appendix C.

To provide a more accurate representation of U.S. financial flows, several categories of activities have been expanded from those in the UNFCCC guidance, and two new categories have been added. The new category *Support for FCCC Participation* refers to activities where the United States has supported developing and transition economies to participate in international meetings, discussions, and training events. *Crosscutting Activities* refers to activities and programs that cannot be easily listed under a single category. Many of these "crosscutting" activities, for example, simultaneously provide both mitigation and adaptation benefits.

It is important to note that U.S. funding data—collected from hundreds of offices and divisions of over a dozen U.S. government agencies, as well as from numerous other public and private institutions—are difficult to categorize into the list of climate change topics requested in the UNFCCC guidelines. In many instances, U.S.-funded climate change activities could have been included under more than one topic area. For example, U.S. government agencies often label most activities that support industry, transportation, or waste management as "energy." In

---

TABLE 7-1 Financial Contributions to the Global Environment Facility: 1997–2000 (Millions of U.S. Dollars)

Since 1997, the U.S. government has provided $285.8 million to the GEF, which has a number of focal areas, including climate change.

Institution	1997	1998	1999	2000	Total
Global Environment Facility	35	47.5	167.5	35.8	285.8

Note: Information for GEF contributions is based on U.S. annual appropriations by fiscal year (October 1–September 30), which does not directly correspond to the calendar year. For example, for calendar year 1997, the figure used is from fiscal year 1997 (October 1, 1996–September 30, 1997).

TABLE 7-2 Financial Contributions to Multilateral Institutions and Programs (Millions of U.S. Dollars)

The U.S. government provides direct funding to multilateral institutions in support of sustainable economic development, poverty alleviation, and protection of the global environment.

Institution or Program	1997	1998	1999	2000	Total
Multilateral Institutions					
World Bank	700.0	1,034.0	800.0	771.1	3,305.1
International Finance Corporation	6.7	0	0	0	6.7
African Development Bank	0	45.0	128.0	131.1	304.1
Asian Development Bank	113.2	150.0	223.2	90.7	577.1
European Bank for Reconstruction and Development	11.9	35.8	35.8	35.8	119.3
Inter-American Development Bank	25.6	25.6	25.6	25.6	102.4
United Nations Development Program	76.0	93.7	97.4	77.9	345.0
United Nations Environment Program*	11.0	9.0	12.0	10.0	42.0
UN Framework Convention on Climate Change	2.6	3.9	3.8	4.9	15.2
Multilateral Scientific, Technological, and Training Programs					
World Meteorological Organization*	2.0	1.5	2.0	2.0	7.5
Intergovernmental Panel on Climate Change	0.7	1.0	2.7	1.6	6.0

*U.S. total voluntary contributions only from the International Organizations and Programs account.

addition, it is difficult in U.S. government programs to clearly distinguish between forest and biodiversity conservation programs, or between carbon sequestration programs (that apply forest and biodiversity conservation approaches) and adaptation programs (that seek to protect species endangered by changing climatic conditions). Similarly, many agricultural programs simultaneously support vulnerability assessments for climate impacts (i.e., severe weather), flood risk, desertification, drought, water supply, and/or food security.

While new categories have been included, most have been added as subcategories of the original headings provided in the UNFCCC guidelines. In this manner, total figures may be calculated within each main category for direct comparison with other countries' submissions. In addition, total figures may be calculated across regions and sectors. This more detailed representation of U.S.-funded climate change activities should promote more transparent and comprehensive understanding of the kind of support and attention the United States has provided in responding to climate change through technology transfer and development assistance programs.

SUMMARY OF FINANCIAL FLOW INFORMATION FOR 1997–2000

From 1997 to 2000, the United States provided more than $4.1 billion in direct funding to activities in developing and transition economies. This funding included greenhouse gas mitigation in the energy, industrial, and waste management sectors; carbon sequestration through improved forest and biodiversity conservation and sustainable agriculture; activities that address vulnerability and adaptation to climate impacts through improved water supply, disaster preparedness, food security, and research; and other global climate change activities. In the energy and water supply categories, commercial sales from private industry have enabled the transfer of technologies valued at

approximately $3.6 billion.

As shown in Table 7-3 in Appendix C, funding levels varied considerably between different categories. In addition to variations in U.S. government programming practices, this occurred in part because some categories (such as energy, water supply, and waste management) are very capital-intensive, while others (such as forest management or vulnerability assessment) require less capital investment.

In addition to direct funding and commercial sales, the United States provided $954.3 million in indirect funding between 1997 and 2000. This funding contributed to infrastructure projects and technologies that supported greenhouse gas mitigation in the energy sector.

Funding Types

This chapter reports direct support in the form of official development assistance (ODA) and official assistance (OA), grants from foundations and other philanthropic institutions, U.S. government-backed project financing, NGO funds, foreign direct investment (FDI), and commercial sales from private industry.[30] From 1997 to 2000, commercial sales and ODA/OA accounted for the largest share of direct support, followed by loans, foundation grants, FDI, and NGO funding (Figure 7-1). ODA, OA, grants, and to some extent NGO funds were directed to foreign governments, NGOs, and research institutions, as well as to U.S.-based institutions working in developing countries and transition economies.

It is estimated that U.S. FDI comprises the vast majority of funding that goes to climate change-related activities in developing and transition economies. However, because most information about the financing and implementation of private-sector projects is proprietary, very little FDI is reported under Table 7-3. What is

reported generally includes project development and implementation of USIJI energy and land-use mitigation projects. For these particular projects, annual financial contributions have ranged from $9,000 to $1.8 million per project.

U.S. government-based project financing has supported financing for private-sector infrastructure development. Loan amounts typically ranged from $60 million to $123 million per project, often providing a portion of the full project capitalization in conjunction with other funding sources. U.S. commercial sales of climate-friendly

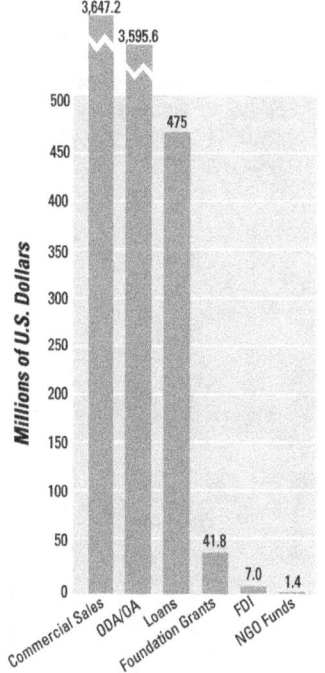

FIGURE 7-1 Commercial Sales and Direct Financial Flows: 1997–2000

From 1997 to 2000, commercial sales and official development assistance/official assistance (ODA/OA) accounted for the largest shares of direct support for activities that address mitigation of and adaptation to climate change.

Note: ODA/OA = official development assistance/official assistance; FDI = foreign direct investment; NGO = nongovernmental organizations.

---

[30] Justification for including commercial sales in this analysis of financial flows is derived from guidance provided in chapter 2 of IPCC 2000: "commercial sales refer to the sale (and corresponding purchase), on commercial terms, of equipment and knowledge."

environmental goods and services capture much of the "hard" technology or equipment exported to developing and transition economies. Annual commercial sales flows have ranged from $2,505 to $75.6 million per transaction.

Indirect financing, which includes risk guarantees, loan guarantees, and investment insurance, has contributed to the development of large private-sector energy infrastructure projects (Figure 7-2). The difference between direct and indirect financing is that the indirect flows do not represent actual transfers of cash, but rather guarantees to financial institutions and companies that the United States will cover the guaranteed amount of the total losses resulting from loan defaults, or other risks to a creditor or company. Indirect

flows typically have ranged from $3.1 million to $200 million per project.

Regional Trends

From 1997 to 2000, the United States provided over $1.1 billion to Asia and the Near East, $2 billion to Latin America and the Caribbean, $390.9 million to sub-Saharan Africa, $276.9 million to Europe and Eurasia, and $275.4 million to other global programs for the direct financing of mitigation, adaptation, and other climate change activities. With commercial sales of technologies and services, the United States provided $1.9 billion to Asia and the Near East, $1.5 billion to Latin America and the Caribbean, $134.0 million to sub-Saharan Africa, and $76.2 million to Europe and Eurasia.

With respect to indirect financing, the United States provided $425.5 million to Asia and the Near East, $467.1 million to Latin America and the Caribbean, and $61.7 million to Europe and Eurasia (Figure 7-3).

Funding has varied across regions in part because of differences between regional development priorities and because of the types of financial resources that have been mobilized for that region. A region's or subregion's development needs, geography, and investment environment often determine the types of climate change mitigation and adaptation projects that the United States funds. In addition, the distribution of the three dominant financial flow types—ODA, loans, and commercial sales—explains the huge

---

FIGURE 7-2 **Indirect Financial Flows in the U.S. Energy Sector: 1997–2000**

Indirect flows, which includes risk guarantees, loan guarantees, and investment insurance, has contributed to the development of large private-sector energy infrastructure projects. Indirect flows represent guarantees to financial institutions and companies that the United States will cover the guaranteed amount of the total losses resulting from loan defaults, or other risks to a creditor or company.

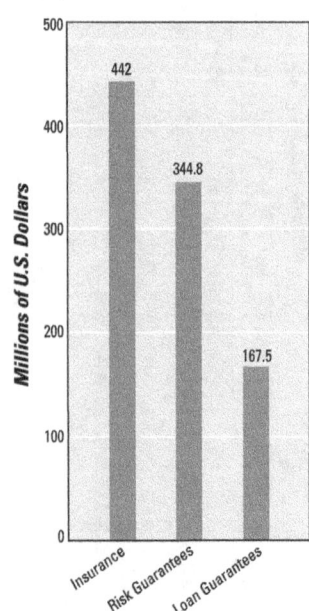

FIGURE 7-3 **Regional and Global Direct, Commercial Sales, and Indirect Financial Flows: 1997–2000**

From 1997 to 2000, the United States provided billions of dollars for mitigation, adaptation, and other climate change activities, specifically: $4.1 billion in direct financing, $3.6 billion for commercial sales of technologies and services, and $943 million in indirect financing.

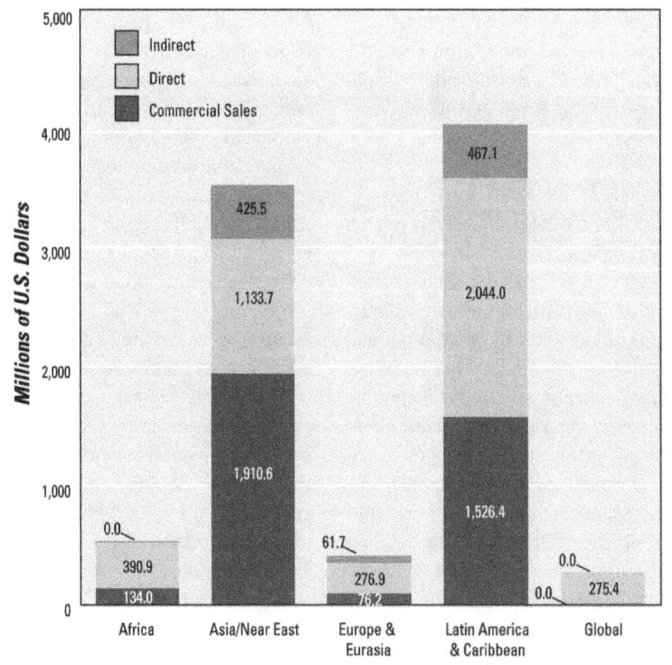

variances in the magnitude of financial flows across regions and across time. In particular, a few loans that supported energy-sector activities far exceeded the relative funding levels provided through ODA, and actually doubled or tripled the baseline flows to a particular region. These activities have tended to be infrequent, one-time loans for single projects in a single country.

For example, from 1997 to 2000 in Asia and the Near East, the United States provided the energy sector $504.5 million in direct financing, $411.2 million in commercial sales, and $425.5 million in indirect financing. For the water supply sector, the United States provided $337.7 million in direct financing and $1.5 billion in commercial sales of relevant equipment and technologies. This funding distribution is representative of the region's experience with water supply constraints and increasing energy demand. In another example, to support forestry-related activities, the United States provided direct financing of $144.3 million to Latin America and the Caribbean, $121.2 million to Africa, and $121.2 million to Asia and the Near East over the same period. These regions boast significant potential for conservation of carbon stocks and other climate-friendly forest and biodiversity conservation opportunities (see Appendix C).

Mitigation Activities

From 1997 to 2000, the United States spent $2.4 billion overall on climate change mitigation in the form of ODA, U.S. government-backed loans, foundation grants, NGO funds, FDI, and commercial sales. The United States also indirectly financed climate change mitigation activities in the amount of $954.3 million. Following the UNFCCC guidance for Table 7-3 (in Appendix C), the mitigation activities reported here include emission-reduction initiatives in the energy,

transportation, forestry, agriculture, waste management, and industrial sectors. To more accurately represent U.S.-supported activities, the forestry sector has been divided into two subcategories: forest conservation and biodiversity conservation (Figure 7-4).

Energy

The majority of U.S. spending on mitigation of climate change from 1997 to 2000 was directed toward energy-related projects, totaling approximately $1 billion in direct financing and $862.4 million in commercial sales.[31] In indirect financing, the United States leveraged $954.3 million for climate-friendly investments, all of which went to the energy sector (see Appendix C). U.S. support for climate technology transfer in this sector has varied widely throughout the world to include complex, large-scale infrastructure investment and development; extensive

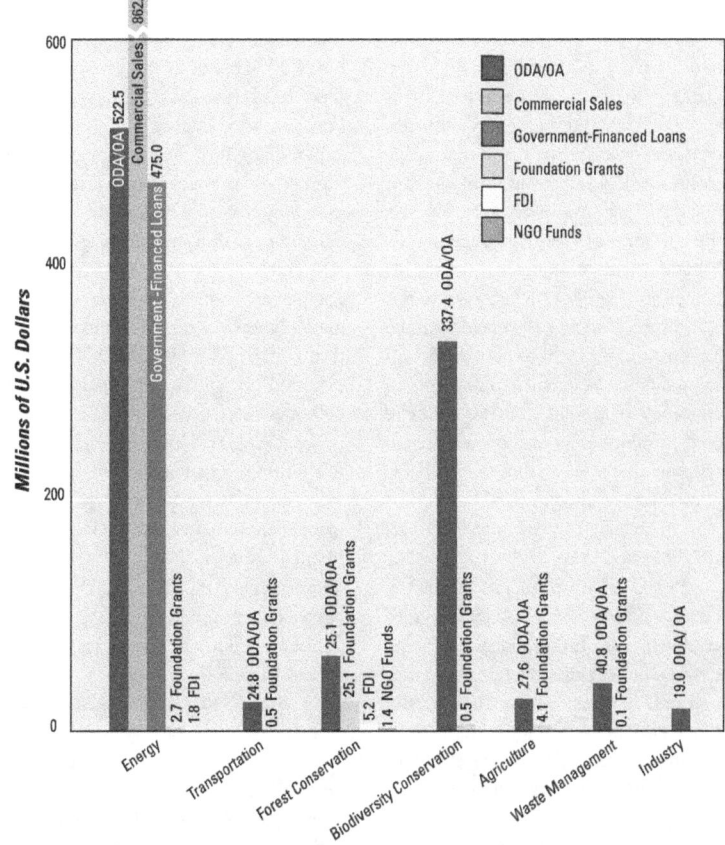

FIGURE 7-4 U.S. Financial Flows by Mitigation Sector and Financial Flow Type: 1997–2000

From 1997 to 2000, the United States directly financed $2.4 billion and indirectly financed $954.3 million for activities to mitigate the effects of climate change.

Note: ODA/OA = official development assistance/official assistance; FDI = foreign direct investment; NGO = nongovernmental organization.

---

[31] In selecting commercial sales transactions applicable to the energy sector, the U.S. limited its query to equipment for heat and energy management and renewable energy plants, as determined by the US–AEP study that examined U.S. environmental exports (US–AEP/USAID 2000). These commodities included (1) photosensitive semiconductor devices/photovoltaic cells and light-emitting diodes; (2) heat-exchange units, nondomestic, nonelectric; (3) electric-generating sets; (4) parts of hydraulic turbines and water wheels, including regulators; (5) hydraulic turbines and water wheels of a power exceeding 10,000 KW; (6) instantaneous or storage water heater, nonelectric; (7) hydraulic turbines and water wheels of a power exceeding 1,000 KW but not exceeding 10,000 KW; and (8) hydraulic turbines and water wheels of a power not exceeding 1,000 KW.

capacity building for power-sector policy and regulatory reform; improvements in the development and propagation of energy-efficiency, renewable-energy, and clean-energy technologies and practices; and conservation practices at the municipal and household levels. U.S. support for this sector has often included overlap with the transportation, industrial, and waste management sectors.

U.S. support for technical assistance and training has contributed to policy reforms and increased energy-efficient operations in the power and industrial sectors. For example, USAID supported a number of significant utility restructuring and regulatory reform activities, including adjustments to energy tariffs and fuel pricing in countries in Asia, the Near East, Europe, and Eurasia. These efforts have largely resulted in improved market efficiency, cost-effective management, and reduced greenhouse gas emissions through the use of innovative technologies, improved management practices, and incentives that increase the efficiency of energy production, distribution, and consumption. Among its numerous energy-efficiency activities worldwide, USAID has worked with the Egyptian government to provide technical assistance to enhance power station efficiency, reduce losses in transmission, and introduce time-of-day metering to regulate the flow of electricity. These efforts have resulted in considerable savings in annual carbon dioxide (CO_2) emissions.

USAID also partnered with the United States Energy Association to establish an International Climate Change Project Fund that provides support to U.S. investor-owned utilities and other energy companies to implement specific projects that mitigate emissions in USAID-assisted countries in Asia, Africa, and Latin America. One of the Fund's projects selected in 2000 is the SENELEC Network Power Generation Efficiency Project in Senegal which, through partnering with the U.S.-based Electrotek Concepts, will increase the efficiency, reliability, and power quality of the primary electricity supply system

operated by this national electric utility. This project is expected to eliminate 315 gigagrams of CO_2 over its 10-year lifetime and reduce fuel imports to Senegal by an estimated 140,000 barrels a year. In Mexico, USAID's Steam and Combustion Efficiency Pilot Project has promoted high-efficiency motors, compressors, pumps, and lighting to demonstrate the linkages between reducing emissions and increasing energy efficiency. In 1999, this effort resulted in a reduction of more than 325 gigagrams of CO_2 emissions.

U.S. support for broader infrastructure financing has also helped advance the use of renewable energy, energy efficiency, and clean energy in developing and transition economies. For example, the Export–Import Bank of the United States has financed combined-cycle plants in Latin America and the Caribbean, Asia, the Near East, Europe, and Eurasia. These plants exhibit high efficiency as they combine the use of natural gas and a low heat rate, which results in lower CO_2 production per kilowatt-hour of generated electricity. Support from U.S. power companies and NGOs financed the pilot phase of a rural solar electrification project in Bolivia, which is expected to avoid 1.3 gigagrams of CO_2 over its 20-year lifetime. In Bulgaria, USAID's Development Credit Authority program provided a partial loan guarantee for United Bank of Bulgaria to enable consumers in Bulgaria to finance municipal energy-efficiency improvements. As a result of this credit enhancement program, USAID leveraged $6.3 million in private capital at a cost of $435,000.

In addition to supporting large projects focused on energy supply, the United States has addressed the demand side of the power sector. For example, EPA has collaborated with authorities in China to reduce energy use by establishing minimum energy-efficiency levels for fluorescent lamp ballasts and room air conditioners. EPA has also worked to increase the energy-efficiency levels of refrigerators. Plans are now underway to strengthen the

Chinese voluntary energy-efficiency label through technical cooperation with the U.S. ENERGY STAR® program—an initiative that promotes energy-efficient solutions for businesses and consumers that save money as well as the environment.

Transportation

From 1997 to 2000, the United States spent approximately $25.3 million in ODA funding on climate-related activities in the transportation sector (see Appendix C). Note that a significant number of U.S. government projects supporting climate-related activities in the transportation sector are counted under "Energy."

U.S. international programs to address climate change through transportation have included efforts to improve engine and fuel efficiency, promote improved transportation management and planning, support alternative transportation systems, and introduce cleaner fuels and alternative-fuel technologies. For example, several USAID programs operating in Egypt, India, the Philippines, and Mexico seek to reduce greenhouse gas emissions from motor vehicles, while also reducing lead, particulates, and smog-forming emissions. The U.S. Department of Transportation's Center for Climate Change and Environmental Forecasting has supported strategic planning, policy research, communication, and outreach, as well as the preliminary assessment of project-specific international emission-trading opportunities in India, China, Indonesia, and Brazil.[32] DOE and the Ministry of Science and Technology of the People's Republic of China have been collaborating on research and development of electric and hybrid electric vehicle technology.

The U.S. Trade and Development Agency (TDA) has financed numerous feasibility studies, orientation visits, and other training and technical

---

[32] Given the current U.S. congressional restrictions on resource allocation for all efforts aimed at implementing Kyoto Protocol provisions, the USDOT/FHWA activities are focused only on raising the level of awareness among the potential domestic and international stakeholders.

assistance activities for railway, mass transit, and transportation system efficiency improvements throughout the developing world. For example, TDA provided $220,000 for a feasibility study of the light rail project on the island of Cebu—the fastest-growing region and second-largest metropolitan area in the Philippines.

Forestry

The United States spent over $439.4 from 1997 to 2000 on climate change activities in the forestry sector (see Appendix C). This funding included traditional forest conservation and management activities, biodiversity conservation, and related natural resource management activities that improved the technical capacity of national and local governments, NGOs, and local communities to manage and conserve forests. The United States has also provided direct investment in protection of natural areas to reduce the rate of loss of, preserve, or increase carbon stock capacity. Overall, the majority of resources expended in this area went toward biodiversity conservation programs.

Forest Conservation. From 1997 to 2000, the U.S. government spent $96.7 million on forest conservation in Central and South America, Africa, Asia, and Europe and Eurasia (Figure 7-5). For example, USAID has addressed rapid deforestation in the Amazon tropical rain forest by funding scientific studies that use satellite imagery to analyze deforestation trends to better understand specific risks from drought, illegal logging, accidental fires, and agriculture practices.

In Mexico, following the 1997 and 1998 wildfire disasters, USAID, the Mexican government, and local NGOs jointly developed a wildfire prevention and land restoration program to mitigate environmental, health, and climate effects from forest fires. USAID helped lead several efforts to adopt policies discouraging slash-and-burn agriculture, improve collaboration between Mexico's federal government and

NGOs, and provide training on fire prevention and wildfire management. As a result, local fire brigades were able to control and extinguish fires much more effectively, and in 1999 Mexico experienced a decrease in the area normally affected by fires. Efforts are underway to assess the amount of carbon potentially sequestered as a result of Mexico's fire restoration efforts.

By working with communities to establish clear boundaries for community management, control agricultural clearing, and implement monitoring plans, USAID facilitated the transfer of over 625,000 hectares (over 15 million acres) of forest to local management in the Philippines. After four years, about 5.5 million hectares of forestland—over 60 percent of the country's open-access forests—are now under community

management. Without such interventions, the country's forest cover would have declined by an estimated 6 percent during the same period.

Through the USIJI program, the U.S.-based NatSource Institutional Energy Brokers, the Costa Rican Ministry of the Environment and Energy, and the Costa Rican National Parks Foundation have begun implementing the Territorial and Financial Consolidation of Costa Rican National Parks and Biological Reserves Project. This "certified tradable offset" project facilitates the transfer of primary forest, secondary forest, and pasture lands that have been declared National Parks or Biological Reserve to the Costa Rican Ministry of Environment and Energy (MINAE). Over its 25-year life, the project is expected to avoid an

---

FIGURE 7-5 Direct, Commercial Sales, and Indirect Financial Flows by Mitigation/Adaptation Sector: 1997–2000

From 1997 to 2000, the majority of U.S. spending on climate change mitigation activities was directed toward energy-related projects, totaling approximately $1 billion in direct financing, $862.4 million in commercial sales, and $954.3 million in indirect financing for climate-friendly investments.

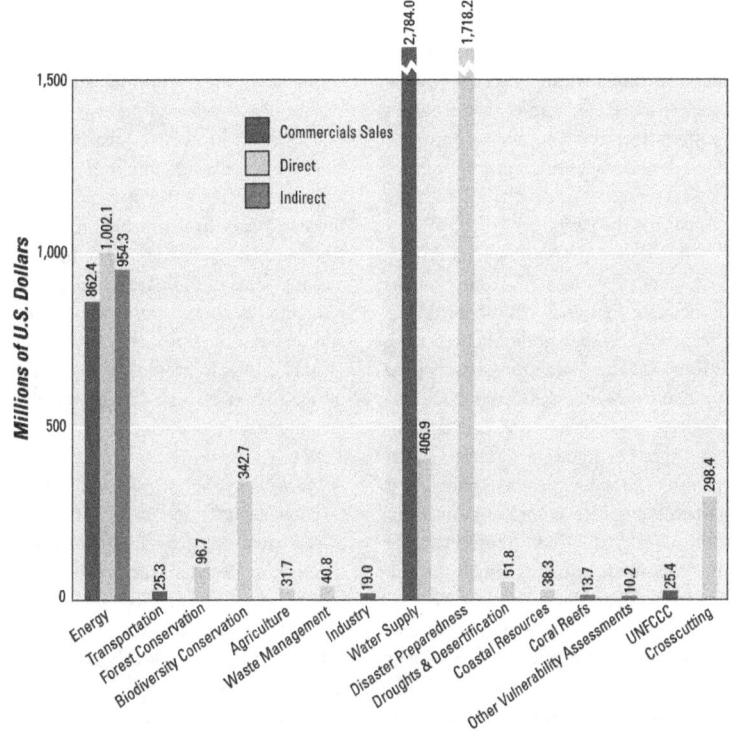

estimated 57 teragrams of CO_2 emissions.

Biodiversity Conservation. From 1997 to 2000, the U.S. government spent $342.7 on biodiversity conservation activities, such as establishing and managing protected areas, providing training in habitat conservation, and promoting sustainable resource management (Figure 7-5). Funding for biodiversity activities has come primarily from USAID, USIJI projects, and private foundations, usually in partnership with international NGOs, research institutions, and host-country governments and organizations.

USAID's Parks-in-Peril program, a partnership with The Nature Conservancy and local NGOs, has become Latin America's largest, most successful site-based conservation effort. Working in 37 protected areas in 15 countries, this program has helped protect over 11 million hectares (more than 271 million acres) of natural forests, of which 6.3 million hectares (more than 155 million acres) contain substantial carbon stocks.

In Bulgaria, USAID's GEF Biodiversity Project has strengthened a network of protected areas, with a specific focus on the Rila and Central Balkan National Parks, totaling 179,622 hectares (over 4 million acres). The project has provided policy development assistance, promoted sustainable economic use of biological resources, and built local capacity to manage the parks.

In similar efforts, the MacArthur Foundation's Ecosytems Conservation Policy grant program has supported initiatives in Nepal and Tibet totaling $100,000. The March for Conservation program has supported coastal zone biodiversity and conservation education in Sri Lanka ($75,000), and Terra Capital Investors Limited's venture capital fund ($1 million) invests in Latin American businesses that involve the sustainable use of natural resources and foster the preservation of biological diversity.

In Guatemala, the home of the Maya Biosphere Reserve and one of the largest tracts of intact tropical forests, USAID has worked to reduce deforestation rates

and promote carbon sequestration. By supporting improved land- and resource-use practices, an improved policy framework, and stronger local institutions through technical assistance, training, and farmer-to-farmer extension networks, this work had led to the protection of approximately 700,000 hectares (more than 17 million acres) in 1999.

USAID's work in Indonesia took steps to protect the West Kalimantan tropical broadleaf forest, where approximately 43,000 hectares (more than 1 million acres) are now under effective management as villagers organize, create maps of, and impose rules on harvesting the natural resources. In 2000, USAID also supported resource valuation studies for communities in Indonesia's Bunaken National Park to demonstrate the relative monetary value per hectare and per family that biologically diverse forests have, as compared with oil palm monoculture forests.

In Madagascar, USAID has sought to preserve biologically diverse carbon stocks and reduce their rate of loss. Working with the National Association for Management of Protected Areas (ANGAP) and the Ministry of Water and Forest (MEF), USAID supported the growth and sound management of Madagascar's Protected Area Network, as well as forests and important biological areas outside of the network. These programs specifically focus on protection and improved management of existing areas of biological importance, reducing slash-and-burn agriculture, and increasing agroforestry and tree nursery efforts to promote reforestation of multiple-use, high-economic-value, or indigenous tree species.

Agriculture

Between 1997 and 2000, the United States spent approximately $31.7 million on climate-related activities in agriculture (see Appendix C). These financial resources have promoted agroforestry, reduced tillage, erosion control, introduction of perennial crops and crop rotation, improved nitrogen

and soil management, use of organic fertilizers, and improved management of agrochemicals.

In Uganda and Madagascar, for example, USAID has supported sustainable farming systems and agroforestry to improve agricultural output while enhancing the carbon storage potential in soils and crops. In Kenya, the Ford Foundation has supported Winrock International's Institute for Agricultural Development to strengthen associations of women professionals in agriculture and the environment in East Africa. The Institute has enhanced food security and environmental conservation by preparing women for leadership roles in agricultural and environment-related sciences.

In Chiapas, Mexico, a ground-breaking partnership between Starbucks Coffee and Conservation International (CI) begun in 1998 has promoted cultivation that incorporates biodiversity protection and environmentally sustainable agricultural practices. Under the partnership, CI assists farmers in the El Triúnfo Biosphere Reserve, in the Sierra Madre de Chiapas, to produce coffee under the shade of the forest canopy using practices that avoid the need to clear forested lands.

Waste Management

The United States spent over $40.8 million from 1997 to 2000 on activities supporting greenhouse gas mitigation in the waste management sector (see Appendix C).[33] These activities primarily addressed the development and implementation of waste-to-energy programs involving the recovery of greenhouse gases, such as methane from solid waste disposal facilities. For example, US–AEP and Conservation Services Group (CSG) Energy Services jointly implemented an energy-efficiency technology and pollution prevention project in India in partnership with several universities and India's

---

[33] Financial information on some waste management initiatives was not available, especially with regard to private-sector activities. Note, a considerable number of industrial-sector activities have been included under "Energy," above.

Thane-Belapur Industries Association. The partners will use the CSG grant to assess the potential of selected landfill methane-recovery sites to mitigate greenhouse gas emissions.

Under USIJI, the regional Argentinean government agency, Coordinación Ecológica Area Metropolitana Sociedad del Estado (CEAMSE) and U.S.-based Pacific Energy Systems, Inc., have developed a landfill gas management project in Greater Buenos Aires, where up to 5 million tons of waste are deposited annually. Studies initiated under the project estimate that capturing and combusting 70 percent of the gas generated from the waste in the CEAMSE landfills could result in an annual net emission reduction of 4 teragrams of CO_2 equivalent. Further reductions could be achieved as the gas is used to displace combustion of more carbon-intensive fossil fuels.[34]

Industry

Between 1997 and 2000, the United States spent more than $19 million on climate-friendly activities in the industrial sector (see Appendix C).[35] These activities have improved industrial energy efficiency, environmental management systems, process efficiency, and waste-to-energy programs, particularly in energy-intensive industries.

In Mexico, for example, USAID and DOE have collaborated to develop greenhouse gas emission benchmarks for key industries, as well as energy-efficiency initiatives in the public sector. These efforts have demonstrated that investments in resource management systems are both technically and economically sound, paying for themselves through energy and other savings within a few years. In the Philippines, USAID supported the adoption of ISO 14000 certification, a voluntary system that promotes environmental management improvements in production practices at a Ford Motor Company plant and throughout its chain of 38 suppliers. In Chennai, India, USAID worked with a starch manufacturing company in the Salem District of Tamil Nadu to recover methane emissions from its tapioca-processing effluents. A USAID-commissioned study found that the 800 manufacturing facilities of Salem produce enough methane to generate about 80 MW of power, compelling the local chamber of commerce to implement a demonstration project in 1998 with USAID assistance to convert the recaptured methane for fuel use.

Other U.S. government facilitation of climate-friendly industrial development has involved the transfer of U.S. equipment and technical expertise. In 1997, the U.S Trade and Development Agency provided a $600,000 grant to the Ukrainian Ministry of Coal to study the feasibility of the production of coalbed methane and utilization of gases to generate electric power in the Donetsk Basin. The U.S. firm International Coal Bed Methane Group (composed of Black Warrior Methane and E.L. Lassister) carried out the study. U.S. exports to the project consisted of drilling and completion equipment, drilling rigs, service rigs, combustion power turbines, logging and geophysical equipment, and engineering and legal services. In 2000, the Department of Commerce, through its International Clean Energy Initiative, began promoting the transfer of U.S.-developed waste recovery technology to developing countries. A trade mission to China involved the participation of the Asian American Coal Company, which has developed technology that captures coalbed methane for conversion to natural gas.[36]

Adaptation Activities

From 1997 to 2000, the United States spent over $5 billion on climate change vulnerability and adaptation activities. These activities, funded mostly by commercial sales and ODA, are presented in Appendix C under the categories provided by the UNFCCC guidance: capacity building, coastal zone management, and other vulnerability assessments. However, to more accurately represent the numerous adaptation activities the United States has supported that are relevant to climate change, the following subcategories were created under capacity building: water supply, disaster preparedness and response, and drought and desertification. Under coastal zone management, the following two categories were created: coastal resources and coral reef protection.

Capacity Building

From 1997 to 2000, the United States provided $4.9 billion in funding for climate change activities in the broad category of capacity building. The major sources of funding for capacity building came from commercial sales for much of the technology transferred in the "water supply" subcategory, while ODA and foundation grants funded disaster preparedness and response programs and droughts and desertification programs.

Water Supply. Between 1997 and 2000, the United States spent approximately $406.9 million in direct financing for water supply programs primarily directed at the development and improvement of water supply and wastewater treatment infrastructure.[37] Hard technologies transferred through commercial sales amounted to approximately $2.8 billion[38] (Figure 7-6). Following are some examples of this financial and technical assistance.

---

[34] USIJI Project Descriptions–CD.

[35] A considerable number of industrial-sector activities have been included under "Energy," above.

[36] ITA Web site.

[37] IPCC Working Group II included water supply as a capacity-building category in IPCC 2001a, based on the integrated water resource management approaches identified for adapting to climate change impacts in the hydrology and water resources sector .

[38] In selecting commercial sales relevant to the water supply sector, the United States limited its query to wastewater treatment equipment, an IPCC-determined supply-side option for adapting to climate change impacts in the hydrology and water resources sector (IPCC 2001a, p. 220). Based on the methodology of a US–AEP study that examined U.S. environmental exports (US–AEP/USAID 2000), the United States chose to include sales of the following types of commodities: (1) mats, matting, and screens of vegetable plaiting materials; (2) rotary positive displacement pumps; (3) centrifugal pumps; (4) filtering or purifying machinery and apparatus for water; (5) filtering or purifying machinery and apparatus for liquids; and (6) machines for mixing, kneading, crushing, grinding, etc.

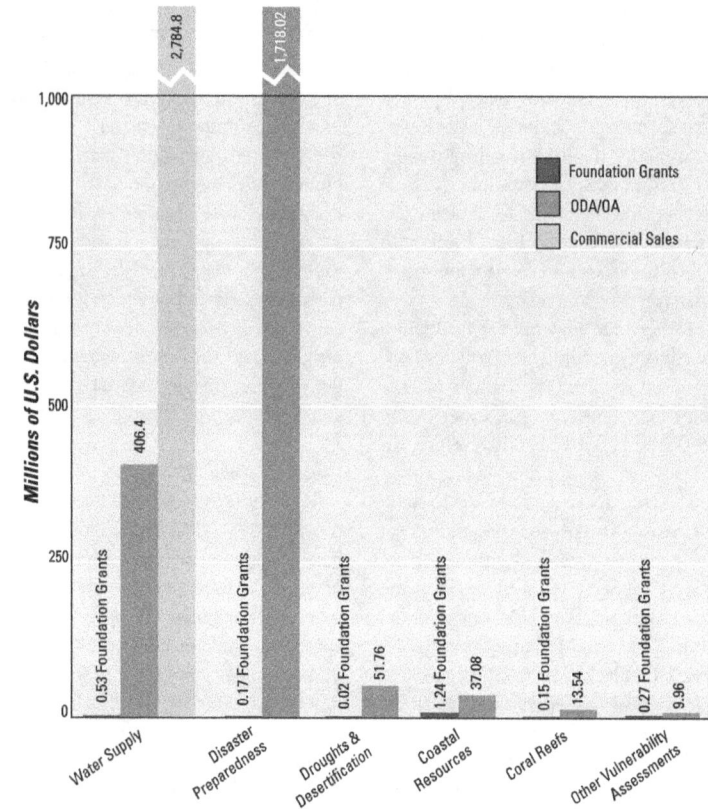

FIGURE 7-6 Financial Flows by Adaptation Sector and Financial Flow Type: 1997–2000

From 1997 to 2000, the United States spent over $5 billion on climate change vulnerability and adaptation activities. These activities were funded primarily by commercial sales and official development assistance/official assistance (ODA/OA).

United States recognizes that prevention, reduction, and preparedness are important factors in reducing the large-scale devastation that disasters can have on vulnerable populations. As a result, the United States has provided extensive assistance for recovery from natural disasters around the world.

Severe weather disasters. In May 1999, the U.S. Congress appropriated $621 million under the Emergency Supplemental Appropriations Act, primarily to support the reconstruction of the Dominican Republic and Haiti, which were devastated in late 1998 by Hurricane Georges. This funding also assisted Central America's recovery from Hurricane Mitch, which struck on the scale of a storm seen only once in 100–200 years. These funds were later extended for reconstruction in the Bahamas and the Caribbean, which were struck by Hurricanes Floyd and Lenny in 1999.

After surveying the extensive damage caused by Hurricane Mitch, the United States announced the $11 million Central American Mitigation Initiative. This project aims to reduce the impacts of natural disasters by building national capacity in Central American countries to forecast, monitor, and prevent those disasters. In the wake of Hurricane Mitch, the United States initiated a multi-agency effort to strengthen worldwide climate-related disaster preparedness and mitigation, with particular emphasis on Mexico and Central America.

In a joint effort, a group of U.S. government agencies[39] implemented a variety of disaster preparedness and relief programs for hurricane-related impacts throughout Latin America. These programs have included, for example, the development of more

In mid-1998, USAID provided emergency assistance to the Zai Water Treatment Plant in Jordan, which is the source of drinking water for 40 percent of Amman's population. In coordination with Japan and Germany, USAID has funded efforts to expand and upgrade the plant to reduce the likelihood of future water crises, and is funding the rehabilitation of 27 contaminated springs and wells. In Egypt, USAID is continuing work on the rehabilitation and expansion of the southern portion of Cairo's Rod El Farag water treatment plant, a $97.4-million project. As a result of this work, four million Cairo residents now benefit from a more reliable and safer water supply service.

In 1997, the U.S. Trade and Development Agency (TDA) provided $168,500 to FMI International to conduct a feasibility study for the development of a wastewater treatment plant in northeastern Estonia. In another example, at the request of the Royal Thai government, TDA provided $40,000 for an orientation visit for 16 Thai officials interested in U.S. flood control technology in 1997. That same year, TDA also granted $367,000 for a feasibility study on water-loss reduction for the city of Curitiba in Parana, Brazil.

Disaster Preparedness and Response. Between 1997 and 2000, the United States spent $1.7 billion on climate-related disaster preparedness, mitigation, and relief (see Appendix C). The

---

[39] USAID, NOAA, USDA, USGS, EPA, Federal Emergency Management Agency (FEMA), the Department of the Interior (DOI), Department of Health and Human Services (HHS), Department of Transportation (DOT), Department of Housing and Urban Development (HUD), the Peace Corps, Export–Import Bank (Ex–Im Bank), Overseas Private Investment Corporation (OPIC), Department of State, and the General Accounting Office (GAO).

resilient infrastructure, climate forecasting and warning systems, and various forms of humanitarian aid. USAID helped establish a training and technical assistance program to develop adaptation plans for extreme climatic events in the region, supported watershed rehabilitation through a transnational watershed program, and helped install stream gauges and early-warning systems in Honduras.

To continue addressing connected disaster risks in the Caribbean region, USAID recently initiated the Caribbean Disaster Mitigation Project. Implemented by the Organization of American States' Unit of Sustainable Development and Environment, this $5-million, six-year project promotes the adoption of natural disaster preparedness and loss-reduction practices by both the public and the private sectors through regional, national, and local activities. These activities target six major themes: (1) community-based preparedness, (2) hazard assessment and mapping, (3) hazard-resistant building practices, (4) vulnerability and risk audits for lifeline facilities, (5) promotion of hazard mitigation within the property insurance industry, and (6) incorporation of hazard mitigation into post-disaster recovery. To date, pilot projects have been implemented in 11 Caribbean countries.

Similarly, the Federal Emergency Management Agency (FEMA) has a long history of interaction with foreign governments to help them more effectively respond to and prevent disasters, including expert exchanges and "train-the-trainer" courses. FEMA recently established pilot projects for building disaster-resistant communities with Argentina, the Dominican Republic, El Salvador, Guatemala, Haiti, Honduras, and Nicaragua and expanded civil emergency planning work through NATO partners to include East European nations.

Watershed management. In continued efforts to reduce severe weather risks in Central America, USAID has undertaken activities in the transboundary Río Lempa watershed, shared by Guatemala, Honduras, and El Salvador. The adaptation strategy for the Río Lempa has focused on three components: the National Weather Service River Forecast Center,[40] capacity building on the operation and maintenance of the forecast system, and the development of a geographic information system and watershed disaster mitigation plan to mitigate the impacts of extreme events. The watershed disaster mitigation plan includes identification of vulnerable populations, flood-prone areas, areas at risk of landslides, the location of shelters, and road networks for delivery of supplies. The program facilitated a tri-national agreement to mitigate the impacts of transnational disasters in the Lempa Watershed, with the goal of exporting the lessons learned from the Río Lempa to other transnational watersheds in the region.

Flood preparedness and response. The United States has also provided flood preparedness and response support to developing countries around the world, both in terms of disaster relief and in planning and mitigating future risks. Among the many catastrophic floods that occurred between 1997 and 2000, the United States has helped victims and communities in over a dozen developing countries around the world.

In 1999, USAID's Office of Foreign Disaster Assistance (OFDA) announced $3 million in funding to assist relief efforts related to massive flooding, landslides, and mudslides in Venezuela, which killed an unknown number of people and displaced many more. The same year, the U.S. Geological Survey provided follow-on disaster planning assistance to produce hazard maps for future response to and recovery from disastrous flood and landslides in Venezuela. In addition, USAID funded the provision of emergency relief supplies to flood victims in Mozambique, South Africa, and Zimbabwe in response to severe flooding in southern Africa in 1999.

In 1998, OFDA provided funds for emergency housing, clothing, mosquito nets, and cooking utensils to Vietnam after heavy rains and severe flooding devastated the country. To minimize future flood risks, in 2000, USAID started supporting efforts to map flood plains and determine where people should avoid building their homes in the future. These efforts included locating emergency shelters and determining evacuation routes to be used during future flooding.

In 1998, floods struck Kinshasa, Democratic Republic of the Congo, affecting an estimated 100,000 people. After the emergency, OFDA designed a project to reduce the population's vulnerability to future floods. With OFDA funding, Catholic Relief Services built 17 small check-dams from locally available materials, cleaned drainage canals, and reseeded degraded watershed areas to improve soil and moisture retention. When torrential rains again struck Kinshasa in February 1999, there were no injuries, no displaced residents, and no damaged homes in the project area. This successful project enabled the residents of Kinshasa—where monthly household incomes are less than $70—to avoid a repeat of the $7.7 million in economic losses they suffered in 1998.

Climate forecasting and research. Climate, meteorological, and hydrological forecasting has played an increasingly important role in warning developing country populations of pending severe storm risks, as well as better informing them of long-term disaster mitigation and response efforts. Under NOAA's National Weather Service, the United States has regularly provided developing countries with meteorological and hydrological forecasts and prediction models; floods, droughts, and river flow predictions; tropical cyclone/hurricane forecasts for the Western Hemisphere; global aviation hazardous weather forecasts; high-sea forecasts for the North Atlantic and North Pacific; and meteorological training programs for countries throughout Central America, the

---

[40] The NWSRFS was developed by NOAA and is being implemented by NOAA, the U.S. Geological Survey, and the System for Central American Integration.

Caribbean, and Africa. NOAA also provides research and response activities to prepare for severe impacts expected from extremes of climate variability, climate forecast research and applications, predictions related to El Niño phenomena, support for a scientific network, and capacity building in Africa, Latin America and Caribbean, South-east Asia, and the South Pacific. During the disastrous flooding in Mozambique in March and April of 2000, for example, NOAA provided real-time weather forecasts to the affected regions as well as to international response and relief agencies.

In cooperation with 26 countries, NASA implemented the Pacific Rim II airborne campaign in the southern and western portions of the Pacific Rim region. The campaign resulted in the deployment of research aircraft and remote-sensing instrumentation for collecting data that will enable scientists to better assess local environmental conditions and natural hazards to enhance disaster management and mitigation practices in Pacific Rim countries. Similarly, NOAA has implemented the Pan American Climate Studies Sounding Network (PACS SONET) for extended monitoring of climate variability over the Americas. This project enhances understanding of low-level atmospheric circulation features within monsoonal North and South America, provides a means of validating numerical model simulations, and establishes a long-term, upper-air observing system for climate prediction and research.

In another climate-hydrological forecasting effort, USAID and NOAA have cooperated to provide snow-monitoring and river-forecasting assistance to Central Asian Hydrometeorological Services, known as *Glavgidromets*. This effort will download imagery over Central Asia from NOAA's polar-orbiting satellites. The imagery will be used by the *Glavgidromets* to monitor the extent of the snowpack in the Himalayan Mountains, which is the source of most of the water that flows through the Amu Darya and Syr Darya rivers.

Numerous partners, including USAID

and NOAA, created the Radio and Internet Technology for Communication of Hydro-Meteorological and Climate Related Information (RANET) program. The program consists of information and applications networks in southern Africa, the Greater Horn of Africa, and West Africa. These networks provide regular seasonal climate forecast information and work directly with users to reduce climate-related vulnerability. RANET will make information, translated into appropriate local languages, directly available to farm-level users through wind-up radio.

Droughts and Desertification. From 1997 to 2000, the United States spent approximately $51.8 million on activities that address droughts and desertification (see Appendix C). These activities are often implemented in connection with the U.S. government's foreign disaster response programs, although a number of long-term adaptation initiatives have also been supported. They include weather forecasting, drought prediction, hazard mapping, and research, technical assistance, and capacity building. Through NOAA, for example, the U.S. government has provided vegetation stress and drought prediction information to China, Georgia, Kazakhstan, Morocco, and Poland, and technical assistance to China and Tajikistan for estimating drought intensity and duration.

The U.S. Department of Agriculture, the U.S. Geological Survey, the GEF, and the government of Kazakhstan have begun implementing the pilot phase of the Kazakhstan Dryland Management Project. The project's objective is to conserve, rehabilitate, and sustainably use natural resources in the marginal cereal-growing area of the Shetsky Raion of northern Karaganda Oblast, Kazakhstan. This project works with communities and the Kazakh government to (1) develop alternative land uses and rehabilitate ecosystems for conservation of plant and animal bio-diversity; (2) develop a coherent framework and national capacity to monitor carbon sequestration; and (3) build public

capacity and develop a replication strategy so that project activities can be adopted in other similar areas of Kazakhstan and other Central Asian countries.

In the drought-prone Bie province of Angola, USAID has funded Africare, a private voluntary organization, to distribute 339 metric tons of seeds and 55,000 farming tools to 27,500 internally displaced people. In 2000 in Afghanistan, USAID/OFDA provided immediate drought relief measures through Save the Children to engage in drought-related activities, with a focus on maternal and child care.

The United States provides much support for food security through foreign agriculture programs and climate monitoring systems. For example, the Famine Early Warning System (FEWS) was started in 1985 and is funded at approximately $6 million a year to provide decision makers with the information they need to effectively respond to drought and food insecurity. Working in 17 drought-prone countries across Sub-Saharan Africa, FEWS analyzes remote-sensing data and ground-based meteorological, crop, and rangeland observations by field staff to track the progress of the rainy seasons in semi-arid regions of Africa and to identify early indications of potential famine. Other factors affecting local food availability and access are also carefully evaluated to identify vulnerable population groups requiring assistance. These assessments are continuously updated and disseminated to provide host-country governments and other decision makers with the most timely and accurate information available. Overall, FEWS activities strengthen the capacities of public and private institutions to monitor and respond to drought, the principal impact of climate variability in Sahelian Africa. By helping to anticipate potential famine conditions and lessen vulnerability, FEWS has helped save lives, while also promoting a more efficient use of limited financial resources.

USDA provides a number of additional food security activities around the world, including:

- the West Africa Regional Food Security Project, which provides information on vulnerable populations, food balances, food needs, food aid, and commercial import requirements;
- development of agricultural and natural hazard profiles for selected African countries to assist in mitigation, response, and rehabilitation;
- direct food aid in response to drought-related famine in Ethiopia and the Horn of Africa;
- disaster management and logistics support for African desert locust response; and
- collaboration with Central American countries to develop strategies to overcome soil erosion, manage water quality, and resolve food safety problems resulting from Hurricanes Mitch and Georges.

Coastal Zone Management

From 1997 to 2000, the United States provided about $52 million in ODA and foundation grants for climate change adaptation activities supporting coastal zone management. These activities included efforts to address coastal resources, sea level rise, severe weather and storm surges, risks to ecosystems (such as rising seawater temperatures), and protection of coral reefs.

Coastal Resources. Adaptation activities addressing coastal resources fall under the broad categories of integrated coastal management (ICM); coastal zone management and planning; conservation of critical coastal habitats and ecosystems (such as coral reefs, mangrove forests, and sand dunes) to maintain vital ecosystem functions; protection of coastal areas from storm surge and sea level rise; reduction of coastal erosion to limit future displacements of settlements and industries; development of guidelines for best coastal development practices and resource use; and the dissemination of best practices for coastal planning and capacity building. The United States financed $38.3 million in coastal resources activities between 1997 and 2000 (Figure 7-5).

The United States has implemented a number of ICM programs in several countries around the world. In 1985, USAID initiated the Coastal Resources Management program and again renewed this program in 2000 as part of a new $32-million commitment for coastal zone management programs worldwide. The CRM project is now funded at approximately $6 million a year and has operated in Mexico, Ecuador, Jamaica, the Dominican Republic, El Salvador, Kenya, Tanzania, Egypt, Thailand, Indonesia, and the Philippines. CRM projects largely promote improved governance, public participation, and stewardship toward the management of multi-sectoral activities within the coastal zone and surrounding watershed—helping to address a variety of climate-related threats to coastal and marine biodiversity and resource-dependent communities (USAID 2001b).

In addition to providing extensive technical assistance and research addressing coastal zone management needs, USAID's Coastal Resources Management program has helped generate a number of significant practical tools, such as coastal maps, program performance management guidelines, community coastal zone management strategies, national ICM policies, and best management guidelines in such areas as aquaculture, mariculture, and tourism development. The program has also promoted outreach mechanisms about best practices through reports, publications, journals, CD-ROMs, e-mail list servers, Web sites, and training and communications publications.

Coral Reefs and Other Marine Resources. Between 1997 and 2000, the United States supported the protection of coral reefs and other marine resources through the creation of marine sanctuaries, the introduction of sustainable fishing practices and coastal zone management, and research on coral reef habitats and climate risks in the amount of $13.7 million (see Appendix C). For example, community-based marine sanctuaries in the

Philippines and South Pacific have proven to be effective in conserving coral reef ecosystems, as well as increasing fish biomass and production. Efforts have been underway to reproduce these successful conservation areas in Indonesia under USAID's ICM project in North Sulawesi. These community-based marine sanctuaries are small areas of subtidal marine environment, primarily coral reef habitat, where all extractive and destructive activities are permanently prohibited. They were developed with the widespread support and participation of the local community and government, were established by formal village ordinance, and are managed by community groups.

USAID has implemented a number of programs involving site preservation for marine-protected areas. For instance, it provided support for the implementation of a new Galapagos Special Law to establish a marine park and has begun funding a Bering Sea Marine Ecoregion Conservation program.

In related efforts, the MacArthur Foundation provided $105,000 between 1998 and 2000 to establish a coral reef monitoring program with the Hong Kong University of Science and Technology. This project will provide important information to international conservation efforts about the health of coral reefs and risks to their survival.

Other Vulnerability Assessments

U.S. funding between 1997 and 2000 on vulnerability assessments and studies associated with adaptation to the impacts of climate change amounted to approximately $10.2 million (see Appendix C). Much of this funding went toward the U.S. Country Studies Program (CSP) to help developing countries assess their unique vulnerabilities to long- and short-term climate impacts, their adaptation options for addressing those risks, and their contributions to global greenhouse gas emissions. Since its inception, the CSP has helped 56 countries build the human and institutional capacities necessary to assess their vulnerability to climate change.

NOAA has focused on reducing the vulnerability of coastal populations to hazardous weather. Since 1997, it has developed a community-based vulnerability assessment methodology to aid local hazard mitigation planning and has begun working with the Organization of American States to provide training on vulnerability assessment to Caribbean countries.

Other Global Climate Change Activities

To account for those activities that did not easily fit within the mitigation and adaptation categories provided by the guidance for Appendix C of this chapter, two additional categories were created: UNFCCC participation and crosscutting activities. Both categories are relevant to implementation of the UNFCCC. Between 1997 and 2000, the United States spent approximately $323.8 million on "other global climate change activities."

UNFCCC Participation

The United States spent approximately $25.4 million between 1997 and 2000 to promote meaningful participation in the UNFCCC process by developing and transition economies (see Appendix C). USAID alone implemented over 70 capacity-building activities designed to strengthen participation in the Convention in 1999. This included promoting efforts to integrate climate change into national development strategies; establishing emission inventories; developing national climate change action plans; promoting procedures for receiving, evaluating, and approving joint implementation proposals; and establishing baselines for linking greenhouse gas emissions to economic growth.

For example, through its Climate Change Center in Ukraine, established in 1999, USAID provided support to the Ukrainian government to establish national administrative structures, develop a national climate change inventory program, and prepare investment projects. USAID assistance in Mexico supported the national government's establishment of an Interagency Commission on Global Climate Change. In connection with those efforts, the Mexican Congress considered a global climate change bill outlining how Mexico could integrate climate change considerations into national strategic, energy, and sustainable development goals.

Crosscutting Activities

The United States spent over $298.4 million on crosscutting climate change activities in developing and transition economies from 1997 to 2000 (Figure 7-5). Many of these activities have simultaneously addressed climate change mitigation and/or adaptation issues. For example, the Rockefeller Foundation awarded the Pacific Environment and Resources Center a $300,000 grant in 2000 to address threats to critical marine and forest ecosystems in the Russian Far East. Similarly, many USAID activities contributed to mitigation of, and adaptation to, climate change.

Chapter 8
Research and Systematic Observation

The United States leads the world in research on climate and other global environmental changes, spending approximately $1.7 billion annually on its focused climate change research programs. This contribution is roughly half of the world's focused climate change research expenditures, three times more than the next largest contributor, and larger than the combined contributions of Japan and all 15 nations of the European Union (Figure 8-1).

Most of this research is coordinated through the U.S. Global Change Research Program (USGCRP). Definition of the program began in the late 1980s, and Congress codified the program in the Global Change Research Act of 1990. The USGCRP was created as a high-priority, national research program to:

- address key uncertainties about changes in the Earth's global environment, both natural and human-induced;

FIGURE 8-1 Research Expenditures by Country: 1999–2000

The United States is responsible for roughly half of the world's focused climate change research expenditures—three times more than the next-largest contributor, and larger than the contributions of Japan and all 15 nations of the European Union combined.

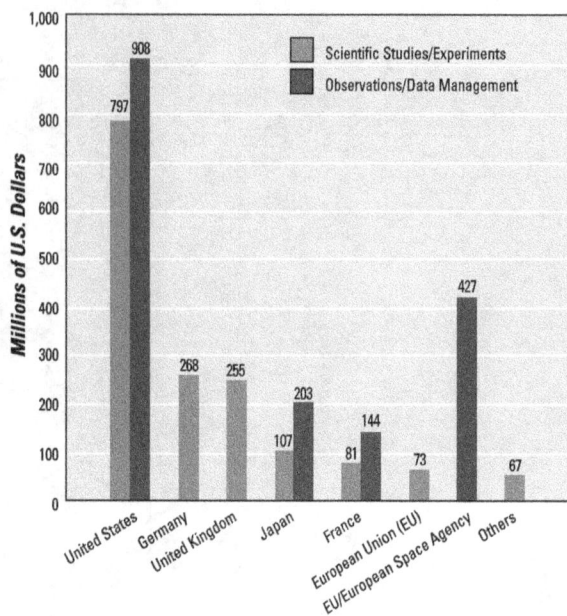

Note: Contributions by the United Kingdom and Germany to the European Space Agency (ESA) are included in the ESA observations total. No data are included from Australia, Brazil, India, Indonesia, Italy, Korea, Mexico, the People's Republic of China, Poland, Russia, Spain, Taiwan, and the Nordic Council. Inclusion of these could raise the totals by 10–20 percent.

Source: IGFA 2000.

- monitor, understand, and predict global change; and
- provide a sound scientific basis for national and international decision making.

The program builds on research undertaken over previous decades by independent researchers and programs. Today the USGCRP facilitates coordination across eleven federal departments and agencies with active global change programs. This distributed structure enables the program to draw on the missions, resources, and expertise of both research and mission-oriented agencies as it works to reduce uncertainties and develop useful applications of global change research. Participants include the Departments of Agriculture, Commerce (National Oceanic and Atmospheric Administration), Defense, Energy, Health and Human Services (National Institutes of Health), Interior (U.S. Geological Survey), and Transportation; the U.S. Environmental Protection Agency; the National Aeronautics and Space Administration; the National Science Foundation; and the Smithsonian Institution. The Office of Science and Technology Policy and the Office of Management and Budget provide oversight on behalf of the Executive Office of the President.

Despite the intensive U.S. investment in climate change science over the past decade, numerous gaps remain in our understanding. President Bush directed a Cabinet-level review of climate policy, including the state of science. As an input to this review, the U.S. National Academy of Sciences (NAS) prepared a report on *Climate Change Science: An Analysis of Some Key Questions* (NRC 2001a). This report was released in June 2001 and reached a number of findings regarding uncertainties and gaps in our knowledge that impede policymaking.[1] The report states:

> Because there is considerable uncertainty in current understanding of how the climate system varies naturally and reacts to emissions of greenhouse gases and aerosols, current estimates of the magnitude of future warming should be regarded as tentative and subject to future adjustments (either upward or downward). Reducing the wide range of uncertainty inherent in current model predictions of global climate change will require major advances in understanding and modeling of both (1) the factors that determine atmospheric concentrations of greenhouse gases and aerosols, and (2) the so-called "feedbacks" that determine the sensitivity of the climate system to a prescribed increase in greenhouse gases. There is also a pressing need for a global system designed for monitoring climate.

With respect to specific areas of knowledge, the NAS report concluded that greenhouse gases are accumulating in the Earth's atmosphere as a result of human activities, causing surface air temperatures and subsurface ocean temperatures to rise (see Appendix D). The changes observed over the last several decades are likely to result mostly from human activities, but some significant part of these changes is also a reflection of natural variability. Human-induced warming and associated sea level rise are expected to continue through the 21st century. Computer model simulations and basic physical reasoning suggest secondary effects, including potential changes in rainfall rates and in the susceptibility of semi-arid regions to drought. The impacts of

---

[1] The National Academy of Science report (NRC 2001a) generally agreed with the assessment of human-caused climate change presented in the recent IPCC Working Group I scientific report (IPCC 2001d), but sought to articulate more clearly the level of confidence that can be ascribed to those assessments and the caveats that need to be attached to them (see Appendix D).

these changes will be critically dependent on the magnitude of the effect, the rate at which it occurs, secular trends in technology that affect society's adaptability and vulnerability, and specific measures taken to adapt to or reduce vulnerability to climate change.

Accordingly, the NAS found that reducing the wide range of uncertainty inherent in current approaches to projecting global climate change and its effects on human beings and ecosystems will require major advances in understanding and modeling. To ensure that policies are informed by the best science, the United States is working aggressively to advance the science of climate and global change. In June 2001, President Bush announced the U.S. Climate Change Research Initiative, which is focused on reducing key areas of uncertainty in climate change science.

RESEARCH

U.S. research focuses on the full range of global change issues. The U.S. Congress, in the Global Change Research Act of 1990 (Public Law 101-606), directs the implementation of a program aimed at "understanding and responding to global change, including cumulative effects of human activities and natural processes on the environment." The Act defines global change as "changes in the global environment (including alterations in climate, land productivity, oceans or other water resources, atmospheric chemistry, and ecological systems) that may alter the capacity of the Earth to sustain life." This perspective recognizes the profound socioeconomic and ecological implications of global environmental change.

The USGCRP focuses on sets of interacting changes in the coupled human–environment system, which is undergoing change at a pace unprecedented in human history. These changes are occurring on many time and spatial scales, and many feedbacks and interdependencies link them. These numerous and various forces complicate efforts to understand the interactions of human and natural systems and how they may affect the capacity of the Earth to sustain life over the long term. Indeed, the interactions between changes in external (solar) forcing, human activities, and the intrinsic variability of the Earth's atmosphere, hydrosphere, and biosphere make understanding and projecting atmospheric and oceanic circulation, global energy and water cycles, and biogeochemical cycling among the most demanding scientific challenges.

U.S. Climate Change Research Initiative

On June 11, 2001, President Bush announced the establishment of the U.S. Climate Change Research Initiative to study areas of uncertainty and identify priority areas for investment in climate change science. He directed the Secretary of Commerce to work with other agencies to set priorities for additional investments in climate change research and to fully fund high-priority research areas that are underfunded or need to be accelerated. The definition of this new initiative is underway. It will improve the integration of scientific knowledge, including measures of uncertainty, into effective decision support systems.

Ongoing Broader Agenda for U.S. Research

The Climate Change Research Initiative will take place in the context of the broader global change research program that is ongoing in federal agencies. The USGCRP provides a framework and coordination mechanism for the continuing study of all of the complex, interrelated global change aspects in the NAS recommendations that are not addressed by the initiative.

The USGCRP is engaged in a continuing process to review its objectives and structure so that it can help government, the private sector, and communities to make informed management decisions regarding global environmental changes in light of persistent uncertainties. This will require the program to continue fundamental research to address crucial uncertainties about how human activities are changing the Earth's climate and environment. This program will need to continue developing increasingly detailed projections of how natural variability and human-induced environmental change interact and affect conditions on global to regional scales, and how we can manage natural resources in the future. Scientific understanding and data will need to be applied to tools useful for reducing risks and seizing opportunities resulting from global change.

The program will build on decades of scientific progress and will take advantage of the development of powerful advances in computing, remote sensing, environmental monitoring, and data and information technologies. Through additional focused investment in observations, scientific studies, and modeling, the USGCRP will seek to reduce uncertainties in the understanding of some of the most basic questions. The science needed to accomplish this ambitious objective is organized into the six research elements presented in Table 8-1, each of which focuses on topics crucial to projecting change and understanding its potential importance.

The USGCRP will also work with its partners to transition scientific knowledge to applications in resource management, disaster preparedness, planning for growth and infrastructure, and environmental and health assessment, among other areas. Partnerships among research programs, operational entities, and actors in the private sector and in federal, state, and local governments will be essential for the success of this effort. It will also require significant levels of cooperation and new management techniques to permit co-production of knowledge and deliverables across agencies and stakeholders.

National Climate Change Technology Initiative

The United States is further committed to improving climate change technology research and development, enhancing basic research, strengthening applied research through public–private partnerships, developing

TABLE 8-1 Fundamental Climate Change Research Needs

To support informed decision making, the U.S. Global Change Research Program is addressing uncertainties about how human activities are changing the Earth's climate and environment. The six research elements in this table focus on topics essential to projecting climate change and understanding its potential importance:

Research Uncertainty	USGCRP Research Focus
Atmospheric Composition • How do human activities and natural phenomena change the composition of the global atmosphere? • How do these changes influence climate, ozone, ultraviolet radiation, pollutant exposure, ecosystems, and human health?	• Processes affecting the recovery of the stratospheric ozone layer. • Properties and distribution of greenhouse gases and aerosols. • Long-range transport of pollutants and implications for air quality. • Integrated assessments of the effects of these changes for the nation and the world.
Climate Variability and Change • How do changes in the Earth system that result from natural processes and human activities affect the climate elements that are important to human and natural systems, especially temperature, precipitation, clouds, winds, and extreme events?	• Predictions of seasonal-to-decadal climate variations (e.g., the El Niño–Southern Oscillation). • Detection and attribution of human-induced change. • Projections of long-term climate change. • Potential for changes in extreme events at regional-to-local scales. • Possibility of abrupt climate change. • How to improve the effectiveness of interactions between producers and users of climate forecast information.
Carbon Cycle • How large and variable are the reservoirs and transfers of carbon within the Earth system? • How might carbon sources and sinks change and be managed in the future?	• North American and ocean carbon sources and sinks. • Impacts of land-use changes and resource management practices on carbon sources and sinks. • Future atmospheric carbon dioxide and methane concentrations and changes in land-based and marine carbon sinks. • Periodic reporting (starting in 2010) on the global distribution of carbon sources and sinks and how they are changing.
Global Water Cycle • How do human activities and natural processes that affect climate variability influence the distribution and quality of water within the Earth system? • To what extent are these changes predictable? • How will these changes affect climate, the cycling of carbon and other nutrients, and other environmental properties?	• Trends in the intensity of the water cycle and the causes of these changes (including feedback effects of clouds on the water and energy budgets, as well as the global climate system). • Predictions of precipitation and evaporation on time scales of months to years and longer. • Models of physical and biological processes and human demands and institutional processes, to facilitate efficient management of water resources. • Research supporting reports on the state of the global water cycle and national water resources.
Terrestrial and Marine Ecosystems • How do natural and human-induced changes in the environment interact to affect ecosystems (from natural to intensively managed), their ability to provide natural resources and commodities, and their influence on regional and global climate?	• Structure and function of ecosystems, including cycling of nutrients and how they interact with the carbon cycle. • Key processes that link ecosystems with climate. • Vulnerability of ecosystems to global change. • Options for enhancing resilience and sustaining ecosystem goods and services. • Scientific underpinning for improved interactions with resource managers.
Changes in Land Use and Land Cover • What processes determine land cover and land use at local, regional, and global scales? • How will land use and land cover evolve over time scales of 10–50 years?	• Identifying the human drivers of changes in land use and cover. • Monitoring, measuring, and mapping land use and land cover and managing data systems. • Developing projections of land-cover and land-use changes under various assumptions about climate, demographic, economic, and technological trends. • Integrating information about land use, land management, and land cover into other research elements.

improved technologies for measuring and monitoring gross and net greenhouse gas emissions, and supporting demonstration projects for new technologies.

Enhanced Carbon Technologies

The United States has committed to a number of projects to develop enhanced carbon technologies for capturing, storing, and sequestering carbon. Two contracts signed on July 11, 2001, solidified partnerships with The Nature Conservancy and with an international team of energy companies.

The Nature Conservancy Project. The Department of Energy will work in partnership with The Nature Conservancy and such companies as General Motors Corporation and American Electric Power to study how carbon dioxide can be stored more effectively by changing land-use practices and by investing in forestry projects. Using newly developed aerial and satellite-based technology, researchers will study forestry projects in Brazil and Belize to determine their carbon sequestration potential. Researchers will also test new software models that predict how carbon is sequestered by soil and vegetation at sites in the United States and abroad. The United States will provide $1.7 million of the $2 million cost of the three-year project.

International Team of Energy Companies. The Department of Energy will also collaborate with nine energy companies from four nations to develop breakthrough technologies to reduce the cost of capturing carbon dioxide from fossil fuel combustion and safely storing it underground. The nine energy companies are: BP–Amoco, Shell, Chevron, Texaco, Pan Canadian (Canada), Suncor Energy (Canada), ENI (Italy), Statoil Forskningssenter (Norway), and Norsk Hydro ASA (Norway). The U.S. government's contribution of $5 million will leverage an international commitment that will total more than $25 million over the next three years, including funding from the European Union, Norway's Klimatek Program, and the nine industry partners.

Human Effects on and Responses to Environmental Changes

In an effort to identify strategies to enhance the resilience of human systems to climate change, the U.S. Global Change Research Program continues to support research both on human activities that influence environmental change from local and regional to global scales and on how human systems prepare for and respond to environmental changes. An expanding research area will focus on analyses of the regional impacts of climate change on human systems and how improved information about climate change impacts can help decision makers in the public and private sectors.

Recent Accomplishments

Following are some recent USGCRP accomplishments in human dimensions and socioeconomic analyses:

- The U.S. Environmental Protection Agency (EPA) and the National Oceanic and Atmospheric Administration (NOAA) have established ongoing regional research and assessment projects in six U.S. regions to study the effects of climate variability and change on natural and human systems. These projects have been highly successful in analyzing the regional context of global change impacts, fostering relationships between scientists and stakeholders in the regions, and determining how research can meet stakeholders' needs for water-resource planning, fisheries management, ranching, and other climate-sensitive resource management issues.
- The U.S. Department of Transportation (DOT) established a center to identify effective ways to reduce the transportation sector's emissions and to help prepare the nation for the impacts of climate change. As part of its research efforts, the center will investigate how climate change could affect transportation infrastructure.
- Interdisciplinary investigations of human responses to seasonal and yearly swings in climate are highlighting the effects of market forces, access to resources, institutional flexibility, impacts across state boundaries, and the role of local culture and experience on the likelihood that individuals and institutions will use improved scientific information.

International Research Cooperation

The Working Group on International Research and Cooperation provides international affairs support for the USGCRP. The working group has representatives from interested government agencies and departments and acts as a forum to keep them informed on international global change research and funding issues. It addresses interagency support for international global change research programs and coordination, and infrastructure funding for such organizations as the Asia–Pacific Network for Global Change Research, the Inter-American Institute for Global Change Research, the International Human Dimensions Programme, the International Geosphere-Biosphere Programme, the World Climate Research Programme, and the Global Change System for Analysis, Research, and Training. The working group also addresses concerns raised by international nongovernmental global change organizations, such as free and open data exchange. These organizations include the International Group of Funding Agencies for Global Change Research and the Arctic Ocean Sciences Board.

The USGCRP contributes to and benefits from international research efforts to improve understanding of climate change on regional and global scales. USGCRP-supported scientists coordinate many of their programs with those of their counterparts in other countries, providing essential inputs to the increasingly complex models that enable scientists to improve analysis and prediction of climate change. Following are some examples of recent, ongoing, and planned climate change research and related activities in which USGCRP-supported scientists are heavily involved and for which international cooperation, participation, and support are especially important.

U.S.–Japan Cooperation in Global Change Research

During 2000, the United States and Japan co-sponsored a series of scientific workshops to identify important climate change research problems of mutual interest and to recommend how scientists from the two countries might constructively address them. Conducted under the auspices of the U.S.–Japan Agreement on Cooperation in Research and Development in Science and Technology, these workshops are managed on the U.S. side by the Working Group on International Research and Cooperation of the federal interagency Subcommittee on Global Change

The workshops developed recommendations to study the health impacts of climate change, in particular the impacts of greater and longer-lasting exposures to higher temperatures interacting with different air pollutants. A workshop on monsoon systems identified a number of cooperative bilateral and multilateral activities for the two countries to undertake. In 2001, Japan hosted the ninth workshop in this series, entitled Carbon Cycle Management in Terrestrial Ecosystems. The workshops have stimulated cooperation between Japanese and U.S. scientists and have led to numerous follow-up activities, including more focused planning workshops, data exchanges, and collaborative projects.

Climate and Societal Interactions

NOAA's Climate and Societal Interactions Program supports Regional Climate Outlook Fora, pilot application projects, workshops, training sessions, capacity building, and technical assistance for better understanding of climate variability and extreme events and for improving prediction and forecasting capability and data management, in Africa, Latin America, the Caribbean, Southeast Asia, and the Pacific. The Climate Information Project is develop-

ing a new program—Radio and Internet for the Communication of Hydro-Meteorological and Climate Information—to provide training to meteorological services worldwide on the use and production of radio and multimedia content in conjunction with digital satellite communication. This effort is being led by NOAA and involves a number of international partners, including the U.S. Agency for International Development; the World Bank; the World Meteorological Organization; the Inter-American Institute for Global Change Research; the Global Change System for Analysis, Research, and Training; and the Asia–Pacific Network for Global Change Research.

Eastern Pacific Investigation of Climate Processes

The Department of Commerce, through NOAA, and the National Science Foundation are bringing together more than 100 scientists from the United States, Mexico, Chile, and Peru to cooperate in the Eastern Pacific Investigation of Climate (EPIC). EPIC's scientific objectives are to observe and understand: (1) ocean–atmosphere processes in the equatorial and northeastern Pacific portions of the Inter-Tropical Convergence Zone (ITCZ); and (2) the properties of cloud decks in the trade wind and cross-equatorial flow regime and their interactions with the ocean below.

The project will study stratus cloud decks located off the west coast of South America, a region of cool sea-surface temperatures located along the equator in the eastern Pacific Ocean, and a region of intense precipitation located in the eastern Pacific north of the equator. All three of these phenomena interact to control the climate of the Southwest United States and Central and South America.

Studies of Global Ocean Ecosystem (GLOBEC) Dynamics

Scientists and research vessels from Germany, the United Kingdom, and the United States are conducting a closely coordinated major GLOBEC

field study on krill near the West Antarctic Peninsula. Krill are an essential component of the Southern Ocean food web and a commercially important species. Their predators—including sea birds, seals, and whales—depend on this food resource for survival. Sea ice plays an essential role as a habitat for krill (which feed beneath the ice) and their predators. Since evidence suggests that interannual variation in the extent of sea ice affects the abundance of krill, improving understanding of the role of climate factors affecting sea ice will comprise a critical component of the Southern Ocean GLOBEC program.

IGBP Open Science Conference

The International Geosphere–Biosphere Programme (IGBP) convened an open science conference in July 2001 in Amsterdam. A major objective of this conference was to present the latest results of climate change research at a series of levels: research conducted through the individual IGBP core projects and research integrated across these projects; research that has been integrated between the IGBP and the World Climate Research Programme, the International Human Dimensions Programme, Diversitas, and the Global Change System for Analysis, Research and Training, and other regional programs; and individual research projects on which these integrated efforts are based. The conference also identified new approaches to the study of the complex planetary system in which human activities are closely linked to natural processes.

International Group of Funding Agencies

The International Group of Funding Agencies (IGFA) is a forum through which national agencies that fund research on global change identify issues of mutual interest and ways to address them through coordinated national and, when appropriate, international actions. IGFA's focus is not on the funding of single projects, which is still a matter of national procedures; instead, it coordinates the support for the programs

Climate, Ecosystems, and Infectious Disease: Key Findings

The USGCRP, six federal agencies, and the Electric Power Research Institute sponsored a study completed in 2001 by the U.S. National Research Council's Committee on Climate, Ecosystems, Infectious Disease, and Human Health, entitled *Under the Weather: Climate, Ecosystems, and Infectious Disease* (NRC 2001b). Following are the Committee's key findings related to linkages between climate and infectious diseases.

Weather fluctuations and seasonal-to-interannual climate variability influence many infectious diseases. The characteristic geographic distributions and seasonal variations of many infectious diseases are *prima facie* evidence of linkages to weather and climate. Studies have shown that such factors as temperature, precipitation, and humidity affect the life cycles of many disease pathogens and vectors (both directly, and indirectly through ecological changes) and thus can affect the timing and intensity of disease outbreaks. However, disease incidence is also affected by such factors as sanitation and public health services, population density and demographics, land-use changes, and travel patterns. The importance of climate relative to these other variables must be evaluated in the context of each situation.

Observational and modeling studies must be interpreted cautiously. Although numerous studies have shown an association between climatic variations and disease incidence, they are not able to fully account for the complex web of causation that underlies disease dynamics. Thus, they may not be reliable indicators of future changes. Likewise, a variety of models have been developed to simulate the effects of climatic changes on the incidence of such diseases as malaria, dengue, and cholera. While these models are useful heuristic tools for testing hypotheses and carrying out sensitivity analyses, they are not necessarily intended to serve as predictive tools, and often do not include such processes as physical/biological feedbacks and human adaptation. Thus, caution must be exercised in using these models to create scenarios of future disease incidence and to provide a basis for early warnings and policy decisions.

The potential disease impacts of global climate change remain highly uncertain. Changes in regional climate patterns caused by long-term global warming could affect the potential geographic range of many infectious diseases. However, if the climate of some regions becomes more suitable for transmission of disease agents, human behavioral adaptations and public health interventions could serve to mitigate many adverse impacts. Basic public health protections, such as adequate housing and sanitation, as well as new vaccines and drugs, may limit the future distribution and impact of some infectious diseases, regardless of climate-associated changes. These protections, however, depend upon maintaining strong public health programs and ensuring vaccine and drug access in the poorer countries of the world.

Climate change may affect the evolution and emergence of infectious diseases. The potential impacts of climate change on the evolution and emergence of infectious disease agents create another important but highly uncertain risk. Ecosystem instabilities brought about by climate change and concurrent stresses, such as land-use changes, species dislocation, and increasing global travel, potentially influence the genetics of pathogenic microbes through mutation and horizontal gene transfer, and could give rise to new interactions among hosts and disease agents.

There are potential pitfalls in extrapolating climate and disease relationships from one spatial/temporal scale to another. The relationships between climate and infectious disease are often highly dependent upon local-scale parameters, and it is not always possible to extrapolate these relationships meaningfully to broader spatial scales. Likewise, disease impacts of seasonal-to-interannual climate variability may not always provide a useful analog for the impacts of long-term climate change. Ecological responses on the time scale of an El Niño event, for example, may be significantly different from the ecological responses and social adaptation expected under long-term climate change. Also, long-term climate change may influence regional climate variability patterns, hence limiting the predictive power of current observations.

Recent technological advances will aid efforts to improve modeling of infectious disease epidemiology. Rapid advances being made in several disparate scientific disciplines may spawn radically new techniques for modeling infectious disease epidemiology. These include advances in sequencing of microbial genes, satellite-based remote sensing of ecological conditions, the development of geographic information system (GIS) analytical techniques, and increases in inexpensive computational power. Such techniques will make it possible to analyze the evolution and distribution of microbes and their relationship to different ecological niches, and may dramatically improve our abilities to quantify the disease impacts of climatic and ecological changes.

themselves (Secretariats, International Project Offices, etc.). IGFA facilitates international climate change research by bringing the perspective of national funding agencies to strategic research planning and implementation. At its October 2000 meeting, most IGFA member nations reported increases in funding for climate change research, initiation and deployment of new national programs, and establishment of some new research centers.

Diversitas

Diversitas was established in 1991 as an umbrella program to coordinate a broad research effort in the biodiversity sciences at the global level. The program has played an important role at the interface between science and policy by building a partnership with the Convention on Biological Diversity. Diversitas has signed a Memorandum of Understanding with the Secretariat of the Convention and has provided input to its Subsidiary Body on Scientific, Technical and Technological Advice. Among the issues that IGFA considered at its 2001 plenary meeting in Stockholm was the development of a new implementation strategy for Diversitas. Countries, via IGFA, have committed funds to help strengthen the international infrastructure for biodiversity research through Diversitas according to the model of the other partner global change programs.

International Paleoclimate Research

An international team of researchers from the United States, Germany, and Russia is investigating El'gygytgyn Lake in northeastern Siberia, just north of the Arctic Circle. This crater was formed 3.6 million years ago by a meteorite impact. Its sediments hold the promise of revealing the evolution of Arctic climate a full one million years before the first major glaciation of the Northern Hemisphere. In addition, through an international consortium of researchers, the Nyanza Project team, involving scientists from the United States, Europe, and four countries in Africa, is studying climate variability, as well as environmental and ecological change, through the entire episode of human evolution. As part of this project, a unique 2,000-year-old annually resolved record of atmospheric circulation and dynamics, revealing El Niño–Southern Oscillation and solar cycles, has been recovered from sediments in Lake Tanganyika, the second deepest lake on the planet.

SYSTEMATIC OBSERVATION

Long-term, high-quality observations of the global environmental system are essential for defining the current state of the Earth's system, and its history and variability. This task requires both space- and surface-based observation systems. The term "climate observations" can encompass a broad range of environmental observations, including:

* routine weather observations, which, collected over a long enough period, can be used to help describe a region's climatology;
* observations collected as part of research investigations to elucidate chemical, dynamic, biological, or radiative processes that contribute to maintaining climate patterns or to their variability;
* highly precise, continuous observations of climate system variables collected for the express purpose of documenting long-term (decadal-to-centennial) change; and
* observations of climate proxies, collected to extend the instrumental climate record to remote regions and back in time to provide information on climate change for millennial and longer time scales.

The various federal agencies involved in observing climate through space-based and ground-based activities provide many long-term observations. Space-based systems have the unique advantage of obtaining global spatial coverage, particularly over the vast expanses of the oceans, sparsely populated land areas (e.g., deserts, mountains, forests, and polar regions), and the mid and upper troposphere and stratosphere. They provide unique measurements of solar output; the

Earth's radiation budget; vegetation cover; ocean biomass productivity; atmospheric ozone; stratospheric water vapor and aerosols; greenhouse gas distributions; sea level and ocean interior; ocean surface conditions; winds, weather, and tropical precipitation; and other variables.

Satellite observations alone are not sufficient. *In-situ* observations are required for the measurement of parameters that cannot be estimated from space platforms (e.g., biodiversity, ground water, carbon sequestration at the root zone, and subsurface ocean parameters). *In-situ* observations also provide long time-series of observations required for the detection and diagnosis of global change, such as surface temperature, precipitation and water resources, weather and other natural hazards, the emission or discharge of pollutants, and the impacts of multiple stresses on the environment due to human and natural causes. To meet the need for the documentation of global changes on a long-term basis, the United States integrates observations from both research and operational systems. The goal of the U.S. observation and monitoring program is to ensure a long-term, high-quality record of the state of the Earth system, its natural variability, and changes that occur.

Since 1998, Parties to the United Nations Framework Convention on Climate Change (UNFCCC) have noted with concern the mounting evidence of a decline in the global observing capability and have urged Parties to undertake programs of systematic observations and to strengthen the collection, exchange, and use of environmental data and information. It has long been recognized that the range of global observations needed to understand and monitor Earth processes contributing to climate and to assess the impact of human activities cannot be satisfied by a single program, agency, or country. The United States supports the need to improve global observing systems for climate and to exchange information on national plans and programs that contribute to the global capacity in this area.

Documentation of U.S. Climate Observations

As part of its continuing contributions to systematic observations in support of climate monitoring, the United States forwarded *The U.S. Detailed National Report on Systematic Observations for Climate* to the UNFCCC Secretariat on September 6, 2001 (U.S. DOC/NOAA 2001c). Because this was the U.S. government's first attempt to document all U.S. contributions to global climate observations, a wide net was cast to include information on observations that fell into each of the following categories: (1) *in-situ* atmospheric observations; (2) *in-situ* oceanographic observations; (3) *in-situ* terrestrial observations; (4) satellite-based observations, which by their nature cut across the atmospheric, oceanographic, and terrestrial domains; and (5) data and information management related to systematic observations. The report attempted to cover all relevant observation systems and is representative of the larger U.S. effort to collect environmental data.

Material for the report was developed by a U.S. interagency Global Climate Observing System (GCOS) coordination group comprised of representatives from the following federal agencies: (1) the U.S. Department of Agriculture's Natural Resources Conservation Service and U.S. Forest Service; (2) three line offices of the U.S. Department of Commerce's National Oceanic and Atmospheric Administration; (3) the U.S. Department of Energy's Office of Science; (4) the U.S. Environmental Protection Agency; (5) the U.S. Department of the Interior's U.S. Geological Survey; (6) the National Aeronautics and Space Administration; (7) the U.S. Department of Transportation's Federal Aviation Administration; (8) the National Science Foundation; (9) the U.S. Naval Oceanographic Office; (10) the U.S. Army Corps of Engineers; and (11) the U.S. Air Force. The report was coordinated with the U.S. Global Change Research Program.

In-situ Climate Observation

The United States supports a broad network of global atmospheric, ocean, and terrestrial observation systems.

Atmospheric Observation

The United States supports 75 stations in the GCOS Surface Network (GSN), 20 stations in the GCOS Upper Air Network (GUAN), and 4 stations in the Global Atmospheric Watch (GAW). These stations are distributed geographically as prescribed in the GCOS and GAW network designs. The data (metadata and observations) from these stations are shared according to GCOS and GAW protocols. The GSN and GUAN stations are part of a larger network, which was developed for purposes other than climate monitoring. Nonetheless, the stations fully meet the GCOS requirements.

The United States has no comprehensive system designed to observe climate change and climate variability. Basically, U.S. sustained observing systems provide data principally for nonclimatic purposes, such as predicting weather, advising the public, and managing resources. In addition, U.S. research-observing systems collect data for climate purposes, but are often oriented toward gathering data for climate process studies or other research programs, rather than climate monitoring. They are usually limited in their spatial and temporal extent. Because the U.S. climate record is based upon a combination of existing operational and research programs, it may not be "ideal" from a long-term climate monitoring perspective. Nevertheless, these observing systems collectively provide voluminous and significant information about the spatial and temporal variability of U.S. climate and contribute to the international climate observing effort as well. The atmospheric section in the main body of the detailed national report examines *in-situ* climate monitoring involving systems from the surface, upper air, and atmospheric deposition domains (U.S. DOC/NOAA 2001c).

Ocean Observation

The Global Ocean Observing System (GOOS) requirements are the same as the GCOS requirements. Both are based on the Ocean Observing System Development Panel Report (OOSDP 1995). Like GCOS, the GOOS is based on a number of *in-situ* and space-based observing components. The United States supports the Integrated Global Ocean Observing System's surface and marine observations through a variety of components, including fixed and surface-drifting buoys, subsurface floats, and volunteer observing ships. It also supports the Global Sea Level Observing System through a network of sea-level tidal gauges. The United States currently provides satellite coverage of the global oceans for sea-surface temperatures, surface elevation, ocean surface winds, sea ice, ocean color, and other climate variables. These satellite activities are coordinated internationally through the Committee on Earth Observation Satellites.

Terrestrial Observation

For terrestrial observations, the requirements for climate observations were developed jointly between GCOS and the Global Terrestrial Observing System (GTOS) through the Terrestrial Observations Panel for Climate (WMO 1997). GCOS and GTOS have identified permafrost thermal state and permafrost active layer as key variables for monitoring the state of the cryosphere. GCOS approved the development of a globally comprehensive permafrost-monitoring network to detect temporal changes in the solid earth component of the cryosphere. As such, the Global Terrestrial Network for Permafrost (GTN-P) is quite new and still very much in the developmental stage. The International Permafrost Association has the responsibility for managing and implementing the GTN-P.

U.S. contributions to the GTN-P network are provided by the Department of the Interior and the National Science Foundation, through grants to various universities. All the U.S.

GTN-P stations are located in Alaska. The active layer thickness is currently being monitored at 27 sites. Forty-eight bore holes exist in Alaska where permafrost thermal state can be determined. Of these, 4 are classified as *surface* (0–10 m) sites, 1 is *shallow* (10–25 m), 22 are *intermediate depth* (25–125 m), and 21 are *deep bore holes* (>125 m). The U.S. contribution to the GTN-P network comes from short-term (three- to five-year) research projects.

The United States operates a long-term "benchmark" glacier program to intensively monitor climate, glacier motion, glacier mass balance, glacier geometry, and stream runoff at a few select sites. The data collected are used to understand glacier-related hydrologic processes and improve the quantitative prediction of water resources, glacier-related hazards, and the consequences of climate change.

The approach has been to establish long-term, mass-balance monitoring programs at three widely spaced U.S. glacier basins that clearly sample different climate-glacier-runoff regimes. The three glacier basins are South Cascade Glacier in Washington State, and Gulkana and Wolverine Glaciers in Alaska. Mass-balance data are available beginning in 1959 for the South Cascade Glacier, and beginning in 1966 for the Gulkana and Wolverine Glaciers.

The AmeriFLUX network endeavors to establish an infrastructure for guiding, collecting, synthesizing, and disseminating long-term measurements of CO_2, water, and energy exchange from a variety of ecosystems. Its objectives are to collect critical new information to help define the current global CO_2 budget, enable improved projections of future concentrations of atmospheric CO_2, and enhance the understanding of carbon fluxes, net ecosystem production, and carbon sequestration in the terrestrial biosphere.

The terrestrial section of the detailed U.S. report examines *in-situ* climate monitoring and discusses, in addition to the GTN-P, the Global Terrestrial Network for Glaciers (GTN-G), and the

AmeriFLUX programs, stream-flow and surface-water gauging, ground-water monitoring, snow and soil monitoring, the U.S. paleoclimatology program, ecological observation networks, fire-weather observation stations, as well as global, national, and regional land cover characterization. The United States contributes to all of these activities.

Satellite Observation Programs

Space-based, remote-sensing observations of the atmosphere–ocean–land system have evolved substantially since the early 1970s, when the first operational weather satellite systems were launched. Over the last decade satellites have proven their capability to accurately monitor nearly all aspects of the total Earth system on a global basis—a capability unmatched by ground-based systems, which are limited to land areas and cover only about 30 percent of the planetary surface.

Currently, satellite systems monitor the evolution and impacts of El Niño, weather phenomena, natural hazards, and extreme events, such as floods and droughts; vegetation cycles; the ozone hole; solar fluctuations; changes in snow cover, sea ice and ice sheets, ocean surface temperatures, and biological activity; coastal zones and algal blooms; deforestation and forest fires; urban development; volcanic activity; tectonic plate motions; and other climate-related information. These various observations are used extensively in real-time decision making and in the strategic planning and management of industrial, economic, and natural resources. Examples include weather and climate forecasting, agriculture, transportation, energy and water resource management, urban planning, forestry, fisheries, and early warning systems for natural disasters and human health impacts.

The GCOS planning process addressed satellite requirements for climate. In so doing, it identified an extensive suite of variables that should be observed and monitored from space (WMO 1995). In addition, GCOS plans specified that instrument calibra-

tion and validation be performed to ensure that the resulting space-based observations meet climate requirements for accuracy, continuity, and low bias.

The current generation of U.S. research satellite instruments exceeds the GCOS requirements for the absolute calibration of sensors—something that was lacking in the early satellite platforms used for real-time operational purposes. Several of the historical data series from operational satellites have been reprocessed using substantially improved retrieval algorithms to provide good-quality global data products for use in climate change research and applications.

NPOESS Program

Improving the on-board capabilities for calibration on operational satellites will be one of the objectives in the development of the National Polar-orbiting Operational Environmental Satellite System (NPOESS) program. Prior to the launch of NPOESS in 2008, an NPOESS Preparatory Project (NPP) satellite will be launched in the 2005 time frame as a bridge mission between the NASA Earth Observing Satellites (EOS) program and NPOESS.

The mission of NPP is to demonstrate advanced technology for atmospheric sounding, and to provide ongoing observations after EOS-Terra and EOS-Aqua. It will supply data on atmospheric and sea-surface temperatures, humidity soundings, land and ocean biological productivity, and cloud and aerosol properties. NPP will also provide early instrument and system-level testing and early user evaluation of NPOESS data products, such as algorithms, and will identify opportunities for instrument calibration. The information and lessons learned from NPP will help reduce instrument risk and will enable design modifications in time to ensure NPOESS launch readiness.

U.S. Environmental Satellite Program

A number of U.S. satellite operational and research missions form the basis of a robust national remote-

sensing program that fully supports the requirements of GCOS (U.S. DOC/NOAA 2001c). These include instruments on the Geostationary Operational Environmental Satellites (GOES) and Polar Operational Environmental Satellites (POES), the series of Earth Observing Satellites (EOS), the Landsats 5 and 7, the Total Ozone Mapping Spectrometer satellite, and the TOPEX/Poseidon satellite measuring sea-surface height, winds, and waves. Additional satellite missions in support of GCOS include (1) the Active Cavity Radiometer Irradiance Monitor for measuring solar irradiance; (2) EOS-Terra; (3) QuickSCAT; (4) the Sea-viewing Wide Field-of-view Sensor (SeaWiFS) for studying ocean productivity; (5) the Shuttle Radar Topography Mission; and (6) the Tropical Rainfall Measuring Mission for measuring rainfall, clouds, sea-surface temperature, radiation, and lightning.

Defense Meteorological Satellite Program

The Defense Meteorological Satellite Program (DMSP) is a Department of Defense program run by the Air Force Space and Missile Systems Center. The program designs, builds, launches, and maintains several near-polar-orbiting, sun-synchronous satellites, which monitor the meteorological, oceanographic, and solar–terrestrial physics environments. DMSP satellites are in a near-polar, sun-synchronous orbit. Each satellite crosses any point on the Earth up to two times a day, thus providing nearly complete global coverage of clouds approximately every six hours.

Integrated Global Observing Strategy

The United States cooperates on an international basis with a number of coordinating bodies. The Integrated Global Observing Strategy (IGOS) is a strategic planning process covering major satellite- and surface-based systems for global environmental observations of the atmosphere, oceans, and land, that provides a framework for decisions and resource allocations by individual funding agencies. IGOS assesses Earth-observing requirements, evaluates capabilities of current and planned observing systems, and has begun (at least among the space agencies) to obtain commitments to address these gaps.

An IGOS Ocean Theme is in the implementation phase under leadership from GOOS. An analysis of requirements, gaps, and recommendations for priority observations is underway for integrated global carbon observations as well as integrated global atmospheric chemistry observations. Similar analyses, recommendations, and commitments are also being explored for geological and geophysical hazards, coasts and coral reefs, and the global water cycle.

Operational Weather Satellites

Operational weather satellites are internationally coordinated through the Coordination Group for Meteorological Satellites, of which the World Meteorological Organization is a member and major beneficiary, along with five other satellite agency members. The primary body for policy and technical issues of common interest related to the whole spectrum of Earth observation satellite missions is the Committee on Earth Observation Satellites (CEOS). CEOS has 22 space agency members, including both research and operational satellite agencies, with funding and program responsibilities for a satellite Earth observation program currently operating or in the later stages of system development. CEOS encourages compatibility among space-borne Earth-observing systems through coordination in mission planning; promotion of full and nondiscriminatory data access; setting of data

product standards; and development of data products, services, applications, and policies.

Global Change Data and Information System

Global environmental concerns are an overriding justification for the unrestricted international exchange of GCOS data and products for peaceful, noncommercial, global scientific, and applications purposes. As such, GCOS developed an overarching data policy that endorses the full and open sharing and exchange of GCOS-relevant data and products for all GCOS users at the lowest possible cost. The United States recognizes and subscribes to this data policy.

Achieving the goals of the U.S. climate observing program requires multidisciplinary analysis of data and information to an extent never before attempted. This includes the analysis of interlinked environmental changes that occur on multiple temporal and spatial scales, which is very challenging both technically and intellectually. For example, many types of satellite and *in-situ* observations at multiple scales need to be integrated with models, and the results need to be presented in understandable ways to all levels of the research community, decision makers, and the public. Additionally, very large volumes of data from a wide variety of sources and results from many different investigations need to be readily accessible to scientists and other stakeholders in usable forms that can be integrated.

Various U.S. agencies have engaged in extensive development of interagency data and information processes to foster better integration and accessi-

bility of data- and discipline-specific information. The Global Change Data and Information System (GCDIS) has been developed to facilitate this goal. GCDIS currently provides a gateway for access to more than 70 federally funded sources of data, both governmental and academic. During the last decade, significant strides have been made in creating seamless connections between diverse data sets and sources, as well as enhancing its ability to search across the full complement of data sources. While the Internet has facilitated this effort, the provision of data and information in forms needed for cross-disciplinary analyses remains a challenge.

The U.S. government's position (as evidenced by its support of the "10 Principles of Climate Observations" and of the U.S. climate research community [NRC 1999]) is that high standards must be met for a particular set of observations to serve the purpose of monitoring the climate system to detect long-term change. In general, the observing programs and resulting data sets described here have not yet fully met these principles. This shortfall stems from two main factors: (1) the principles were articulated only within the past decade (Karl et al. 1995), long after the initiation of most of our long-term observing systems; and (2) more recent observing programs typically do not have climate monitoring as their prime function.

The U.S. systematic climate observing effort will continue to improve and enhance understanding of the climate system. A full copy of the *The U.S. Detailed National Report on Systematic Observations for Climate* (U.S. DOC/NOAA 2001c) can be found at http://www.eis.noaa.gov/gcos/soc_long.pdf.

Chapter 9
Education, Training, and Outreach

Over the last three years, U.S. climate change outreach and education efforts have evolved significantly. Early outreach efforts, which focused primarily on the research and academic community, have helped to expand climate change research activity and have resulted in a robust research agenda that has resolved many scientific uncertainties about global warming. Scientists and decision makers worldwide have used the findings of U.S. research projects. More recent outreach efforts have moved beyond the research community, focusing on public constituencies who may be adversely affected by the impacts of climate change. These constituencies will have the ultimate responsibility to help solve the climate change problem by supporting innovative, cost-effective solutions at the grassroots level.

Federal efforts to increase public education and training on global climate change issues are designed to increase understanding of the Earth's complex climate system. This improved understanding will enable decision makers and people potentially at risk from the impacts of climate change to more accurately interpret complex scientific information and make better decisions about how to reduce their risks.

Federal outreach and educational activities are performed under several U.S. mandates, including the Global Change Research Act of 1990, the National Climate Program Act, the Clean Air Act Amendments of 1990, and the Environmental and Education Act of 1990. Federal programs often rely on noneducational programs to simultaneously meet legislative mandates on climate education and U.S. science, policy, and outreach goals.

In addition to outreach conducted at the federal level, a growing movement of nongovernmental outreach efforts has proven to be very effective in engaging the U.S. public and industry on the climate change issue. Most outside groups work independently of government funding in their climate change research and outreach efforts, although some nongovernmental programs are funded in part by the federal government. Many nongovernmental organizations (NGOs) enjoy tax-exempt status, which permits them to receive private support and reduce costs to donors. An extensive list of NGOs conducting climate change outreach and education initiatives may be found at http://www.epa.gov/global-warming/ links/org_links.html.

Industry is also playing an increasing role in climate change outreach and education. Many corporations have worked extensively with federal government partnership programs to resolve climate change issues. These companies spend millions of dollars to promote their climate change investments and viewpoints to consumers and other industries, and most disseminate information about climate change to their customers and the public.

More recent outreach and education efforts, both within government and by NGOs and industry, have encouraged many activities that adapt to a changing climate or that reduce greenhouse gas emissions. Because of these efforts, more citizens understand the issue with a higher level of sophistication. And as people are becoming more familiar with the problem, they are also beginning to appreciate the impacts of society's actions on the climate system.

This chapter presents a sample of current U.S. education and outreach efforts that are building the foundation for broad action to reduce risks from climate change. Because a comprehensive treatment of NGO efforts is beyond the scope of this chapter, it focuses on new and updated governmental activities since the previous National Communication.

U.S. GLOBAL CLIMATE RESEARCH PROGRAM EDUCATION AND OUTREACH

Sponsored by the U.S. Global Change Research Program (USGCRP), the U.S. national assessment of the potential consequences of climate variability and change (NAST 2000 and 2001) has provided an important opportunity to reach out to the many interested parties, or stakeholders, about the potential significance for them of future changes in climate.

Regional Outreach

The National Assessment began in 1997 and 1998 with 20 regional workshops across the country. Each initiated a discussion among the stakeholders, scientific community, and other interested parties about the potential importance of climate change and the types of potential consequences and response options, all in the context of other stresses and trends influencing the region. On average, about 150 people participated in each workshop. There was extensive outreach to local media, drawn in part by the frequent participation of high-level government officials. Halfway through this effort, a National Forum convened in Washington, D.C., attracted about 400 participants, from Cabinet officials to some ranchers who had never traveled outside of the central U.S.

Moving from the workshop phase to an assessment phase, the USGCRP organized a range of activities that involved assessment teams drawn from the research and stakeholder communities. While sponsored by and working with government agencies, these teams were based largely in the academic community to broaden participation and enhance their independence and credibility. To focus analysis on the issues identified in the regional workshops, 16 of these assessment teams had a regional focus. Each team established an advisory and outreach framework that was used for the preparation of each assessment report. The reports are being distributed widely within each region, and outreach activities include workshops, presentations, and the media. USGCRP agencies are continuing to sponsor many of these regional activities as a way of strengthening the dialogue with the public about the potential consequences and significance of climate change, and the anticipatory actions that will be needed.

National Outreach

The USGCRP also sponsored five national sectoral studies covering climate change's potential consequences for agriculture, forests, human health, water resources, and coastal areas and marine resources (NAAG 2001, NFAG 2001, NHAG 2000 and 2001, NWAG 2000, NCAG 2000). The five broadly based teams organized outreach activities ranging from presentations at scientific and special-interest meetings to full workshops and special issues of journals. Each team is now issuing its report, distributing information widely to the public.

The National Assessment Synthesis Team (NAST) was created as an independent federal advisory committee to integrate the findings and significance of the five sectoral studies. The NAST was composed of representatives from academia, government, industry, and NGOs. Through a series of open meetings, followed by a very extensive open review process, the NAST prepared both an overview report that summarizes the findings (NAST 2001a) and a foundation report that provides more complete documentation (NAST 2001b). Both reports are being widely circulated. They are available on the Internet, and copies are being sent to every state and to major U.S. libraries.

The USGCRP is also using other outreach tools to increase public understanding of the potential consequences of climate change. USGCRP's Web site (http://www.usgcrp.gov) helps connect scientists, students and their teachers, government officials, and the general public to accurate and useful information on global change. Also, the newsletter *Acclimations* provides regular information to a broad audience about the national assessment (USGCRP 1998–2000).

The USGCRP is sponsoring the preparation of curriculum materials based on the national assessment. These materials will be made widely available to teachers over the Web, updating the various types of materials made available during the mid-1990s by a number of federal agencies. Through these mechanisms, the national assessment has directly involved several thousand individuals, while reaching out to many thousands more through the reports and the media.

FEDERAL AGENCY EDUCATION INITIATIVES

Climate change education at the primary and secondary (K–12) and university levels has grown considerably over the past three years. The growth of the Internet has allowed educators throughout the country to use on-line educational global change resources. Federal government programs have supported numerous initiatives, ranging from online educational programs to research support. This section and Table 9-1 present a sampling of these initiatives.

Department of Energy

The Department of Energy (DOE) sponsors several programs that support advanced global change research.

Global Change Education Program

DOE's Global Change Education Program continues to support three coordinated components aimed at pro-viding research and educational support to postdoctoral scientists, graduate students, faculty, and undergraduates at minority colleges and universities: the Summer Undergraduate Research Experience, the Graduate Research Environmental Fellowships, and the Significant Opportunities in Atmospheric Research and Science program.

Oak Ridge Institute for Science and Education

The Science/Engineering Education Division at the Oak Ridge Institute for Science and Education continues to develop and administer collaborative research appointments, graduate and postgraduate fellowships, scholarships, and other programs that capitalize on the resources of federal facilities across the nation and the national academic community. The aim is to enhance the quality of scientific and technical education and literacy, thereby increasing the number of graduates in science and engineering fields, particularly those related to energy and the environment.

National Aeronautics and Space Administration

From helping design K–12 curricula to teacher training, NASA is heavily involved in education initiatives related to Earth science.

Earth System Science Education Program

Sponsored by NASA through the Universities Space Research Association, this program supports the development of curricula in Earth System Science and Global Change at 44 participating colleges and universities. The program's Web site provides educational resources for undergraduates.

Earth Science Enterprise

Every year tens of thousands of students and teachers participate in NASA's Earth Science Enterprise program. The program attempts to improve people's understanding of the natural processes that govern the global environment and to assess the effects of human activities on these processes. It is expected to yield better weather forecasts, tools for managing agriculture and forests, and information for commercial fishers and coastal planners. Ultimately, the program will improve our ability to predict how climate will change.

While the program's ostensible goal is scientific understanding, its ultimate product is education in its broadest form. The Earth Science Enterprise has formulated education programs that focus on teacher preparation, curriculum and student support, support for informal education and public communication, and professional training. Its Earth System Science Fellowship program encourages student research, modeling, and analysis in support of the USGCRP. More than 500 Ph.D. and M.S. fellowships have been awarded since the program's inception in 1990.

Partnerships

Partnerships allow agencies with similar goals to combine resources and expertise to serve the interests of educators and students.

Climate Change Partnership Education Program

The Environmental Protection Agency (EPA), NASA, and NOAA initiated a partnership outreach program for broadcast meteorologists on climate change impacts and science. They formed the partnership in response to broadcasters' requests for educational materials that they could use in their community outreach and education activities, particularly during school visits. The resulting *Climate Change Presentation Kit* CD-ROM includes fact sheets that can be downloaded, printed, and distributed to audiences who have varying levels of scientific literacy, a complete PowerPoint slide presentation that can be shown from a computer or printed as overhead transparencies, science experiments and games for classroom use, contact names and phone numbers for additional scientific information, and links to informative Web sites (U.S. EPA, NASA, and NOAA 1999).

TABLE 9-1 U.S. Government On-line Climate Change Educational Resources

Resource	Description	Web Site
Department of Energy		
Energy Efficiency and Renewable Energy Kids' Site	A wealth of information on types of renewable energy.	http://www.eren.doe.gov/kids/
Energy Information Administration Kids' Page	Interactive Web page with energy information, activities, and resources.	http://www.eia.doe.gov/kids/
Fossil Energy—Education Main Page	An introduction to fossil fuels for students.	http://www.fe.doe.gov/education/main.html
Environmental Protection Agency		
Global Warming Site	Information for general audiences about the science of climate change, its impacts, green-house gas emissions, and mitigation actions.	http://www.epa.gov/globalwarming
Global Warming Kids' Site	Overview of global warming and climate science; includes interactive games.	http://www.epa.gov/globalwarming/kids/index.html
GLOBE Program		
GLOBE Program Home Page	Interactive science and education site for participants in the GLOBE program, grades K–12.	http://www.globe.gov
National Aeronautics and Space Administration		
Educational Links	List of Earth science educational links.	http://eospso.gsfc.nasa.gov/eos_homepage/education.html
Teaching Earth Science Site	Resources and information for Earth science educators for elementary through university levels.	http://www.earth.nasa.gov/education/index.html
For Kids Only Site	From NASA's Earth science Enterprise, contains a wealth of Earth science information, teacher resources, and interactive games.	http://kids.earth.nasa.gov/
National Oceanic and Atmospheric Administration		
CLIMGRAPH	Educational graphics on global climate change and the greenhouse effect.	http://www.fsl.noaa.gov/~osborn/CLIMGRAPH2.html
Specially for Students—Climate Change and Our Planet	List of NOAA's climate change-related sites tailored for kids.	http://www.education.noaa.gov/sclimate.html
Specially for Teachers	List of NOAA's climate change-related sites tailored for educators.	http://www.education.noaa.gov/tclimate.html
A Paleo Perspective on Global Warming	For general audiences, a site to help teach the importance of paleoclimate research and its relation to global warming.	http://www.ngdc.noaa.gov/paleo/globalwarming/home.html
U.S. Global Change Research Information Office		
Global Change and Environmental Education Resources	List of global change and environmental education on-line resources.	http://gcrio.org/edu/educ.html
GCRIO Home Page	Data and information on climate change research, adaptation/mitigation strategies and technologies.	http://gcrio.org
Common Questions About Climate Change	Intended for general audiences.	http://www.gcrio.org/ipcc/qa/cover.html
Global Warming and Climate Change	Brochure explaining the issue for general audiences.	http://gcrio.org/gwcc/toc.html
U.S. Global Change Research Program		
USGCRP Home Page	Global change information for students and educators.	http://www.usgcrp.gov/
U.S. Geological Survey		
Global Change Teacher Packet	An introduction and five activities for classroom use.	http://mac.usgs.gov/mac/isb/pubs/teachers-packets/globalchange/globalhtml/guide.html
Global Change Educational Activities	Information about global change for grades 4–6.	http://www.usgs.gov/education/learnweb/GC.html

GLOBE Program

Administered by NOAA, NASA, NSF, and EPA, the Global Learning and Observations to Benefit the Environment (GLOBE) program continues to bring together students, educators, and scientists throughout the world to monitor the global environment. The program aims to increase environmental awareness and to improve student achievement in science and mathematics. GLOBE's worldwide network has expanded to represent more than 10,000 K–12 schools in over 95 countries. These students make scientific observations at or near their schools in the areas of atmosphere, hydrology, biology, and soils, and report their findings to the network.

FEDERAL AGENCY OUTREACH

Federal agencies provide the public, state and local governments, industry, and private groups with information about national and global climate change research and risk assessments, U.S. mitigation activities, and policy developments. Agencies work on outreach efforts independently and in partnership with other federal agencies, NGOs, and industry. Although outreach activities may vary from agency to agency, most of them share the common goal of increasing awareness about the potential risks climate change poses to the environment and society. Current outreach encourages constituencies to participate in existing federal voluntary programs that promote climate change mitigation and adaptation activities.

Department of Energy

DOE supports numerous initiatives focused on increasing energy efficiency and reducing greenhouse gas emissions.

Carbon Dioxide Information Analysis Center

The Carbon Dioxide Information Analysis Center (CDIAC), which includes the World Data Center for Atmospheric Trace Gases, is DOE's primary center for global change data

and information analysis. CDIAC responds to data and information requests from users from all over the world who are concerned about the greenhouse effect and global climate change. CDIAC's data holdings include historical records of the concentrations of carbon dioxide and other radiatively active gases in the atmosphere; the role of the terrestrial biosphere and the oceans in the biogeochemical cycles of greenhouse gases; emissions of carbon dioxide to the atmosphere; long-term climate trends; the effects of elevated carbon dioxide on vegetation; and the vulnerability of coastal areas to rising sea level.

National Institute for Global Environmental Change

The National Institute for Global Environmental Change conducts research on global climate change in six U.S. regions: Great Plains, Midwest, Northeast, South Central, Southeast, and West. The Institute integrates and synthesizes information to help decision makers and communities better respond to the effects of climate change.

Each region has a "host institution," a prominent university that appoints a Regional Director who acts in an administrative capacity. Regional centers develop their own research programs by soliciting proposals from scholars throughout the nation. These programs must focus on areas important to global environmental change and must meet DOE's research priorities and the following criteria:
- Improve scientific understanding of global environmental and climate change issues.
- Reduce uncertainties surrounding key environmental and climate change science.
- Create experimental or observation programs to enhance the understanding of regional- or ecosystem-scale processes contributing to global change.
- Improve decision-making tools for resolving global environmental and climate change issues.

- Build education and training opportunities and develop new curriculum materials to increase the flow of talented scholars into global environmental change research areas.
- Focus contributions to public education on the subject of global climate change and other energy-related environmental risks.

Regional Roundtables

DOE held roundtable meetings with various segments of the energy industry to discuss implementing its planned energy partnership programs for energy efficiency. Workshop participants were asked to advise DOE's Office of Energy Efficiency and Renewable Energy about how to improve the quality of the individual program implementation plans, as well as the overall package of initiatives. Attendees represented manufacturers, builders, utility executives, engineers, and others who offered a variety of perspectives on the programs. These meetings were instrumental in shaping the final energy partnership programs, and many of the participants' suggestions were incorporated into the revised implementation plans.

Environmental Protection Agency

Following are some examples of EPA's numerous climate change outreach and education initiatives.

Business/Industry Outreach

EPA has taken various steps to engage business and industry on climate change-related issues. For example, EPA, the Risk and Insurance Management Society, Inc., the Federal Emergency Management Agency, NOAA, DOE, and the National Renewable Energy Laboratory co-sponsored a climate change and insurance roundtable in March 2000 to share information and ideas about the risks that climate change poses to the insurance industry and society. The roundtable provided insurance and financial executives with information about climate science and policy information. It also explored alternative risk management tools as a way to mitigate and adapt to the impacts of climate change. EPA also partnered with DOE to produce the publication *U.S. Insurance Industry Perspectives on Global Climate Change* (Mills et al. 2001).

Global Warming Site

Provided as a public service in support of EPA's mission to protect human health and the natural environment, the Global Warming Site strives to present accurate information on climate change and global warming in a way that is accessible and meaningful to all parts of society. The site is broken down into four main sections: climate (science), emissions, impacts, and actions. Updated daily to reflect the latest peer-reviewed science and policy information, the site contains over 2,000 content pages, as well as hundreds of official documents and publications. During 2001 the site averaged several hundred thousand page hits per month.

Outdoor/Wildlife Outreach

Since 1997 EPA has conducted climate change outreach activities for the outdoor recreation and wildlife enthusiast community. EPA staff have attended conferences and conventions of such diverse groups as Ducks Unlimited, the Izaak Walton League, the Wildlife Management Institute, the Federation of Fly Fishers, the National Association of Interpretation, and America Outdoors, distributing information about climate change science and impacts as they relate to the interests of each community. EPA has given presentations and conducted workshops at conventions and has contributed articles to the various groups' newsletters and magazines. To convey the vulnerabilities of specific recreational activities to the impacts of climate change, EPA has also developed targeted brochures and educational kits for use with the outdoor enthusiast audience. In 2002 EPA plans to release a toolkit for leaders of hunting and angling organizations to use with their constituencies.

Sea Level Rise Outreach

To meet U.S. obligations under the Framework Convention on Climate Change for taking measures to adapt to climate change, EPA supports a number of activities that encourage timely measures in anticipation of sea level rise. For example, EPA's continual recommendations to state and local governments to consider sea level rise within their ongoing initiatives has resulted in four states' passing regulations that ensure the inland migration of wetlands as sea level rises. A planning scenario mapping project is working with coastal planners to develop county-scale maps that illustrate where people are likely to hold back the sea and which areas are likely to flood. To stimulate dialogue within communities about how to prepare for sea level rise, EPA is developing brochures that explain the risks of sea level rise and also include the county-scale maps. Additionally, an outreach program to sand and gravel companies—who supply the fill material needed to elevate areas as the sea rises—is getting underway in one coastal state.

State and Local Climate Change Program

States and localities can play a significant role in promoting the reduction of greenhouse gases if they have the tools they need for assessing climate change issues in their daily decision making. By providing them with guidance and technical information about climate change, local air quality, and the health and economic benefits of reducing greenhouse gas emissions, EPA's State and Local Climate Change Program is enhancing the ability of state and local decision makers to comprehensively address their environmental and economic goals.

The program provides a variety of technical and outreach or education services and products related to clean air and climate change issues, including:
- assistance for states to analyze the co-benefits of mitigating greenhouse gases, developing and updating emission inventories, and assessing the impacts of climate change policies on state economies;
- new tools and models that build understanding of the broader benefits of climate protection and better integrate multi-emission reductions, as well as multi-goal (e.g., energy efficiency and renewable energy) strategies in state implementation plans submitted to EPA;
- capacity-building outreach through EPA's Web site, an electronic "listserv," and case studies;
- a best-practices clearinghouse to promote multi-emission reduction strategies, energy efficiency, sustainability, clean energy, and other greenhouse gas mitigation measures;
- information on state and local legislative activities related to greenhouse gases;
- state forest carbon data; and
- additional enhanced opportunities to promote state and local efforts, including creating success stories for wide dissemination and replication.

In 2000 the program distributed over 4,200 CD-ROM outreach kits to state and local leaders, providing information on voluntary strategies for reducing greenhouse gases. The kits are helping states and communities save money, improve air quality, lower risks to human health, and reduce traffic congestion, among other benefits. Their slide show on climate change is suitable for presentations to community groups, business organizations, and others. They also include more than 100 information sheets on climate change science, its potential impacts on each state, and technologies and policies that lower greenhouse gas emissions.

National Aeronautics and Space Administration

NASA's well-established outreach activities are designed to draw public and press attention to its work in the climate change arena.

Workshops for Journalists

NASA's co-sponsored workshops on global climate change provide science reporters with basic tutorials, information on major scientific advances, access to international science leaders, and opportunities to visit major scientific facilities. In 1999 NASA hosted its first Global Change Workshop for journalists in concert with the American Geophysical Union.

Media Directory for Global Change Experts

Published biennially, NASA's *Earth Observing System Global Change Media Directory* provides journalists with a ready source of international expertise on global climate change science and policy (NASA 2001). The directory contains contact information for more than 300 science experts available to the media in climate change, natural hazards, ozone, water resources, global warming, and many other areas. It is available on-line and is searchable by topic, name, affiliation, or location.

Earth Observatory On-Line Newsroom

NASA's on-line newsroom for journalists features the latest news on Earth science research released from all NASA centers and more than 80 universities participating in NASA's Earth programs through sponsored research. Resources updated weekly include media announcements, summaries of headline news, listings of newly published research, a searchable directory of experts, and selected writers' guides.

National Park Service

As the guardian of the world's finest system of national parks, the National Park Service applies innovative techniques to reach out to and actively involve diverse audiences in preserving and restoring our nation's parks. Following are some examples of the Park Service's increased support of education on global warming and environmental stewardship.

Environmental Leadership Program

As part of its Environmental Leadership Program, the Park Service has turned Utah's Zion National Park visitor center into a model environmentally sus-

tainable facility. The new center incorporates passive solar design to reduce overall energy consumption and uses only 80 percent of the energy required for other national park visitor centers. The center also receives 30 percent of its total electricity needs from solar power. Through an innovative transportation agreement with the nearby town of Bonneville, visitors can reduce fuel consumption by parking in town and riding alternative-fueled buses to the park.

Green Energy Parks

Green Energy Parks focuses on conserving energy and incorporating renewable-energy resources into the national park system to save money in park operations, as well as to promote more environmentally friendly facilities. The Park Service educates its visitors about its sustainable environment efforts through a combination of sign- age, brochures, and fact sheets. For example, at Lake Meade, Nevada, the Park Service has turned the park entrance tollbooth into a state-of-the-art, renewable-energy facility that is powered solely by the building's photovoltaic roof panels. Road signs describe the facility to drivers and explain the technology's environmental benefits.

National Oceanic and Atmospheric Administration

Several NOAA offices are significantly contributing to climate change and weather-related research education and public outreach efforts.

National Climatic Data Center

NOAA's National Climatic Data Center maintains a vast database of weather-related information used by specialists in meteorology, insurance, and agriculture and by various business sectors. The center provides information through special reports and its Web site.

The National Climatic Prediction Center

NOAA's National Climatic Prediction Center recently developed climate outlook products to help farmers, businesses, and the public better plan for extreme weather events related

to variations in climate. The new products are available on the center's Expert Assessment Web page at http://www.cpc.ncep.noaa.gov/products/expert_assessment/. They include drought, hurricane, and winter outlooks, along with an El Niño–Southern Oscillation advisories and threat assessments. The center also maintains a climate educational Web site.

National Geophysical Data Center

NOAA's National Geophysical Data Center's primary mission is data management. The center plays a leading role in the nation's research into the environment, while providing public domain data to a wide group of users. It features a Web site on paleoclimate at http://www.ngdc.noaa.gov/paleo/global warming/home.html, which was developed both to help educate, inform, and highlight the importance of paleoclimate research and to illustrate how paleoclimate research relates to global warming and other important issues of climate variability and change.

Office of Global Programs

NOAA's Office of Global Programs (OGP) released the fourth of its *Reports to the Nation* series in 1997. The reports offer educators and the public a clear understanding of complex atmospheric phenomena, such as El Niño, the ozone layer, and climate change. Through a grant to the Lamont–Doherty Earth Observatory, OGP produced a public fact sheet on the North Atlantic Oscillation. OGP also created a special climate Web page to make NOAA's climate information more accessible to the general public.

During the 1997–98 El Niño and 1998–99 La Niña, OGP and the National Climatic Prediction Center worked closely with the Federal Emergency Management Agency, state agencies, and the press to educate the public about seasonal climate variability, the importance of advisories of El Niño–Southern Oscillations and other seasonal and decadal oscillations to our daily lives, and the need to prepare for related extreme weather events.

Smithsonian Institution

Every year the Smithsonian Institution's exhibits educate millions of U.S. and foreign visitors about many areas of science, including global warming.

Understanding the Forecast: Global Warming

Originally shown at New York's American Museum of Natural History, this exhibit was updated by the Smithsonian in the summer of 1997 at the National Museum of Natural History in Washington, D.C. Nearly 443,000 visitors passed through the exhibit that summer, and many more viewed it on its nationwide tour. The exhibit's interactive displays provided information on climate change science and explained the connections between our daily use of electricity, gasoline, and consumer products and greenhouse gas emissions. The displays also demonstrated how we can reduce our individual contributions to greenhouse gas emissions.

Under the Sun: An Outdoor Exhibition of Light

Tens of thousands of visitors viewed the Cooper Hewitt's outdoor solar energy exhibit in the gardens of the museum's Andrew Carnegie mansion in New York City. The Smithsonian later sent the exhibit on tour to other cities, including a summer stay in the gardens behind the Smithsonian's castle on the Mall in Washington, D.C. The exhibit demonstrated how solar energy systems can meet architectural and design preferences, while providing energy that reduces pollution and greenhouse gas emissions. The exhibit script paid special attention to helping visitors understand how energy consumption is linked to global warming. Both federal agencies and private industry partners helped fund the exhibit.

Forces of Change

The Smithsonian is working on an exhibit that examines the geological, environmental, and cultural processes that have shaped and continue to change our world. It consists of a permanent exhibit hall at the Smithsonian's National Museum of Natural History, traveling exhibitions, publications, interactive computer products, and public programs, including a lecture series and electronic classroom courses. Opened in the summer of 2001, the exhibit is expected to be seen by six million museum visitors annually. Its outreach programs and materials will reach additional millions throughout the nation. The exhibit's supporters include NASA, the W.K. Kellogg Foundation, USDA, the Mobil Foundation, Inc., the American Farmland Trust, EPA, and the U.S. Global Change Research Program.

Global Links

As part of its Forces of Change program, the Smithsonian is developing the Global Links exhibit, designed to tell a series of global climate change stories. The first story will explore El Niño and its possible links to global warming. The second story will examine greenhouse gases and the ozone hole. An EPA grant has supported preliminary planning of the Global Links exhibit.

Antarctica Exhibit

The National Museum of Natural History is seeking funding for an exhibit that explores how research in Antarctica allows us to learn more about global climate change in the past and to improve predictions for future change. The exhibit is scheduled to open in June 2003.

Partnerships

Government organizations with joint interests in climate change have formed partnerships to educate the public about climate change and to offer suggestions for how individuals and communities can help reduce its risks. Following are some examples.

It All Adds Up to Cleaner Air

This collaborative effort of the U.S. Department of Transportation and EPA is informing the public about the connections between their transportation choices, traffic congestion, and air pollution. The program emphasizes simple, convenient actions people can take that can improve air quality when practiced on a wide scale.

Outdoor Interpreter's Tool Kit

EPA led a partnership effort with the National Park Service, the U.S. Fish and Wildlife Service, and NOAA to develop a climate change educational toolkit CD-ROM for park wildlife interpreters (U.S. EPA and NPS 2001). The kit provides interpreters with fact sheets and presentation materials that investigate the links between climate change and changes to habitat, ecosystems, wildlife, and our national parks. The partnership also produced a climate change video that will inform park visitors about climate change and its impacts on national parks. Released early in 2002, the kit includes other outreach materials, such as Park Service climate change bookmarks.

Reporter's Guide on Climate Change

Supported by NOAA and DOE, the nonprofit National Safety Council's Environmental Health Center produced a second-edition guide for journalists on climate change in 2000 (NSC 2000). *Reporting on Climate Change: Understanding the Science* is part of a series of reporters' guides designed to enhance public understanding of the significant environmental health risks and challenges facing modern society. Based on the findings of the 1995 Intergovernmental Panel on Climate Change assessment report, the guide explains major global warming issues in detail, as well as broader strategies for successful science reporting, interaction with the scientific community, and understanding scientific reporting methods. The guide also contains a glossary and list of public and private information sources and Web links.

Appendix A
Emission Trends

TABLE 10 EMISSIONS TRENDS (CO₂)
(Sheet 1 of 5)

United States of America
1999
Submission 2001

GREENHOUSE GAS SOURCE AND SINK CATEGORIES	Base year[1]	1990	1991	1992	1993	1994	1995	1996	1997	1998	1999
					(Gg)						
1. Energy	0.00	4,840,609.87	4,788,679.21	4,887,348.69	4,996,714.56	5,085,608.10	5,134,849.78	5,315,958.31	5,386,939.02	5,397,690.27	5,664,788.74
A. Fuel Combustion (Sectoral Approach)	0.00	4,835,688.49	4,782,408.64	4,881,064.68	4,986,699.51	5,078,407.35	5,121,262.89	5,302,960.54	5,374,912.89	5,386,761.75	5,453,088.14
1. Energy Industries		1,757,344.40	1,736,958.54	1,735,395.83	1,793,606.14	1,811,882.58	1,810,564.91	1,880,288.04	1,953,514.41	2,010,670.18	1,953,352.50
2. Manufacturing Industries and Construction		1,023,471.15	1,007,631.47	1,064,905.86	1,062,395.51	1,090,871.96	1,101,048.15	1,140,592.33	1,141,145.94	1,113,318.51	1,155,609.76
3. Transport		1,422,584.88	1,386,224.43	1,425,797.26	1,456,126.29	1,499,862.42	1,537,307.47	1,571,795.55	1,588,137.82	1,615,078.66	1,677,713.82
4. Other Sectors		549,373.48	560,839.92	570,245.56	588,643.73	580,616.65	586,246.09	623,239.13	608,568.63	558,376.01	577,133.24
5. Other		82,914.58	90,754.29	84,720.16	85,927.83	93,173.74	86,096.26	87,045.49	83,546.09	89,318.39	89,278.82
B. Fugitive Emissions from Fuels	0.00	5,121.37	6,270.57	6,276.01	10,015.05	10,200.75	13,586.90	12,997.77	12,026.14	10,838.53	11,700.60
1. Solid Fuels		0.00	0.00	0.00	0.00	0.00	0.00	0.00	0.00	0.00	0.00
2. Oil and Natural Gas		5,121.37	6,270.57	6,276.01	10,015.05	10,200.75	13,586.90	12,997.77	12,026.14	10,838.53	11,700.60
2. Industrial Processes	0.00	54,565.48	53,318.67	53,655.59	55,263.81	58,199.44	61,895.21	63,272.84	66,063.41	66,984.48	67,401.42
A. Mineral Products		53,765.48	52,478.67	52,773.59	54,352.09	57,301.87	60,926.88	62,132.51	64,769.82	65,571.09	65,829.18
B. Chemical Industry		800.00	840.00	882.00	911.72	897.57	968.33	1,140.33	1,293.59	1,413.39	1,572.24
C. Metal Production		0.00	0.00	0.00	0.00	0.00	0.00	0.00	0.00	0.00	0.00
D. Other Production		NE	NE	NE	NE	NE	NE	NE	NE	NE	NE
E. Production of Halocarbons and SF₆											
F. Consumption of Halocarbons and SF₆		0.00	0.00	0.00	0.00	0.00	0.00	0.00	0.00	0.00	0.00
G. Other		0.00	0.00	0.00	0.00	0.00	0.00	0.00	0.00	0.00	0.00
3. Solvent and Other Product Use	0.00	0.00	0.00	0.00	0.00	0.00	0.00	0.00	0.00	0.00	0.00
4. Agriculture	0.00	0.00	0.00	0.00	0.00	0.00	0.00	0.00	0.00	0.00	0.00
A. Enteric Fermentation		0.00	0.00	0.00	0.00	0.00	0.00	0.00	0.00	0.00	0.00
B. Manure Management		0.00	0.00	0.00	0.00	0.00	0.00	0.00	0.00	0.00	0.00
C. Rice Cultivation		0.00	0.00	0.00	0.00	0.00	0.00	0.00	0.00	0.00	0.00
D. Agricultural Soils[2]		NA	NA	NA	NA	NA	NA	NA	NA	NA	NA
E. Prescribed Burning of Savannas		0.00	0.00	0.00	0.00	0.00	0.00	0.00	0.00	0.00	0.00
F. Field Burning of Agricultural Residues		0.00	0.00	0.00	0.00	0.00	0.00	0.00	0.00	0.00	0.00
G. Other		0.00	0.00	0.00	0.00	0.00	0.00	0.00	0.00	0.00	0.00
5. Land-Use Change and Forestry[3]	0.00	-1,059,900.00	-1,046,800.00	-1,297,466.67	-1,024,200.00	-1,028,590.00	-1,019,600.00	-1,021,400.00	-981,900.00	-983,400.00	-990,400.00
A. Changes in Forest and Other Woody Biomass Stocks		-750,200.00	-738,100.00	-711,300.00	-712,300.00	-719,000.00	-711,600.00	-716,000.00	-718,700.00	-712,500.00	-720,900.00
B. Forest and Grassland Conversion		0.00	0.00	0.00	0.00	0.00	0.00	0.00	0.00	0.00	0.00
C. Abandonment of Managed Lands		0.00	0.00	0.00	0.00	0.00	0.00	0.00	0.00	0.00	0.00
D. CO₂ Emissions and Removals from Soil		-291,900.00	-291,200.00	-569,066.67	-296,600.00	-295,900.00	-295,400.00	-295,400.00	-253,800.00	-262,100.00	-261,800.00
E. Other		-17,800.00	-17,500.00	-17,100.00	-15,300.00	-13,600.00	-12,000.00	-10,000.00	-9,400.00	-8,800.00	-7,700.00
6. Waste	0.00	17,571.80	19,225.29	19,933.61	20,952.94	22,009.61	23,065.35	23,968.46	25,674.27	25,144.74	25,959.73
A. Solid Waste Disposal on Land		0.00	0.00	0.00	0.00	0.00	0.00	0.00	0.00	0.00	0.00
B. Waste-water Handling		0.00	0.00	0.00	0.00	0.00	0.00	0.00	0.00	0.00	0.00
C. Waste Incineration		17,571.80	19,225.29	19,933.61	20,952.94	22,009.61	23,065.35	23,968.46	25,674.27	25,144.74	25,959.73
D. Other		0.00	0.00	0.00	0.00	0.00	0.00	0.00	0.00	0.00	0.00
7. Other (please specify)	0.00	NA	NA	NA	NA	NA	NA	NA	NA	NA	NA
Total Emissions/Removals with LUCF[4]	0.00	3,853,847.14	3,814,423.18	3,663,463.21	4,048,731.30	4,140,317.14	4,200,810.34	4,381,799.61	4,496,776.71	4,506,329.49	4,567,749.89
Total Emissions without LUCF[4]	0.00	4,912,947.14	4,861,223.18	4,960,929.88	5,072,931.30	5,168,817.14	5,219,810.34	5,403,199.61	5,478,676.71	5,489,729.49	5,558,149.89
Memo Items:											
International Bankers	0.00	114,000.92	120,019.82	109,964.95	99,885.75	98,016.54	101,014.03	102,197.42	109,788.01	112,771.27	107,345.46
Aviation		46,728.43	46,681.66	47,142.96	47,615.43	48,327.04	51,093.40	52,135.04	55,899.42	54,987.80	60,969.72
Marine		67,272.49	73,337.36	62,821.98	52,270.33	49,689.50	49,920.63	50,062.38	53,888.59	57,783.48	46,375.74
Multilateral Operations		NE	NE	NE	NE	NE	NE	NE	NE	NE	NE
CO₂ Emissions from Biomass		180,562.60	179,318.10	188,276.17	183,975.48	191,666.84	200,488.33	202,117.50	194,316.62	194,762.26	234,062.51

(1) Fill in the base year adopted by the Party under the Convention, if different from 1990.

(2) See footnote 4 to Summary 1.A of this common reporting format.

(3) Take the net emissions as reported in Summary 1.A of this common reporting format. Please note that for the purposes of reporting, the signs for uptake are always (-) and for emissions (+).

(4) The information in these rows is requested to facilitate comparison of data, since Parties differ in the way they report CO₂ emissions and removals from Land-Use Change and Forestry.

TABLE 10 EMISSIONS TRENDS (CH₄)
(Sheet 2 of 5)

United States of America
1999
Submission 2001

GREENHOUSE GAS SOURCE AND SINK CATEGORIES	Base year[1]	1990	1991	1992	1993	1994	1995	1996	1997	1998	1999
					(Gg)						
Total Emissions	**0.00**	**30,688.55**	**30,605.64**	**30,883.05**	**30,389.09**	**30,784.20**	**30,977.73**	**30,378.63**	**30,095.69**	**29,754.00**	**29,504.02**
1. Energy	**0.00**	**11,890.79**	**11,770.17**	**11,686.86**	**11,248.36**	**11,140.66**	**11,284.29**	**11,095.51**	**10,867.82**	**10,669.40**	**10,388.16**
A. Fuel Combustion (Sectoral Approach)	0.00	640.57	644.37	661.22	637.84	637.61	654.05	658.96	611.01	580.00	601.03
1. Energy Industries		22.98	22.73	22.32	23.21	23.39	22.70	23.47	24.53	26.12	25.34
2. Manufacturing Industries and Construction		128.52	126.12	130.14	132.28	137.29	140.64	143.32	145.30	143.83	156.99
3. Transport		237.42	234.84	237.24	236.22	234.98	231.98	228.49	224.73	218.65	214.72
4. Other Sectors		251.65	260.67	271.52	246.14	241.96	258.72	263.66	216.45	191.39	203.98
5. Other		0.00	0.00	0.00	0.00	0.00	0.00	0.00	0.00	0.00	0.00
B. Fugitive Emissions from Fuels	0.00	11,250.22	11,125.80	11,025.64	10,610.52	10,503.04	10,630.24	10,436.55	10,256.81	10,089.40	9,787.14
1. Solid Fuels		4,183.70	3,975.39	3,834.90	3,355.84	3,389.85	3,550.02	3,301.04	3,274.14	3,167.64	2,944.20
2. Oil and Natural Gas		7,066.52	7,150.40	7,190.73	7,254.68	7,113.19	7,080.22	7,135.52	6,982.67	6,921.76	6,842.94
2. Industrial Processes	**0.00**	**56.76**	**57.62**	**60.97**	**66.63**	**71.25**	**72.41**	**75.86**	**77.30**	**77.66**	**80.12**
A. Mineral Products		0.00	0.00	0.00	0.00	0.00	0.00	0.00	0.00	0.00	0.00
B. Chemical Industry		56.76	57.62	60.97	66.63	71.25	72.41	75.86	77.30	77.66	80.12
C. Metal Production		0.00	0.00	0.00	0.00	0.00	0.00	0.00	0.00	0.00	0.00
D. Other Production		0.00	0.00	0.00	0.00	0.00	0.00	0.00	0.00	0.00	0.00
E. Production of Halocarbons and SF₆											
F. Consumption of Halocarbons and SF₆											
G. Other		0.00	0.00	0.00	0.00	0.00	0.00	0.00	0.00	0.00	0.00
3. Solvent and Other Product Use	**0.00**	**0.00**	**0.00**	**0.00**	**0.00**	**0.00**	**0.00**	**0.00**	**0.00**	**0.00**	**0.00**
4. Agriculture	**0.00**	**7,862.31**	**7,881.17**	**8,101.50**	**7,935.85**	**8,407.71**	**8,445.64**	**8,205.28**	**8,207.84**	**8,258.84**	**8,232.39**
A. Enteric Fermentation		6,166.13	6,143.04	6,288.80	6,159.82	6,447.30	6,491.60	6,295.47	6,171.80	6,072.14	6,057.20
B. Manure Management		1,256.28	1,310.57	1,326.20	1,337.28	1,445.19	1,477.44	1,463.04	1,552.70	1,676.61	1,637.82
C. Rice Cultivation		414.44	403.90	458.52	416.41	485.33	452.18	418.77	454.78	480.52	509.19
D. Agricultural Soils		0.00	0.00	0.00	0.00	0.00	0.00	0.00	0.00	0.00	0.00
E. Prescribed Burning of Savannas		0.00	0.00	0.00	0.00	0.00	0.00	0.00	0.00	0.00	0.00
F. Field Burning of Agricultural Residues		25.46	23.67	27.97	22.34	29.89	24.42	28.00	28.56	29.56	28.18
G. Other		0.00	0.00	0.00	0.00	0.00	0.00	0.00	0.00	0.00	0.00
5. Land-Use Change and Forestry	**0.00**	**0.00**	**0.00**	**0.00**	**0.00**	**0.00**	**0.00**	**0.00**	**0.00**	**0.00**	**0.00**
A. Changes in Forest and Other Woody Biomass Stocks		0.00	0.00	0.00	0.00	0.00	0.00	0.00	0.00	0.00	0.00
B. Forest and Grassland Conversion		0.00	0.00	0.00	0.00	0.00	0.00	0.00	0.00	0.00	0.00
C. Abandonment of Managed Lands		0.00	0.00	0.00	0.00	0.00	0.00	0.00	0.00	0.00	0.00
D. CO₂ Emissions and Removals from Soil		0.00	0.00	0.00	0.00	0.00	0.00	0.00	0.00	0.00	0.00
E. Other		0.00	0.00	0.00	0.00	0.00	0.00	0.00	0.00	0.00	0.00
6. Waste	**0.00**	**10,878.69**	**10,896.68**	**11,033.72**	**11,138.26**	**11,164.58**	**11,175.39**	**11,001.97**	**10,942.73**	**10,748.11**	**10,803.34**
A. Solid Waste Disposal on Land		10,345.89	10,358.17	10,489.01	10,587.63	10,608.52	10,614.01	10,435.38	10,370.69	10,170.77	10,220.79
B. Waste-water Handling		532.80	538.52	544.72	550.63	556.07	561.38	566.60	572.04	577.33	582.56
C. Waste Incineration		0.00	0.00	0.00	0.00	0.00	0.00	0.00	0.00	0.00	0.00
D. Other		0.00	0.00	0.00	0.00	0.00	0.00	0.00	0.00	0.00	0.00
7. Other (please specify)	**0.00**	**0.00**	**0.00**	**0.00**	**NA**	**0.00**	**0.00**	**0.00**	**0.00**	**0.00**	**0.00**
Memo Items:											
International Bunkers	**0.00**	**1.91**	**1.96**	**1.88**	**1.79**	**1.79**	**1.87**	**1.90**	**2.04**	**2.05**	**2.11**
Aviation		1.28	1.28	1.30	1.31	1.33	1.41	1.44	1.54	1.52	1.68
Marine		0.62	0.68	0.58	0.48	0.46	0.46	0.46	0.50	0.53	0.43
Multilateral Operations		NE	NE	NE	NE	NE	NE	NE	NE	NE	NE
CO₂ Emissions from Biomass											

TABLE 10 EMISSIONS TRENDS (N₂O)
(Sheet 3 of 5)

United States of America
1999
Submission 2001

GREENHOUSE GAS SOURCE AND SINK CATEGORIES	Base year[1]	1990	1991	1992	1993	1994	1995	1996	1997	1998	1999
						(Gg)						
Total Emissions	**0.00**	**1,280.23**	**1,304.18**	**1,339.98**	**1,341.59**	**1,429.59**	**1,393.58**	**1,424.61**	**1,432.57**	**1,399.00**	**1,395.45**
1. Energy	**0.00**	**219.05**	**229.10**	**243.14**	**253.11**	**259.94**	**261.58**	**258.61**	**258.73**	**255.77**	**255.01**
A. Fuel Combustion (Sectoral Approach)	0.00	219.05	229.10	243.14	253.11	259.94	261.58	258.61	258.73	255.77	255.01
1. Energy Industries		23.82	23.77	24.03	24.97	25.19	25.15	26.53	27.37	27.98	27.89
2. Manufacturing Industries and Construction		15.52	15.04	15.57	15.59	16.28	16.37	16.70	17.03	16.92	18.79
3. Transport		175.15	185.61	198.73	208.06	214.05	215.42	210.61	210.22	207.17	204.42
4. Other Sectors		4.56	4.68	4.82	4.50	4.42	4.64	4.78	4.11	3.69	3.92
5. Other		0.00	0.00	0.00	0.00	0.00	0.00	0.00	0.00	0.00	0.00
B. Fugitive Emissions from Fuels	0.00	0.00	0.00	0.00	0.00	0.00	0.00	0.00	0.00	0.00	0.00
1. Solid Fuels		0.00	0.00	0.00	0.00	0.00	0.00	0.00	0.00	0.00	0.00
2. Oil and Natural Gas		0.00	0.00	0.00	0.00	0.00	0.00	0.00	0.00	0.00	0.00
2. Industrial Processes	**0.00**	**116.60**	**119.46**	**115.91**	**121.36**	**128.71**	**129.73**	**133.87**	**123.66**	**90.79**	**94.23**
A. Mineral Products		0.00	0.00	0.00	0.00	0.00	0.00	0.00	0.00	0.00	0.00
B. Chemical Industry		116.60	119.46	115.91	121.36	128.71	129.73	133.87	123.66	90.79	94.23
C. Metal Production		0.00	0.00	0.00	0.00	0.00	0.00	0.00	0.00	0.00	0.00
D. Other Production		0.00	0.00	0.00	0.00	0.00	0.00	0.00	0.00	0.00	0.00
E. Production of Halocarbons and SF₆											
F. Consumption of Halocarbons and SF₆											
G. Other		0.00	0.00	0.00	0.00	0.00	0.00	0.00	0.00	0.00	0.00
3. Solvent and Other Product Use	**0.00**	**0.00**	**0.00**	**0.00**	**0.00**	**0.00**	**0.00**	**0.00**	**0.00**	**0.00**	**0.00**
4. Agriculture	**0.00**	**920.64**	**931.29**	**956.22**	**942.07**	**1,014.97**	**974.84**	**1,006.12**	**1,023.84**	**1,025.54**	**1,019.02**
A. Enteric Fermentation		0.00	0.00	0.00	0.00	0.00	0.00	0.00	0.00	0.00	0.00
B. Manure Management		51.77	53.40	52.74	54.02	54.11	53.14	54.46	55.22	55.34	55.43
C. Rice Cultivation		0.00	0.00	0.00	0.00	0.00	0.00	0.00	0.00	0.00	0.00
D. Agricultural Soils		867.70	876.76	902.18	886.98	959.43	920.51	950.32	967.20	968.74	962.20
E. Prescribed Burning of Savannas		0.00	0.00	0.00	0.00	0.00	0.00	0.00	0.00	0.00	0.00
F. Field Burning of Agricultural Residues		1.17	1.13	1.30	1.07	1.43	1.19	1.34	1.42	1.46	1.39
G. Other		0.00	0.00	0.00	0.00	0.00	0.00	0.00	0.00	0.00	0.00
5. Land-Use Change and Forestry	**0.00**	**0.00**	**0.00**	**0.00**	**0.00**	**0.00**	**0.00**	**0.00**	**0.00**	**0.00**	**0.00**
A. Changes in Forest and Other Woody Biomass Stocks		0.00	0.00	0.00	0.00	0.00	0.00	0.00	0.00	0.00	0.00
B. Forest and Grassland Conversion		0.00	0.00	0.00	0.00	0.00	0.00	0.00	0.00	0.00	0.00
C. Abandonment of Managed Lands		0.00	0.00	0.00	0.00	0.00	0.00	0.00	0.00	0.00	0.00
D. CO₂ Emissions and Removals from Soil		0.00	0.00	0.00	0.00	0.00	0.00	0.00	0.00	0.00	0.00
E. Other		0.00	0.00	0.00	0.00	0.00	0.00	0.00	0.00	0.00	0.00
6. Waste	**0.00**	**23.94**	**24.33**	**24.70**	**25.04**	**25.96**	**27.41**	**26.00**	**26.34**	**26.90**	**27.19**
A. Solid Waste Disposal on Land		0.00	0.00	0.00	0.00	0.00	0.00	0.00	0.00	0.00	0.00
B. Waste-water Handling		23.03	23.57	23.83	24.20	25.08	26.52	25.11	25.51	26.14	26.40
C. Waste Incineration		0.92	0.76	0.87	0.84	0.88	0.89	0.89	0.83	0.77	0.80
D. Other		0.00	0.00	0.00	0.00	0.00	0.00	0.00	0.00	0.00	0.00
7. Other (please specify)	**0.00**	**0.00**	**0.00**	**0.00**	**0.00**	**0.00**	**0.00**	**0.00**	**0.00**	**0.00**	**0.00**
Memo Items:											
International Bunkers	**0.00**	**3.13**	**3.28**	**3.04**	**2.80**	**2.76**	**2.85**	**2.89**	**3.10**	**3.17**	**3.08**
Aviation		1.48	1.47	1.49	1.51	1.53	1.62	1.65	1.77	1.74	1.93
Marine		1.66	1.81	1.55	1.29	1.22	1.23	1.23	1.33	1.43	1.14
Multilateral Operations		NA	NA	NA	NA	NA	NA	NA	NA	NA	NE
CO₂ Emissions from Biomass		NE	NE	NE	NE	NE	NE	NE	NE	NE	NE

TABLE 10 EMISSION TRENDS (HFCs, PFCs and SF₆)
(Sheet 4 of 5)

United States of America
1999
Submission 2001

GREENHOUSE GAS SOURCE AND SINK CATEGORIES	Base year[1]	1990	1991	1992	1993	1994	1995	1996	1997	1998	1999
						(Gg)						
Emissions of HFCs[5] - CO_2 equivalent (Gg)	0.00	35,743.33	31,638.33	36,319.02	37,139.33	41,474.69	51,098.74	65,177.41	72,236.56	89,647.65	87,105.19
HFC-23		2.97	2.63	2.97	2.73	2.69	2.32	2.67	2.58	3.44	2.62
HFC-32*		0.00	0.00	0.00	0.00	0.00	0.00	0.00	0.01	0.01	0.02
HFC-41		1.40	0.70	0.29	0.65	5.14	23.80	43.88	50.93	58.65	62.96
HFC-43-10mee		0.00	0.00	0.00	0.00	0.00	0.00	0.00	0.00	0.00	0.00
HFC-125		0.00	0.00	0.24	0.48	0.30	0.48	0.67	0.89	1.12	1.29
HFC-134		0.00	0.00	0.00	0.00	0.00	0.00	0.00	0.00	0.00	0.00
HFC-134a		0.56	0.56	0.63	2.88	6.28	14.35	18.96	23.48	26.85	30.34
HFC-152a		0.00	0.00	0.00	0.00	0.00	0.00	0.00	0.00	0.00	0.00
HFC-143		0.00	0.00	0.00	0.01	0.05	0.11	0.21	0.33	0.49	0.68
HFC-143a		0.00	0.00	0.00	0.00	0.00	0.00	0.00	0.00	0.00	0.00
HFC-227ea		0.00	0.00	0.00	0.00	0.00	0.00	0.00	0.00	0.00	0.00
HFC-236fa		0.00	0.00	0.00	0.00	0.00	0.00	0.00	0.01	0.12	0.21
HFC-245ca		0.00	0.00	0.00	0.00	0.00	0.00	0.00	0.00	0.00	0.00
Emissions of PFCs[5] - CO_2 equivalent (Gg)	0.00	22,158.83	20,113.92	19,157.78	17,664.34	15,694.35	16,744.92	18,615.77	17,843.50	16,927.78	16,843.57
CF₄		2.58	2.31	2.18	1.89	1.56	1.54	1.59	1.49	1.39	1.38
C₂F₆		0.59	0.55	0.54	0.58	0.60	0.74	0.90	0.89	0.86	0.85
C₃F₈		0.00	0.00	0.00	0.00	0.00	0.00	0.00	0.00	0.00	0.00
C₄F₁₀		0.00	0.00	0.00	0.00	0.00	0.00	0.00	0.00	0.00	0.00
c-C₄F₈		0.00	0.00	0.00	0.00	0.00	0.00	0.00	0.00	0.00	0.00
C₅F₁₂		0.00	0.00	0.00	0.00	0.00	0.00	0.00	0.00	0.00	0.00
C₆F₁₄		0.00	0.00	0.00	0.00	0.00	0.00	0.00	0.00	0.00	0.00
Emissions of SF₆[5] - CO_2 equivalent (Gg)	0.00	26,000.00	27,040.00	28,080.00	29,020.00	29,760.00	31,200.00	31,300.00	33,200.00	32,000.00	31,800.00
SF₆		1.09	1.13	1.17	1.21	1.25	1.31	1.31	1.39	1.34	1.33

Chemical	GWP
HFCs	
HFC-23	11700
HFC-32	650
HFC-41	150
HFC-43-10mee	1300
HFC-125	2800
HFC-134	1000
HFC-134a	1300
HFC-152a	140
HFC-143	300
HFC-143a	3800
HFC-227ea	2900
HFC-236fa	6300
HFC-245ca	560
PFCs	
CF₄	6500
C₂F₆	9200
C₃F₈	7000
C₄F₁₀	7000
c-C₄F₈	8700
C₅F₁₂	7500
C₆F₁₄	7400
SF₆	23900

(5) Enter information on the actual emissions. Where estimates are only available for the potential emissions, specify this in a comment to the corresponding cell. Only in this row the emissions are expressed as CO_2 equivalent emissions in order to facilitate data flow among spreadsheets.

* Represents a weighted total of confidential data, which includes emissions of HFC-152a, HFC-227ea, HFC-4310mee, and PFC/PFPEs, the latter being a proxy for a diverse collection of PFCs and perfluoropolyethers (PFPEs) employed for solvent applications.

TABLE 10 EMISSION TRENDS (SUMMARY)
(Sheet 5 of 5)

GREENHOUSE GAS EMISSIONS	Base year[1]	1990	1991	1992	1993	1994	1995	1996	1997	1998	1999
					CO_2 equivalent (Gg)						
Net CO_2 emissions/removals	0.00	3,853,047.14	3,814,423.18	3,663,463.21	4,048,731.30	4,140,317.14	4,200,810.34	4,381,799.61	4,496,776.71	4,506,329.49	4,567,749.89
CO_2 emissions (without LUCF) [6]	0.00	4,912,947.14	4,861,223.18	4,960,929.88	5,072,931.30	5,168,817.14	5,219,810.34	5,403,199.61	5,478,676.71	5,489,729.49	5,558,149.89
CH_4	0.00	644,459.49	642,718.35	648,544.07	638,170.89	646,468.10	650,532.39	637,951.13	632,009.44	624,834.01	619,584.42
N_2O	0.00	396,871.55	404,295.51	415,393.49	415,893.01	443,172.61	432,008.27	441,628.52	444,097.12	433,690.74	432,589.31
HFCs	0.00	35,743.33	31,638.33	36,319.02	37,139.33	41,474.69	51,098.74	65,177.41	72,236.56	89,647.65	87,105.19
PFCs	0.00	22,158.83	20,113.92	19,157.78	17,664.34	15,694.35	16,744.92	18,615.77	17,843.50	16,927.78	16,843.57
SF_6	0.00	26,000.00	27,040.00	28,080.00	29,020.00	29,760.00	31,200.00	31,300.00	33,200.00	32,000.00	31,800.00
Total (with net CO_2 emissions/removals)	0.00	4,978,280.35	4,940,229.29	4,810,957.58	5,186,618.87	5,316,886.90	5,382,394.66	5,576,472.43	5,696,163.31	5,703,429.67	5,755,672.39
Total (without CO_2 from LUCF) [6]	0.00	6,038,180.35	5,987,029.29	6,108,424.25	6,210,818.87	6,345,386.90	6,401,394.66	6,597,872.43	6,678,063.31	6,686,829.67	6,746,072.39

GREENHOUSE GAS SOURCE AND SINK CATEGORIES	Base year[1]	1990	1991	1992	1993	1994	1995	1996	1997	1998	1999
					CO_2 equivalent (Gg)						
1. Energy	0.00	5,158,421.65	5,106,874.47	5,208,139.74	5,311,395.73	5,403,144.43	5,452,911.08	5,629,134.60	5,695,369.18	5,700,945.25	5,761,992.91
2. Industrial Processes	0.00	175,805.88	170,352.27	174,425.72	178,109.52	186,524.87	202,677.06	221,459.27	229,301.96	235,335.22	234,043.92
3. Solvent and Other Product Use	0.00	0.00	0.00	0.00	0.00	0.00	0.00	0.00	0.00	0.00	0.00
4. Agriculture	0.00	450,505.62	454,204.66	466,559.50	458,695.14	491,203.28	479,560.17	484,209.46	489,753.99	491,353.71	488,775.40
5. Land-Use Change and Forestry [7]	0.00	-1,059,990.00	-1,046,800.00	-1,297,466.67	-1,024,200.00	-1,028,500.00	-1,019,000.00	-1,021,400.00	-981,900.00	-983,400.00	-990,400.00
6. Waste	0.00	253,447.20	255,597.89	259,299.29	262,618.49	264,514.32	266,246.34	263,069.11	263,638.19	259,195.50	261,260.16
7. Other	0.00	0.00	0.00	0.00	0.00	0.00	0.00	0.00	0.00	0.00	0.00

[6] The information in these rows is requested to facilitate comparison of data, since Parties differ in the way they report CO_2 emissions and removals from Land-Use Change and Forestry.

[7] Net emissions.

SUMMARY 2 SUMMARY REPORT FOR CO₂ EQUIVALENT EMISSIONS
(Sheet 1 of 1)

United States of America
1999
Submission 2001

GREENHOUSE GAS SOURCE AND SINK CATEGORIES	CO₂ [1]	CH₄	N₂O	HFCs	PFCs	SF₆	Total
	CO₂ equivalent (Gg)						
Total (Net Emissions) [1]	4,567,749.89	619,584.42	432,589.31	87,105.19	16,843.57	31,800.00	5,755,672.39
1. Energy	5,464,788.74	218,151.43	79,052.75				5,761,992.91
A. Fuel Combustion (Sectoral Approach)	5,453,088.14	12,621.56	79,052.75				5,544,762.45
1. Energy Industries	1,953,352.50	532.12	8,645.76				1,962,530.39
2. Manufacturing Industries and Construction	1,155,609.76	3,296.79	5,823.37				1,164,729.92
3. Transport	1,677,713.82	4,509.02	63,369.66				1,745,592.50
4. Other Sectors	577,133.24	4,283.63	1,213.95				582,630.82
5. Other	89,278.82	0.00	0.00				89,278.82
B. Fugitive Emissions from Fuels	11,700.60	205,529.87	0.00				217,230.46
1. Solid Fuels	0.00	61,828.15	0.00				61,828.15
2. Oil and Natural Gas	11,700.60	143,701.71	0.00				155,402.31
2. Industrial Processes	67,401.42	1,682.52	29,211.21	87,105.19	16,843.57	31,800.00	234,043.92
A. Mineral Products	65,829.18	0.00	0.00				65,829.18
B. Chemical Industry	1,572.24	1,682.52	29,211.21	0.00	0.00	0.00	32,465.97
C. Metal Production	0.00	0.00	0.00		10,040.05	6,100.00	16,140.05
D. Other Production	NE						0.00
E. Production of Halocarbons and SF₆				30,400.00	0.00	0.00	30,400.00
F. Consumption of Halocarbons and SF₆				56,705.19	6,803.52	25,700.00	89,208.72
G. Other	0.00	0.00	0.00	0.00	0.00	0.00	0.00
3. Solvent and Other Product Use	0.00		0.00				0.00
4. Agriculture	0.00	172,880.28	315,895.13				488,775.40
A. Enteric Fermentation		127,201.20					127,201.20
B. Manure Management		34,394.30	17,181.77				51,576.07
C. Rice Cultivation		10,692.97					10,692.97
D. Agricultural Soils [2]	IE	0.00	298,281.28				298,281.28
E. Prescribed Burning of Savannas		0.00	0.00				0.00
F. Field Burning of Agricultural Residues		591.81	432.08				1,023.89
G. Other		0.00	0.00				0.00
5. Land-Use Change and Forestry [1]	-990,400.00	0.00	0.00				-990,400.00
6. Waste	25,959.73	226,870.20	8,430.22				261,260.16
A. Solid Waste Disposal on Land	0.00	214,636.53					214,636.53
B. Wastewater Handling		12,233.68	8,183.31				20,416.99
C. Waste Incineration	25,959.73	0.00	246.91				26,206.64
D. Other	0.00	0.00	0.00				0.00
7. Other (please specify)	0.00	0.00	0.00	0.00	0.00	0.00	0.00
	NA	NA	NA	NA	NA	NA	0.00
Memo Items:							
International Bunkers	107,345.46	44.30	953.35				108,343.11
Aviation	60,969.72	35.30	599.03				61,604.05
Marine	46,375.74	9.00	354.32				46,739.06
Multilateral Operations	NE	0.00	0.00				0.00
CO₂ Emissions from Biomass	234,062.51						234,062.51

[1] For CO₂ emissions from Land-Use Change and Forestry the net emissions are to be reported. Please note that for the purposes of reporting, the signs for uptake are always (-) and for emissions (+).

[2] See footnote 4 to Summary 1.A of this common reporting format.

GREENHOUSE GAS SOURCE AND SINK CATEGORIES	CO₂ emissions	CO₂ removals	Net CO₂ emissions / removals	CH₄	N₂O	Total emissions
Land-Use Change and Forestry	CO₂ equivalent (Gg)					
A. Changes in Forest and Other Woody Biomass Stocks	0.00	-720,900.00	-720,900.00			-720,900.00
B. Forest and Grassland Conversion	0.00		0.00	0.00	0.00	0.00
C. Abandonment of Managed Lands	0.00	0.00	0.00			0.00
D. CO₂ Emissions and Removals from Soil	32,300.00	-294,100.00	-261,800.00			-261,800.00
E. Other	0.00	-7,700.00	-7,700.00	0.00	0.00	-7,700.00
Total CO₂ Equivalent Emissions from Land-Use Change and Forestry	32,300.00	-1,022,700.00	-990,400.00	0.00	0.00	-990,400.00

Total CO₂ Equivalent Emissions without Land-Use Change and Forestry [a]	6,746,072.39
Total CO₂ Equivalent Emissions with Land-Use Change and Forestry [a]	5,755,672.39

[a] The information in these rows is requested to facilitate comparison of data, since Parties differ in the way they report emissions and removals from Land-Use Change and Forestry.

Appendix B
Policies and Measures

Energy: Commercial and Residential
Energy: Industrial
Energy: Supply
Transportation
Industry (Non-CO$_2$)
Agriculture
Forestry
Waste Management
Cross-sectoral Policies and Measures

Energy: Commercial and Residential

ENERGY STAR® for the Commercial Market[1]

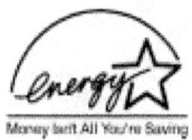

Description: Commercial buildings account for more than 15 percent of total U.S. carbon dioxide emissions. Many commercial buildings could effectively operate with 30 percent less energy if owners invested in energy-efficient products, technologies, and best management practices. ENERGY STAR® in the commercial sector is a partnership program that promotes the improvement of the energy performance of entire buildings.

Objectives: ENERGY STAR® provides information and motivation to decision makers to help them improve the energy perform-ance of their buildings and facilities. The program also provides performance benchmarks, strategies, technical assistance, and recognition.

Greenhouse gas affected: Carbon dioxide.

Type of policy or measure: Voluntary agreement.

Status of implementation: ENERGY STAR® has been underway since 1991 with the introduction of Green Lights. The program developed a strong partnership with large and small businesses and public organizations, such as state and local governments and school systems. The program's strategy has evolved substantially since the 1997 *U.S. Climate Action Report*, with the major program focus now on promoting high-performing (high-efficiency) buildings and providing decision makers throughout an organization with the information they need to undertake effective building improvement projects.

An innovative tool introduced in 1999 allows the benchmarking of building energy performance against the national stock of buildings. This tool is being expanded to represent the major U.S. building types, such as office, school (K–12), retail, and hos-pitality buildings. This national building energy performance rating system also allows for recognizing the highest-performing buildings, which can earn the ENERGY STAR® label. By the end of 2001, the program expects to be working with more than 11 billion square feet of building space across the country and to show over 7,000 rated buildings and more than 1,750 buildings labeled for excellence. EPA estimates that the program avoided 23 teragrams of CO_2 in 2000 and projects reductions of 62 ter-agrams of CO_2 in 2010.

Implementing entities: The partnership is a national program, managed by the Environmental Protection Agency (EPA). Implementing entities include a wide range of building owners and users, such as retailers, healthcare organizations, real estate investors, state and local governments, schools and universities, and small businesses.

Costs of policy or measure: Costs are defined as those monetary expenses necessary for participants' implementation of the program. Participants evaluate the cost-effective opportunities for improved energy performance and upgrade their facilities and operations accordingly. While energy-efficiency improvements require an initial investment, these costs are recovered over a period of time.

Non-GHG mitigation benefits of policy or measure: By reducing energy demand and use, ENERGY STAR® also reduces emis-sions of nitrogen oxides and sulfur dioxide.

Interaction with other policies or measures: By developing established energy performance benchmarks for commercial buildings, ENERGY STAR® in the commercial sector complements other measures at the national level, such as the Department of Energy's (DOE's) Rebuild America.

Contact: Angela Coyle, EPA, Climate Protection Partnerships Division, (202) 564-9719, coyle.angela@epa.gov.

---

[1] Actions 1 and 2 in the 1997 *U.S. Climate Action Report*, continuing.

Commercial Buildings Integration: Updating State Building Codes

Description: This program provides the technical assistance for implementing the energy-efficiency provisions of building codes and applicable standards that affect residential and commercial construction. These efforts involve partnerships with federal agencies, state and local governments, the building industry, financial institutions, utilities, public interest groups, and building owners and users. This measure is supported by Residential Building Codes, which is part of the Residential Buildings Integration program; Commercial Buildings Codes, which is part of the Commercial Buildings Integration Program; and Training and Assistance for Codes, which is part of the Community Energy Program.

Objectives: This program aims to improve the energy efficiency of the nation's new residential and commercial buildings, as well as additions and alterations to existing commercial buildings. Within applicable residential building codes, it incorporates the most technologically feasible, economically justified energy conservation measures. It also provides state and local governments with the technical tools and information they need for adopting, using, and enforcing efficient building codes for residential construction.

Greenhouse gases affected: Carbon dioxide, nitrous oxide, and carbon monoxide.

Type of policy or measure: Regulatory.

Status of implementation: Implemented.

Implementing entities: DOE and state legislatures.

Non-GHG mitigation benefits of policy or measure: This program increases energy efficiency; builds cooperation among stakeholders; shares information between federal and state entities; and educates builders, consumers, and homeowners.

Interaction with other policies or measures: The program complements DOE's efforts to develop and introduce advanced, highly efficient building technologies.

Commercial Buildings Integration:
Partnerships for Commercial Buildings and Facilities

Description: This program develops and demonstrates advanced technologies, controls, and equipment in collaboration with the design and construction community; advances integrated technologies and practices to optimize whole-building energy performance; and helps reduce energy use in commercial multifamily buildings by promoting construction of efficient buildings and their operation near an optimum level of performance. It also performs research on energy-efficient, sustainable, and low-cost building envelope materials and structures. This program is supported by a number of DOE programs: Commercial Buildings R&D, which is part of the Commercial Buildings Integration program, and Analysis Tools and Design Strategies and Building Envelope R&D, which are parts of the Equipment, Materials, and Tools program.

Objectives: This program aims to develop high-performance building design, construction, and operation processes; provide the tools needed for replicating the processes and design strategies for creating high-performance buildings; research new technologies for high-performance buildings; define the criteria and methods for measuring building performance; measure and document building performance in high-profile examples; and develop a fundamental understanding of heat, air, and moisture transfer through building envelopes and insulation materials, and apply the results to develop construction technologies to increase building energy efficiency.

Greenhouse gases affected: Carbon dioxide, nitrous oxide, and carbon monoxide.

Type of policy or measure: Research.

Status of implementation: Implemented.

Implementing entities: Federal government R&D in partnership with the private sector.

Non-GHG mitigation benefits of policy or measure: This program increases energy efficiency, shares information with and educates stakeholders, builds criteria for industry use, and collects useful data. It also has environmental benefits not related to greenhouse gases.

ENERGY STAR® for the Residential Market[2]

Description: Residential buildings account for over 18 percent of total U.S. carbon dioxide emissions. Many homes could use 30 percent less energy if owners purchased efficient technologies, incorporated efficiency into home improvement projects, or demanded an efficient home when buying a new home. EPA's ENERGY STAR® -labeled new homes, and EPA's Home Improvement Toolkit featuring ENERGY STAR®, deliver the information consumers need beyond labels on efficient products and equipment to make these decisions.

Objective: ENERGY STAR® provides information to consumers and homeowners so that they can make sound investments when buying a new home or when undertaking a home improvement project. This includes information on which products to purchase, how to achieve a high-performing home, the current energy performance of a home, and the improved performance that results from improvement projects.

Greenhouse gas affected: Carbon dioxide.

Type of policy or measure: Voluntary agreements and outreach.

Status of implementation: The ENERGY STAR® label for new homes has been available since 1995, building upon the success of the ENERGY STAR® label in a variety of product areas. The ENERGY STAR® program has been underway since 1992, with the introduction of the ENERGY STAR®-labeled computer. The ENERGY STAR® label is now on more than 25,000 U.S. homes that are averaging energy savings of about 35 percent higher than the model energy code. Since the 1997 *U.S. Climate Action Report*, this residential effort has expanded significantly to home improvement projects in the existing homes market. The program now provides guidance for homeowners on designing efficiency into kitchens, additions, and whole-home improvement projects. It offers a Web-based audit tool and a home energy benchmark tool to help homeowners get underway and monitor progress. The program is also working with energy efficiency program partners around the country so that they can use this unbiased information at consumer transaction points to promote energy efficiency. EPA projects the program will avoid 20 teragrams of CO_2 annually by 2010.

Implementing entities: ENERGY STAR® is a national program. EPA implements this effort with partners around the country.

Costs of policy or measure: All costs are recovered over a period of time.

Non-GHG mitigation benefits of policy or measure: By reducing energy demand and use, ENERGY STAR® also reduces nitrogen oxides and sulfur dioxide.

Interaction with other policies or measures: ENERGY STAR® for new homes works closely with DOE's Rebuild America program.

Contact: David Lee, EPA, Climate Protection Partnerships Division, (202) 564-9131, lee.davidf@epa.gov.

---

[2] Part of Action 6 in the 1997 *U.S. Climate Action Report*; continuing.

Community Energy Program:
Rebuild America

Description: Rebuild America connects people, resources, proven ideas, and innovative practices through collaborative partnerships with states, small towns, large metropolitan areas, and Native American tribes, creating a large network of peer communities. The program provides one-stop shopping for information and assistance on how to plan, finance, implement, and manage retrofit projects to improve energy efficiency. Rebuild America supports communities with access to DOE regional offices, state energy offices, national laboratories, utilities, colleges and universities, and nonprofit agencies.

Objective: Rebuild America aims to assist states and communities in developing and implementing environmentally and economically sound activities through smarter energy use.

Greenhouse gases affected: Carbon dioxide, nitrous oxide, and carbon monoxide.

Type of policy or measure: Voluntary, information, and education.

Status of implementation: As of May 2001, Rebuild America had formed 340 partnerships with approximately 550 million square feet of buildings complete or underway in all 50 states and two U.S. territories.

Implementing entities: State and local community partnerships with the federal government.

Non-GHG mitigation benefits of policy or measure: Rebuild America expands knowledge and technology base through education, improves energy efficiency, promotes private–public cooperation and information sharing, creates peer networks, preserves historic buildings, builds new facilities, retrofits existing buildings, stimulates economic development, promotes community development, and avoids urban sprawl.

Interaction with other policy or measure: Rebuild America helps to promote many of the resources made available by other DOE programs, such as Updating State Building Codes and ENERGY STAR®.

Residential Building Integration: Energy Partnerships for Affordable Housing (Building America)

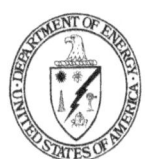

Description: The Residential Buildings Integration program operates Energy Partnerships for Affordable Housing. This new program consolidates the formerly separate systems-engineering programs of Building America, Industrialized Housing, Passive Solar Buildings, Indoor Air Quality, and existing building research into a comprehensive program. Systems-integration research and development activities analyze building components and systems and integrate them so that the overall building performance is greater than the sum of its parts. Building America is a private–public partnership that provides energy solutions for production housing and combines the knowledge and resources of industry leaders with DOE's technical capabilities to act as a catalyst for change in the home building industry.

Objective: This program aims to accelerate the introduction of highly efficient building technologies and practices through research and development of advanced systems for production builders.

Greenhouse gases affected: Carbon dioxide, nitrous oxide, and carbon monoxide.

Type of policy or measure: Voluntary, research, and education.

Status of implementation: Implemented.

Implementing entities: DOE and private industry partners.

Non-GHG mitigation benefits of policy or measure: This program increases energy efficiency, and software and information sharing; incorporates renewable resources and distributed generation; improves builder productivity; reduces construction time; provides new product opportunities to manufacturers and suppliers; and promotes teamwork within the segmented building industry.

ENERGY STAR®-Labeled Products[3]

Description: Many homeowners and businesses could use 30 percent less energy, without sacrificing services or comfort, by investing in energy efficiency. Introduced by EPA in 1992 for computers, the ENERGY STAR® label has been expanded to more than 30 product categories. Since the mid-1990s, EPA has collaborated with DOE, which now has responsibility for certain product categories. The ENERGY STAR® label is now recognized by more than 40 percent of U.S. consumers, who have purchased over 600 million ENERGY STAR® products. The program has developed a strong partnership with business, representing over 1,600 manufacturers with more than 11,000 ENERGY STAR®-labeled products.

Objective: The ENERGY STAR® label is used to distinguish energy-efficient products in the marketplace so that businesses and consumers can easily purchase these products, save money on energy bills, and avoid air pollution.

Greenhouse gas affected: Carbon dioxide.

Type of policy or measure: Voluntary agreement.

Status of implementation: The program's strategy has evolved substantially since the 1997 *U.S. Climate Action Report*, not only with its addition of new products to the ENERGY STAR® family, but also with its expanded outreach to consumers in partnership with their local utility or similar organization. ENERGY STAR® works in partnership with utilities, representing about 50 percent of U.S. energy customers. To date, more than 600 million ENERGY STAR®-labeled products have been purchased. EPA estimates that the program avoided 33 teragrams of CO_2 in 2000 and projects it will reduce 75 teragrams of CO_2 in 2010.

Implementing entities: ENERGY STAR® is a national program. EPA and DOE implement the ENERGY STAR® label on products with partners across the country.

Costs of policy or measure: All costs are recovered over a period of time.

Non-GHG mitigation benefits of policy or measure: By reducing energy demand and use, ENERGY STAR® also reduces nitrogen oxides and sulfur dioxide.

Interaction with other policies or measures: ENERGY STAR® is implemented in concert with the minimum efficiency standards developed by DOE, where those standards exist, such as with household appliances and heating and cooling equipment.

Contact: Rachel Schmeltz, EPA, Climate Protection Partnerships Division, (202) 564-9124, schmeltz.rachel@epa.gov

---

[3] Part of Action 6 in the 1997 *U.S. Climate Action Report;* continuing.

Building Equipment, Materials, and Tools:
Superwindow Collaborative

Description: The Superwindow Collaborative develops commercially viable, advanced electrochromic windows and Superwindows for competing producers. These programs intend to reward industry through market mechanisms for their investments in the research, development, and deployment of energy-efficient windows. In an area that is less suited to national standards and that has a growing international market, significant investments are required to establish a technical basis for performance standards recognized for their scientific excellence. The Superwindow Collaborative is supported by two DOE programs—Building Envelope: Electrochromic Windows and Building Envelope: Superwindows—both of which are part of DOE's Equipment, Materials, and Tools program.

Objectives: The Superwindow Collaborative aims to change windows from net energy loss centers to net energy savers across the United States; to strengthen the market position of U.S. industry in global markets; and to provide building owners cost-effective savings, a more comfortable building climate, and possible productivity improvements.

Greenhouse gases affected: Carbon dioxide, nitrous oxide, and carbon monoxide.

Type of policy or measure: Research.

Status of implementation: Implemented.

Implementing entities: Federal government R&D in partnership with the private sector. The electrochromic participants include two national laboratories and four industrial partners. Supporting research on materials, durability, and energy performance is performed at DOE's national laboratories.

Non-GHG mitigation benefits of policy or measure: The Superwindow Collaborative increases economic competitiveness, energy efficiency, and building climate comfort, and provides possible productivity improvements for buildings.

Building Equipment, Materials, and Tools: Lighting Partnerships

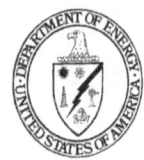

Description: Lighting Partnerships supports research and development in three areas:
- Advanced light sources, consisting of research that is heavily cost-shared with industry to advance lighting technology, with the goal of developing replacements for the inefficient incandescent lamp. The program supports improvements in compact fluorescent lamps and in new lamps using improved incandescent, fluorescent, high-intensity-discharge, and electrode-less technologies.
- Lighting fixtures, controls, and distribution systems consisting of cost-shared research on lighting controls in commercial buildings and light fixtures for advanced light sources, primarily compact fluorescent lamps.
- The impact of lighting on vision, consisting of industry cost-shared research on outdoor lighting.

Objectives: The program aims to develop and accelerate the introduction of advanced lighting technologies and to make solid-state lighting more efficient than conventional sources and more easily integrated into building systems. Additional goals are to develop lighting technologies that last for 20,000 to 100,000 hours and to significantly reduce greenhouse gas emissions from coal-fired power plants.

Greenhouse gases affected: Carbon dioxide, nitrous oxide, and carbon monoxide.

Type of policy or measure: Research.

Status of implementation: Implemented.

Implementing entities: Lighting Partnerships is a federal research and development program that collaborates with manufacturers, utilities, user groups, and trade and professional organizations.

Building Equipment, Materials, and Tools: Partnerships for Commercial Buildings and Facilities

Description: This program develops and demonstrates advanced technologies, controls, and equipment in collaboration with the design and construction community; advances integrated technologies and practices to optimize whole-building energy performance; and helps reduce energy use in commercial multifamily buildings by promoting the construction of efficient buildings and their operation near an optimum level of performance. It also performs research on energy-efficient, sustainable, and low-cost building-envelope materials and structures. The program is supported by a number of DOE programs, including Commercial Buildings R&D, which is part of the Commercial Buildings Integration program; and Analysis Tools and Design Strategies and Building Envelope R&D, which are parts of the Equipment, Materials, and Tools program.

Objectives: This program aims to develop high-performance building design, construction, and operation processes; provide the tools needed for replicating the processes and design strategies for creating high-performance buildings; research new technologies for high-performance buildings; define the criteria and methods for measuring building performance; measure and document building performance in high-profile examples; and develop a fundamental understanding of heat, air, and moisture transfer through building envelopes and insulation materials, and apply the results to develop construction technologies to increase building energy efficiency.

Greenhouse gases affected: Carbon dioxide, nitrous oxide, and carbon monoxide.

Type of policy or measure: Research.

Status of implementation: Implemented.

Implementing entities: Federal government R&D in partnership with the private sector.

Non-GHG mitigation benefits of policy or measure: This program increases energy efficiency, shares information with and educates stakeholders, builds criteria for industry use, and collects useful data. It also has environmental benefits not related to greenhouse gases.

Building Equipment, Materials, and Tools: Collaborative Research and Development

Description: This program researches, develops, and commercializes food display and storage technologies that use less energy and less refrigerant; new, super-efficient electric dryers; low-cost, high-reliability heat pump water heaters; and energy-efficient heating, ventilation, and air conditioning systems. It also brings new products to market and provides an independent third-party evaluation of highly efficient products. The Space Conditioning and Refrigeration program and the Appliances and Emerging Technologies program are part of the Equipment, Materials, and Tools program.

Objectives: This program aims to develop and promote the use of low-cost, energy-efficient equipment, materials, and tools.

Greenhouse gases affected: Carbon dioxide, nitrous oxide, and carbon monoxide.

Type of policy or measure: Research.

Status of implementation: Implemented.

Implementing entities: State and local partnerships with the federal government.

Non-GHG mitigation benefits of policy or measure: This program promotes energy efficiency, evaluates energy-efficient products and accelerates their commercialization, and improves economic competitiveness and data collection.

Residential Appliance Standards

Description: Administered by DOE's Office of Codes and Standards, the Residential Appliance Standards program periodically reviews and updates efficiency standards for most major household appliances.

Objective: The program's standards aim to ensure that American consumers receive a minimum practical energy efficiency for every appliance they buy.

Greenhouse gases affected: Carbon dioxide, nitrous oxide, and carbon monoxide.

Type of policy or measure: Regulatory.

Status of implementation: Implemented.

Implementing entities: DOE and other federal entities. DOE promulgates revised or new regulations, while the Federal Trade Commission prescribes the labeling rules for residential appliances.

Non-GHG mitigation benefits of policy or measure: The program enhances energy security, increases competitiveness and reliability, and improves energy efficiency.

State and Community Assistance: State Energy Program; Weatherization Assistance Program; Community Energy Grants; Information Outreach

Description: Several programs and initiatives support DOE's State and Community Assistance efforts:
- The *State Energy Program* provides a supportive framework with sufficient flexibility to enable the states to address their energy priorities in concert with national priorities, and supports the federal–state partnerships that are crucial to energy policy development and energy technology deployment.
- The *Weatherization Assistance Program* provides cost-effective, energy-efficiency services to low-income constituencies who otherwise could not afford these services and who stand to benefit greatly from the cost savings of energy-efficient technologies.
- The *Community Energy Grants* program provides funding to competitively selected communities to support community-wide energy projects that improve energy efficiency and implement sustainable building design and operation concepts.
- The *Information Outreach* program helps to conceptualize, plan, and implement a systematic approach to marketing and communication objectives and evaluation.

Objectives: The objectives of DOE's State and Community Assistance efforts are based on the combined objectives of its major programs and initiatives:
- *State Energy:* To maximize energy, environmental, and economic benefits through increased collaboration at the federal, state, and community levels; to increase market acceptance of energy-efficient and renewable-energy technologies, practices, and products; and to use innovative approaches to reach market segments and meet policy goals not typically addressed by market-based solutions.
- *Weatherization:* To develop new weatherization technologies, further application of best methods and practices throughout the national weatherization network, leverage and integrate weatherization with other energy efficiency resources, and demonstrate program effectiveness.
- *Community Energy Grants:* To save energy, create jobs, promote growth, and protect the environment.
- *Information Outreach:* To provide technical assistance needed to conduct the various planned activities that will educate target audiences; to follow strategic plan goals and support long-term success in developing energy-efficient systems and processes; to improve technology-transfer and information-exchange processes; and to emphasize partnering with strategic allies, communications, education and training, and information support.

Greenhouse gases affected: Carbon dioxide, nitrous oxide, and carbon monoxide.

Type of policy or measure: Economic, information.

Status of implementation: Implemented.

Implementing entities: States and local communities through partnerships with the federal government. For example, DOE makes grants to states through its Weatherization Assistance Program, which in turn awards grants to local agencies—usually community action agencies or other nonprofit or government organizations—to perform the actual weatherization services.

Non-GHG mitigation benefits of policy or measure: These programs enhance energy efficiency; create jobs and boost economic development; have non-greenhouse gas environmental benefits; increase collaboration at federal, state, and local levels; provide health benefits; educate consumers; promote technology transfer to industry; and provide training.

Heat Island Reduction Initiative[4]

HEAT ISLAND REDUCTION
I N I T I A T I V E

Description: This initiative is a multi-agency effort to work with communities and state and local officials to reduce the impacts of urban heat islands. It promotes common-sense measures, such as planting shade trees, installing reflective roofs, and using light-colored pavements to reduce ambient temperature, ozone pollution, cooling energy demand, and greenhouse gas emissions. This initiative also supports research to quantify the air quality, health, and energy-saving benefits of measures for reducing the impacts of urban heat islands.

Objective: The program's objective is to work with state and local governments to reverse the effects of urban heat islands by encouraging the widespread use of mitigation strategies.

Greenhouse gas affected: Carbon dioxide.

Type of policy or measure: Voluntary, information exchange, and research.

Status of implementation: The program was redesigned in 1997 and is currently ongoing. EPA performs research on the up-front costs, potential savings, and options for reflective surfaces, to assist with implementing measures for reducing the demands of heat islands. In addition, information on the air quality benefits may allow states to incorporate these measures into their air quality plans. Pilot projects have been established in five cities that have agreed to assist with research and work to implement the measures. For example, several cities in Utah have implemented ordinances with measures for reducing the impacts of urban heat islands. And California's state legislature and governor have authorized using over $24 million for measures to reduce peak summer heat island demand for electricity.

Implementing entities: EPA, in partnership with state and local governments.

Costs of policy or measure: Reflective surfaces are generally implemented during new construction or when replacing old materials. While initial costs are comparable between nonreflective and reflective surfaces, cost savings can be expected when evaluating life-cycle costs (as energy savings and reduced maintenance are considered).

Non-GHG mitigation benefits of policy or measure: The program's measures can reduce emissions of volatile organic compounds and nitrogen oxides due to reduced energy use and ambient temperatures. Lower temperatures may also help reduce ozone concentrations due to the heat-dependent reaction that forms this pollutant. In addition, energy savings can be expected from implementing heat-island reduction measures.

Interaction with other policies or measures: The program interacts with the ENERGY STAR® Roofs Program and state implementation plans.

Contact: Niko Dietsch, EPA, Global Programs Division, (202) 564-3479, dietsch.nikolaas@epa.gov.

---

[4] Action 9 in the 1997 *U.S. Climate Action Report;* continuing.

Economic Incentives/Tax Credits

Description: Current law provides a 10 percent business energy investment tax credit for qualifying equipment that uses solar energy to generate electricity, to heat or cool, to provide hot water for use in a structure, or to provide solar process heat. No credit is available for nonbusiness purchases of solar energy equipment. The Administration is proposing a new tax credit for individuals of photovoltaic equipment and solar water-heating systems for use in a dwelling that the individual uses as a residence. Equipment would qualify for the credit only if used exclusively for purposes other than heating swimming pools. An individual would be allowed a cumulative maximum credit of $2,000 per residence for photovoltaic equipment and $2,000 per residence for solar water-heating systems. The credit for solar water-heating equipment would apply only if placed in service after December 31, 2001, and before January 1, 2006, and to photovoltaic systems placed in service after December 31, 2001, and before January 1, 2008.

Objective: This proposed tax credit aims to expand the future market of residential solar energy systems.

Greenhouse gases affected: Carbon dioxide, nitrous oxide, and carbon monoxide.

Type of policy or measure: Economic.

Status of implementation: This measure is in the proposal stage.

Energy: Industrial

Industries of the Future

Description: Industries of the Future creates partnerships among industry, government, and supporting laboratories and institutions to accelerate technology research, development, and deployment. Led by DOE's Office of Industrial Technologies, this strategy is being implemented in nine energy- and waste-intensive industries. Two key elements of the strategy include an industry-driven document outlining each industry's vision for the future, and a technology roadmap to identify the technologies that will be needed to reach that industry's goals.

Objective: This strategy aims to help nine key energy-intensive industries reduce their energy consumption while remaining competitive and economically strong.

Greenhouse gases affected: Carbon dioxide, carbon monoxide, nitrous oxide, methane, and volatile organic compounds.

Type of policy or measure: Voluntary, research, and information.

Status of implementation: Implemented.

Implementing entities: Partnerships among industry, government, and supporting laboratories and institutions.

Non-GHG mitigation benefits of policy or measure: This strategy enhances economic security and energy efficiency, allows for competitive restructuring, has non-GHG environmental benefits, forms cooperative alliances, increases productivity, and disseminates information.

Best Practices Program

Description: This initiative of DOE's Office of Industrial Technologies offers industry the tools to improve plant energy efficiency, enhance environmental performance, and increase productivity. Selected best-of-class large demonstration plants are showcased across the country, while other program activities encourage the replication of these best practices in still larger numbers of large plants.

Objective: Best Practices is designed to change the ways industrial plant managers make decisions affecting energy use by motors and drives, compressed air, steam, combustion systems, and other plant utilities.

Greenhouse gases affected: Carbon dioxide, carbon monoxide, nitrous oxide, methane, and volatile organic compounds.

Type of policy or measure: Voluntary, information.

Status of implementation: Implemented.

Implementing entities: DOE and industrial partners.

Non-GHG mitigation benefits of policy or measure: Best Practices enhances economic security and energy efficiency, has non-greenhouse gas environmental benefits, increases productivity and industry cooperation, and disseminates knowledge.

ENERGY STAR® for Industry (Climate Wise)[5]

Description: Nearly one-third of U.S. carbon dioxide emissions result from industrial activities. The primary source of these emissions is the burning of carbon-based fuels, either on site in manufacturing plants or through the purchase of generated electricity. Recently, ENERGY STAR® and Climate Wise were integrated under ENERGY STAR® to compose a more comprehensive partnership for industrial companies. Through established energy performance benchmarks, strategies for improving energy performance, technical assistance, and recognition for accomplishing reductions in energy, the partnership contributes to a reduction in energy use for the U.S. industrial sector.

Objectives: ENERGY STAR® enables industrial companies to evaluate and cost-effectively reduce their energy use. This reduction, in turn, results in decreased carbon dioxide emissions when carbon-based fuels are the source of that energy.

Greenhouse gas affected: Carbon dioxide.

Types of policy or measure: Voluntary agreement.

Status of implementation: This program has been underway since 1994 with the launch of Climate Wise. In 2000, ENERGY STAR®and the Climate Wise Partnership were integrated to provide the industrial sector with a more comprehensive set of industrial benchmarking and technical assistance tools. The partnership currently has more than 500 industrial partners representing a large share of energy use in the industrial sector. EPA estimates that the program avoided 11 teragrams of CO_2 in emissions in 2000 and projects reductions of 16 teragrams of CO_2 in 2010.

Implementing entities: The partnership is primarily a national program, managed by EPA. State and local governments voluntarily participate by promoting the program to industries within their jurisdictions.

Costs of policy or measure: Costs are defined as the monetary expenses necessary for an industrial participant to implement the program. Participants evaluate the cost-effective opportunities for energy performance and complete adjustments to their operation. While an initial outlay of funds is possible, these costs are recovered over a period of time.

Non-GHG mitigation benefits of policy or measure: The burning of fossil fuels creates airborne pollutants, including nitrogen oxides and sulfur dioxide. By reducing energy demand and use, ENERGY STAR® helps to decrease emissions of these pollutants.

Interaction with other policies or measures: ENERGY STAR® is the only national program that offers industrial companies the ability to evaluate and minimize energy use through established energy performance benchmarks, strategies, and technical assistance. ENERGY STAR® complements programs managed by DOE. DOE oversees partnerships with nine energy-intensive industrial sectors to accelerate technology research, development, and deployment, with a goal of reducing energy use and the environmental impacts of these industries. DOE also manages a program to improve a plant's technical systems, or components of a plant, including the motors, steam, compressed air, combined heat and power, and process heat. ENERGY STAR® complements these programs with a system for evaluating plant-wide energy performance.

Contact: Elizabeth Dutrow, EPA, Climate Protection Partnerships Division, (202) 564-9061, dutrow.elizabeth@epa.gov.

---

[5] Foundation action 9 in the 1997 *U.S. Climate Action Report;* continuing.

Industrial Assessment Centers

Description: Teams of engineering faculty and students from 26 universities around the country conduct free comprehensive energy audits or industrial assessments and provide recommendations to eligible small and medium-sized manufacturers to help them identify opportunities to improve productivity, reduce waste, and save energy.

Objectives: The assessments aim to improve energy efficiency and productivity, minimize waste, and prevent pollution.

Greenhouse gases affected: Carbon dioxide, carbon monoxide, nitrous oxide, methane, and volatile organic compounds.

Type of policy or measure: Voluntary, information, and education.

Status of implementation: Implemented.

Implementing entities: DOE and universities.

Non-GHG mitigation benefits of policy or measure: The program has non-greenhouse gas environmental benefits, improves energy efficiency, economic productivity, and competitiveness; encourages public–private-sector interaction and cooperation and information sharing within industry; provides student educational experience; and collects industry data for industry progress assessments, thereby enabling the quantification of the state of energy, waste, and productivity management in small and medium-sized industries.

Enabling Technologies:
Industrial Materials for the Future

Description: DOE's Industrial Materials for the Future program is the combination of the Advanced Industrial Materials and Continuous Fiber Ceramic Composite programs. The new program focuses on areas that offer major improvements in energy efficiency and emission reductions across all industries.

Objective: Consistent with the mission of DOE's Office of Industrial Technologies, this program's mission is to lead a national effort to research, design, develop, engineer, and test new and improved materials for the Industries of the Future.

Greenhouse gases affected: Carbon dioxide, carbon monoxide, nitrous oxide, methane, and volatile organic compounds.

Types of policy or measure: Research.

Status of implementation: Implemented.

Implementing entities: DOE and industry partners.

Non-GHG mitigation benefits of policy or measure: This program reduces emissions of non-greenhouse gas pollutants; improves energy efficiency, economic productivity, and competitiveness; encourages public–private-sector interaction and cooperation; and facilitates information sharing within industry.

Financial Assistance: NICE[3]
(National Industrial Competitiveness through
Energy, Environment, & Economics)

Description: Sponsored by DOE's Office of Industrial Technologies, NICE[3] is an innovative, cost-sharing grant program that provides funding to state and industry partnerships (large and small businesses) for projects that develop and demonstrate advances in energy efficiency and clean production technologies.

Objectives: The NICE[3] program was authorized to improve the energy efficiency and cost-effectiveness of pollution prevention technologies and processes, including source-reduction and waste-minimization technologies and processes. It also aims to advance the global competitiveness of U.S. industry.

Greenhouse gases affected: Carbon dioxide, carbon monoxide, nitrous oxide, methane, and volatile organic compounds.

Type of policy or measure: Voluntary, research.

Status of implementation: Implemented.

Implementing entities: State agencies, industry, and universities.

Non-GHG mitigation benefits of policy or measure: The NICE[3] program increases economic production, energy efficiency, industry competitiveness, and cooperation between the public and private sectors. It also has non-greenhouse gas environmental benefits.

Energy: Supply
Renewable Energy Commercialization:
Wind; Solar; Geothermal; Biopower

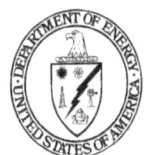

Description: DOE's Office of Power Technologies maintains several programs on individual renewable energy technologies, including wind, solar, geothermal, and biomass. Renewable technologies use naturally occurring energy sources to produce electricity, heat, fuel, or a combination of these energy types.

Objectives: The program aims to develop clean, competitive power technologies; to diversify the nation's energy supply portfolio; to use abundant domestic resources; to help the nation meet its commitments to curb greenhouse gas emissions; and to achieve tax incentives for renewable energy production and use.

Greenhouse gases affected: Carbon dioxide, carbon monoxide, nitrous oxide, methane, and volatile organic compounds.

Type of policy or measure: Research, regulatory.

Status of implementation: Implemented.

Implementing entities: DOE and industry partners.

Non-GHG mitigation benefits of policy or measure: The program enhances the nation's energy and economic security; has non-greenhouse gas environmental benefits; builds energy infrastructure; creates jobs; increases industrial competitiveness and energy reliability; and diversifies the nation's energy portfolio.

Climate Challenge

Description: The Climate Challenge program is a joint, voluntary effort of DOE and the electric utility industry to reduce, avoid, or sequester greenhouse gases. Utilities, in partnership with DOE, developed individual agreements to identify and implement cost-effective activities for reducing greenhouse gas emissions. Electric utility trade associations are active in promoting the program and in developing industry-wide initiatives. Details on the program are available at http://www.eren.doe.gov/climatechallenge/.

Objective: Established as a Foundation Action under the 1993 *Climate Change Action Plan*, Climate Challenge persuaded electric utilities to develop Participation Accords with DOE. These individual agreements identified cost-effective activities for the utility to implement, with the goal of reducing emissions in 2000. Each utility must annually report its results to DOE's Energy Information Administration Voluntary Reporting of Greenhouse Gases Program (http://www.eia.doe.gov/oiaf/1605/frntvrgg.html), consistent with the voluntary reporting of greenhouse gas emission guidelines developed under Section 1605(b) of the Energy Policy Act of 1992. Reductions will continue to be reported beyond 2000.

Greenhouse gases affected: Primarily carbon dioxide, but also other greenhouse gases, such as methane and sulfur hexafluoride. Carbon dioxide activities include both reductions in emissions and increases in carbon sequestration.

Type of policy or measure: DOE and the individual utilities sign Voluntary Participation Accords (or Letters of Participation for smaller utilities), describing the utilities' commitments in the form of specific projects, entity-wide actions, and/or industry-wide initiatives.

Status of implementation: Implemented.

Implementing entities: The program is a joint, voluntary effort between the electric utility industry and the DOE. Parameters of the Climate Challenge program were defined in a 1994 Memorandum of Understanding between DOE and all the national utility trade associations.

Non-GHG mitigation benefits of policy or measure: The reduction in carbon dioxide emissions from Climate Challenge projects often results in a concurrent reduction in sulfur dioxide, oxides of nitrogen, and other emissions associated with fossil fuel combustion. Other projects have reduced landfill requirements by recycling and reusing coal combustion by-products and other materials. Participating utilities have indicated that corporate learning about climate change and mitigation opportunities has been a significant benefit of the program. Climate Challenge has helped shift the thinking of electric utility management and strategic planners to include the mitigation of greenhouse gas emissions into their corporate culture and philosophy.

Interaction with other policies or measures: As a Foundation Action under the 1993 Climate Change Action Plan, Climate Challenge was designed as a platform from which participating utilities could undertake a broad range of activities (individually, through industry-wide initiatives, and through other federal voluntary programs). In addition, Climate Challenge utilities agree to report their results annually to DOE's Energy Information Administration, consistent with the voluntary reporting of greenhouse gas emission guidelines developed under Section 1605(b) of the Energy Policy Act of 1992.

Distributed Energy Resources

Description: This program directs and coordinates a diverse portfolio of research and development, consolidating programs and staff from across DOE's Office of Energy Efficiency and Renewable Energy related to the development and deployment of distributed energy resources (DER). It focuses on technology development and the elimination of regulatory and institutional barriers to the use of DER, including interconnection to the utility grid and environmental siting and permitting. DER partners with industry to apply a wide array of technologies and integration strategies for on-site use, as well as for grid-enhancing systems. Successful deployment of DER technologies affects the industrial, commercial, institutional, and residential sectors of our economy—in effect, all aspects of the energy value chain.

Objectives: This program aims to develop a cleaner, more reliable, and affordable U.S. energy resource portfolio to reduce pollution and greenhouse gas emissions; enhance electric grid operations; boost local economic development; and increase energy and economic efficiency.

Greenhouse gases affected: Carbon dioxide, carbon monoxide, nitrous oxide, methane, and volatile organic compounds.

Type of policy or measure: Research, information, education, and regulatory.

Status of implementation: Implemented.

Implementing entities: DOE, industry.

Non-GHG mitigation benefits of policy or measure: This program has non-greenhouse gas environmental benefits, improves energy reliability, reduces the strain on the electric grid infrastructure, allows energy choices among consumers (creating a more dynamic energy market), and hedges against peak power prices.

High-Temperature Superconductivity

Description: This program investigates the properties of crystalline materials that become free of electrical resistance at the temperature of liquid nitrogen. The lack of resistance makes possible electrical power systems with super-efficient generators, transformers, and transmission cables that reduce energy losses associated with electricity transmission.

Objectives: The next few years may see the beginning of the widespread utilization of superconductivity technologies. This program leads the DOE research and development effort geared toward making this happen. It supports aggressive projects to design advanced electrical applications. The industry-led Second-Generation Wire Development exploits breakthroughs at national laboratories that promise unprecedented current-carrying capacity.

Greenhouse gases affected: Carbon dioxide, carbon monoxide, nitrous oxide, methane, and volatile organic compounds.

Types of policy or measure: Research.

Status of implementation: Implemented.

Implementing entities: DOE, industry.

Non-GHG mitigation benefits of policy or measure: This program has non-greenhouse gas environmental benefits, such as reducing SO_x emissions; improves energy reliability; reduces strain on the electric grid infrastructure; cuts transmission losses by half; and allows electrical equipment to be reduced in size dramatically (which opens more potential site applications).

Hydrogen Program

Description: This program has four strategies to carry out its objective: (1) expand the use of hydrogen in the near term by working with industry, including hydrogen producers, to improve efficiency, lower emissions, and lower the cost of technologies that produce hydrogen from natural gas for distributed filling stations; (2) work with fuel cell manufacturers to develop hydrogen-based electricity storage and generation systems that will enhance the introduction and penetration of distributed, renewable-energy-based utility systems; (3) coordinate with the Department of Defense and DOE's Office of Transportation Technologies to demonstrate safe and cost-effective fueling systems for hydrogen vehicles in urban nonattainment areas and to provide onboard hydrogen storage systems; and (4) work with the national laboratories to lower the cost of technologies that produce hydrogen directly from sunlight and water.

Objective: The program's mission is to enhance and support the development of cost-competitive hydrogen technologies and systems that will reduce the environmental impacts of energy use and enable the penetration of renewable energy into the U.S. energy mix.

Greenhouse gases affected: Carbon dioxide, carbon monoxide, nitrous oxide, methane, and volatile organic compounds.

Type of policy or measure: Research, education.

Status of implementation: Implemented.

Implementing entities: DOE, industry, and national laboratories.

Non-GHG mitigation benefits of policy or measure: This program has non-greenhouse gas environmental benefits, develops new infrastructure, creates jobs, enhances the nation's energy and economic security, diversifies the nation's energy portfolio, and expands our technology base.

Clean Energy Initiative:
Green Power Partnership;
Combined Heat and Power Partnership

Description: Increased economic growth has been fueled in large part by energy produced from fossil fuels, with the unintended consequence of increased air pollution and an increased threat of climate change. EPA's Clean Energy Initiative is designed to reduce greenhouse gas emissions associated with the energy supply sector by promoting available technologies. EPA's strategy includes: (1) increasing corporate and institutional demand for renewable energy, (2) facilitating combined heat and power (CHP) and other clean "distributed generation" technologies in targeted markets, and (3) working with state and local governments to develop policies that favor clean energy.

EPA's Green Power Partnership works with businesses and other institutions to facilitate bulk purchases of renewable energy. This involves setting green power standards, providing recognition, and quantifying the environmental benefits. EPA's CHP Partnership targets candidate sites in key state markets, and provides these facilities with information about the benefits of CHP, as well as technical assistance. The Policy Team produces a database that quantifies the environmental impacts of power generation, along with other policy tools to help reduce the environmental impacts of electricity generation.

Objective: The Clean Energy Initiative will focus on the energy supply sector, as well as industrial, commercial, and residential energy customers. The approach will aim to remove market barriers to the increased penetration of cleaner, more efficient energy supply through education, technical assistance, demonstration, and partnerships.

Greenhouse gas affected: Carbon dioxide.

Type of policy or measure: Voluntary agreement, education, and technical assistance.

Status of implementation: This effort is currently being implemented. EPA will conduct annual reviews of the program's performance at the end of each calendar year. EPA projects the program will reduce greenhouse gas emissions by about 30 teragrams of CO_2 in 2010.

Implementing entity: EPA.

Non-GHG mitigation benefits of policy or measure: This initiative will reduce criteria air pollutants, which contribute to local and regional air quality problems, and will reduce land and water impacts due to the decrease in fossil fuel use.

Interaction with other policies or measures: This initiative requires interaction with ongoing initiatives at DOE, particularly efforts to commercialize renewable energy and CHP technologies. DOE will continue to play the lead role in research and development and performance benchmarking, while EPA will primarily be involved with market transformation activities for these technologies.

Contact: Tom Kerr, EPA, Climate Protection Partnerships Division, (202) 564-0047, kerr.tom@epa.gov.

Nuclear Energy Plant Optimization

Description: The Nuclear Energy Plant Optimization (NEPO) program conducts scientific and engineering research to develop advanced technologies to manage the aging of nuclear plants. The cost-shared program is part of a comprehensive approach to ensure that the United States has the technological capability to produce adequate supplies of baseload electricity while minimizing greenhouse gas emissions and other harmful environmental impacts. Details on the NEPO program are available at http://nuclear.gov.

Objective: The program aims to ensure that current U.S. nuclear power plants can continue to deliver adequate and affordable energy supplies up to and beyond their initial license period by resolving critical issues related to long-term plant aging and by developing advanced technologies for improving plant reliability, availability, and productivity.

Greenhouse gases affected: Carbon dioxide, carbon monoxide, nitrous oxide, methane, and volatile organic compounds.

Type of policy or measure: Research, information.

Status of implementation: DOE began the NEPO program in fiscal year 2000 and continues to initiate cooperative R&D projects, which are identified through input from electric utilities, the Nuclear Regulatory Commission, and other stakeholders.

Implementing entities: The program is a cost-shared partnership between the nuclear industry and the federal government.

Non-GHG mitigation benefits of policy or measure: The NEPO program and other nuclear energy R&D programs conducted by DOE support the goal in the President's *National Energy Policy* of increasing the development and use of nuclear power as non-greenhouse gas-emitting source of electricity for the nation.

Interaction with other policies or measures: Operation of existing nuclear power plants annually avoids emissions of over 150 teragrams of carbon dioxide, five million tons of sulfur dioxide, and 2.4 million tons of nitrogen oxides. Continued operation of existing nuclear plants through their original license term and a 20-year renewed license term would partly mitigate the need to build more baseload power plants.

Development of Next-Generation Nuclear Energy Systems: Nuclear Energy Research Initiative; Generation IV Initiative

Description: DOE's support for next-generation nuclear energy systems comes primarily from two programs: the Nuclear Energy Research Initiative (NERI) and the Generation IV Initiative (Gen-IV). Complete details on the Gen-IV and NERI programs are available at http://nuclear.gov.

Objectives: NERI is funding small-scale research efforts on promising advanced nuclear energy system concepts, in areas that will promote novel next-generation, proliferation-resistant reactor designs, advanced nuclear fuel development, and fundamental nuclear science. In the future, there is likely to be NERI research in the use of nuclear energy to produce hydrogen fuel for fuel cells.

The present focus of Gen-IV is on the preparation of a technology roadmap that will set forth a plan for research, development, and demonstration of the most promising next-generation advanced reactor concepts. These reactor designs hold high potential for meeting the needs for economic, emission-free, sustainable power generation. R&D will be conducted to increase fuel lifetime, recycle used nuclear fuel, establish or improve material compatibility, improve safety performance, reduce system cost, effectively incorporate passive safety features, enhance system reliability, and achieve a high degree of proliferation resistance.

Greenhouse gas affected: Carbon dioxide.

Type of policy or measure: Research, information.

Status of implementation: As ongoing programs, both the NERI and Gen-IV initiatives are under implementation.

Implementing entities: NERI features a cooperative, peer-reviewed selection process to fund researcher-initiated R&D proposals from universities, national laboratories, and industry. The Gen-IV program is an international effort, in which the United States and other member countries of the Generation IV International Forum (GIF) are jointly developing nuclear energy systems that offer advantages in the areas of economics, safety, reliability, and sustainability and that could be deployed commercially by 2030. A major advantage of this arrangement is that funding for the projects is leveraged among the GIF member countries.

Non-GHG mitigation benefits of policy or measure: With the NERI and Gen-IV programs, DOE is addressing issues that will enable the expanded use of nuclear energy. For the longer term, the DOE believes that Gen-IV nuclear energy and fuel cycle technologies can play a vital role in fulfilling the nation's long-term energy needs. Growing concerns for the environment will favor energy sources that can satisfy the need for electricity and other energy-intensive products on a sustainable basis with minimal environmental impact.

Interaction with other policies or measures: The Gen-IV and NERI programs, and other nuclear energy R&D programs conducted by DOE, support the goal stated in the President's *National Energy Policy* of increasing the development and use of nuclear power as a non-greenhouse gas-emitting source of electricity for the nation.

Support Deployment of New Nuclear Power Plants in the United States

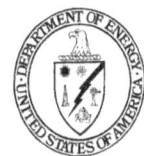

Description: To cope with U.S. near-term needs for nuclear energy, DOE organized a Near-Term Deployment Group (NTDG). The group was tasked with developing a Near-Term Deployment Roadmap ("NTD Roadmap") that would provide conclusions and recommendations to facilitate deployment of new nuclear plants in the United States by 2010. Implementation of these recommendations will be realized through DOE's Nuclear Energy Technologies Program—Nuclear Power 2010.

Objectives: The NTD Roadmap provides DOE and the nuclear industry with the basis for a plan to ensure the availability of near-term nuclear energy options that can be in operation in the United States by 2010. It focuses on making available by 2010 a range of competitive, NRC-certified and/or ready to construct nuclear energy generation options of a range of sizes to meet variations in market need.

The NTD Roadmap identifies the technological, regulatory, and institutional gaps and issues that need to be addressed for new nuclear plants to be deployed in the United States in this time frame. It also identifies specific designs that could be deployed by 2010, along with the actions and resources needed to ensure their availability.

Greenhouse gases affected: Carbon dioxide, carbon monoxide, nitrous oxide, methane, and volatile organic compounds.

Type of policy or measure: Information.

Status of implementation: The NTDG submitted the NTD Roadmap to DOE on October 31, 2001. The Nuclear Energy Research Advisory Committee unanimously endorsed the NTD Roadmap recommendations on November 6, 2001.

Implementing entities: As part of the Nuclear Energy Technologies Program, DOE NE-20 has been in working in collaboration with industry and the Nuclear Regulatory Commission to implement near-term needs identified by the NTDG during fiscal year 2001. Fiscal year 2002 activities include continued DOE/industry cost-shared projects to demonstrate the Early Site Permitting process, support advanced gas-cooled reactor fuel qualification and testing, and conduct preliminary advanced reactor technology R&D recommended in the NTD Roadmap.

Non-GHG mitigation benefits of policy or measure: The deployment of new nuclear power plants could substantially resolve the growing U.S. energy supply deficit. It would also provide for an appropriate and secure energy mix that could help achieve Clean Air Act requirements without harming the U.S. economy.

Interaction with other policies or measures: The NTD Roadmap supports the goal stated in the President's *National Energy Policy* of increasing the development and use of nuclear power as a non-greenhouse gas-emitting source of electricity for the nation.

Carbon Sequestration

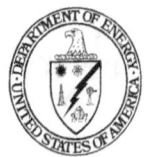

Description: This program develops strategies for the removal of carbon dioxide from man-made emissions or the atmosphere; the safe, essentially permanent storage of carbon dioxide or other carbon compounds; and the reuse of carbon dioxide through chemical or biological conversion to value-added products. The program has five major components: separation and capture, ocean storage, storage in terrestrial ecosystems, storage in geological formations, and conversion and utilization.

Objectives: The primary objectives of the carbon sequestration program are to lower the cost of capturing carbon dioxide, to ensure that the storage of carbon dioxide in geological formations is safe and environmentally secure, and to enhance the productivity and storage of carbon in terrestrial systems.

Greenhouse gas affected: Carbon dioxide.

Type of policy or measure: Research.

Status of implementation: Terrestrial sequestration is underway, and field experiments in geological sequestration are imminent.

Implementing entities: Federal government R&D in partnership with private sector.

Non-GHG mitigation benefits of policy or measure: This program increases the production of oil and natural gas (geological sequestration), reclaims poorly managed lands, and prevents soil erosion and stream sedimentation (terrestrial sequestration).

Hydropower Program

Description: DOE's Hydropower Program develops, conducts, and coordinates hydropower research and development with industry and other federal agencies. Hydropower is a mature technology and has long provided a significant contribution to the national energy supply. Hydropower research today centers on boosting the efficiency of existing hydropower facilities, including incremental hydropower gains. In addition, the program works on developing advanced turbines that reduce fish mortality, use improved sensor technology to understand conditions inside operating turbines, improve compliance with federal water quality standards, and reduce greenhouse gas emissions.

Objective: This program aims to improve the technical, societal, and environmental benefits of hydropower.

Greenhouse gases affected: Carbon dioxide, carbon monoxide, nitrous oxide, methane, and volatile organic compounds.

Type of policy or measure: Research, information.

Status of implementation: Implemented.

Implementing entities: DOE, other federal agencies, and industry partners. DOE's Office of Biopower and Hydropower Technologies administers the program through the DOE Idaho Operations Office.

Non-GHG mitigation benefits of policy or measure: This program has non-greenhouse gas environmental benefits, improves power reliability, and increases the nation's energy security.

International Programs

Description: DOE's International Programs fall under the Office of Technology Access, which promotes exports of renewable energy and energy-efficient products and services and facilitates private-sector infrastructure development to support the delivery and maintenance of these technologies worldwide. The office also provides these same information and technical assistance services to Native Americans on a government-to-government basis.

Objectives: The International Programs aim to service DOE's many Memoranda of Understanding on international energy issues, provide diplomatic and technical assistance to the White House and State Department, and establish a framework to assist Native American governments.

Greenhouse gases affected: Carbon dioxide, carbon monoxide, nitrous oxide, methane, and volatile organic compounds.

Type of policy or measure: Outreach, education.

Status of implementation: Implemented.

Implementing entities: DOE, industry, other government agencies, and international government and nongovernment agencies.

Non-GHG mitigation benefits of policy or measure: This program has non-greenhouse gas environmental benefits, such as reducing SO_x emissions; improves energy reliability; and educates the public internationally on the benefits of energy efficiency and renewable energy and all the benefits associated with overall energy efficiency and renewable energies.

Economic Incentives/Tax Credits

Description: Current law provides taxpayers a 1.5 cent-per-kilowatt-hour (adjusted for inflation after 1992) tax credit for electricity produced from wind, "closed-loop" biomass, and poultry waste. Biomass refers to trees, crops, and agricultural wastes used to produce power, fuels, or chemicals. The electricity must be sold to an unrelated third party, and the credit applies to the first 10 years of production. The current tax credit covers facilities placed in service before January 1, 2002, after which it expires. The new proposal would:

- Extend for three years the 1.5 cent-per-kilowatt-hour biomass credit for facilities placed in service before July 1, 2005.
- Expand the definition of eligible biomass to include certain forest-related resources and agricultural and other sources for facilities placed in service before January 1, 2002. Electricity produced at such facilities from newly eligible sources would be eligible for the credit only from January 1, 2002, through December 31, 2004. The credit for electricity from newly eligible sources would be computed at a rate equal to 60 percent of the generally applicable rate. And the credit for electricity produced from newly eligible biomass co-fired in coal plants would be computed at a rate equal to 30 percent of the generally applicable rate.
- In the case of a wind or biomass facility operated by a lessee, the proposal would permit the lessee, rather than the owner, to claim the credit. This rule would apply to production under the leases entered into after the date on which the proposal is enacted.

Objective: These tax credits aim to accelerate the market penetration of wind- and biomass-based electric generators.

Greenhouse gases affected: Carbon dioxide, nitrous oxide, and carbon monoxide.

Type of policy or measure: Economic.

Status of implementation: These tax credits are in the proposal stage.

Transportation

FreedomCAR Research Partnership

Description: This partnership seeks to substantially improve vehicle fuel efficiency and reduce carbon emissions associated with cars, light trucks, and sport-utility vehicles. FreedomCAR focuses on the long-term, high-risk research needed to achieve a vision of emission- and petroleum-free passenger vehicles, without sacrificing freedom of mobility and freedom of vehicle choice.

Objective: FreedomCAR's mission is to develop a technology and fuel that will reduce consumption of petroleum-based fuel and reduce carbon emissions.

Greenhouse gases affected: Carbon dioxide and other vehicle-related criteria pollutants.

Type of policy or measure: Research and development.

Status of implementation: Adopted.

Implementing entities: This partnership is between DOE and the U.S. Council for Automotive Research (USCAR). Other U.S. government agencies, including EPA and the Department of Transportation (DOT), will participate through related advances in their own programs. The government will seek a cooperative relationship with suppliers and other companies conducting substantial automotive research and development activities in the United States.

Non-GHG mitigation benefits of policy or measure: The maturation of fuel cell technologies for transportation is a major focus of FreedomCAR. Fuel cell vehicles will be free of petroleum, criteria pollutants, and carbon dioxide emissions.

Interaction with other policies or measures: The new partnership supersedes and builds upon the successes of the Partnership for a New Generation of Vehicles (PNGV), which began in 1993. However, FreedomCAR is different in scope and breadth. It shifts government research to more fundamental, higher-risk activities, with applicability to multiple-passenger vehicle models and special emphasis on development of transportation fuel cells and related hydrogen fuel infrastructure.

The transition to a hydrogen fuel cell-powered energy system requires significant investment in order to successfully overcome critical remaining barriers. Since considerable time will be required before fuel cells in transportation become a reality, FreedomCAR also continues support for other technologies that have the potential in the interim to dramatically reduce oil consumption and environmental impacts, and/or are applicable to both fuel cell and hybrid approaches—e.g., batteries, electronics, and motors.

Vehicle Systems R&D

Description: DOE's Office of Heavy Vehicle Technologies works with its industry partners and their suppliers to research and develop technologies that make heavy vehicles more energy efficient and able to use alternative fuels, while reducing vehicle emissions.

Objective: This program aims to encourage optimum performance and efficiency in trucks and other heavy vehicles.

Greenhouse gases affected: Carbon dioxide, carbon monoxide, nitrous oxide, methane, and volatile organic compounds.

Type of policy or measure: Research, information.

Status of implementation: Implemented.

Implementing entities: DOE and national laboratories.

Non-GHG mitigation benefits of policy or measure: This program increases energy and national security, boosts energy efficiency, reduces reliance on foreign energy sources, supports the economy through more efficient transportation of goods, and improves safety through advanced truck materials.

Clean Cities

Description: DOE's Clean Cities program supports public–private partnerships that deploy alternative-fuel vehicles (AFVs) and build supporting infrastructure, including community networks. Clean Cities works directly with local businesses and governments, guiding them through each step in the process of building the foundation for a vibrant local organization, including goal-setting, coalition-building, and securing commitments. Current and potential members of the Clean Cities network also help each other by sharing local innovations, addressing and relaying obstacles they encounter in pursuing alternative-fuel programs, and exchanging "do's" and "don'ts," based on experiences in these programs. Clean Cities continually pioneers innovations and aspires to make strides nationally as well as locally.

Objective: By encouraging AFV use, Clean Cities aims to help cities enhance their energy security and air quality.

Greenhouse gases affected: Carbon dioxide, carbon monoxide, nitrous oxide, methane, and volatile organic compounds.

Type of policy or measure: Voluntary, information.

Status of implementation: Implemented.

Implementing entities: DOE, local stakeholders, and local governments.

Non-GHG mitigation benefits of policy or measure: Clean Cities increases energy efficiency, promotes private–public cooperation and information sharing, provides answers to complex issues, builds a network of contacts, and educates the public.

Biofuels Program

Description: Sponsored by DOE's Office of Fuels Development, the Biofuels Program researches, develops, demonstrates, and facilitates the commercialization of biomass-based, environmentally sound, cost-competitive U.S. technologies to develop clean fuels for transportation, leading to the establishment of a major biofuels industry. The program is currently pursuing the development of conversion technologies for bioethanol and biodiesel fuels. It encourages the use of biomass sources, such as wastepaper and wood residues, to serve near-term niche markets as a bridging strategy to position the biofuels industry for the long-term bulk fuel markets. To meet these ends, the program focuses on researching and developing integrated biofuels systems; creating strategic partnerships with U.S. industry and other stakeholders; and improving the program's operations through well-defined metrics, communication, and coordination with stakeholders and customers.

Objective: The Biofuels Program aims to encourage the large-scale use of environmentally sound, cost-competitive, biomass-based transportation fuels.

Greenhouse gases affected: Carbon dioxide, carbon monoxide, nitrous oxide, methane, and volatile organic compounds.

Type of policy or measure: Research, information.

Status of implementation: Implemented.

Implementing entities: DOE, national laboratories and private-sector partners (industry, individuals, and research organizations).

Non-GHG mitigation benefits of policy or measure: The Biofuels Program increases energy efficiency; reduces reliance on foreign energy sources; promotes the industry internationally, the commercialization of bio-based products, and renewable resources; creates jobs; and provides a larger market for agricultural goods.

Commuter Options Programs: Commuter Choice Leadership Initiative; Parking Cash-Out; Transit Check; Telecommute Initiative; Others[6]

Descriptions:[7] EPA sponsors a number of voluntary commuter initiatives to reduce emissions of greenhouse gases and criteria pollutants from the transportation sector:

- The *Commuter Choice Leadership Initiative* is a voluntary employer-adopted program that helps to increase commuter flexibility by expanding mode options, arranging flexible scheduling, and offering work location choices. EPA provides a variety of technical support measures and recognition. Commuter Choice has also been implemented for workers at all federal agencies.
- *Parking Cash-Out* is a benefit in which employers offer employees the option to receive taxable income in lieu of a free or subsidized parking space at work. A similar set of tax law changes allows employers to offer nontaxable transit/vanpool benefits, currently up to $100 monthly.
- The *National Environmental Policy Institute* (NEPI) initiated an incentive-based pilot Telecommuting Initiative that provides employers with tradable criteria pollutant emission credits for reducing vehicle miles traveled from telecommuting workers and is working to include greenhouse gases. Given rapid technological advances, telecommuting offers substantial opportunity to reduce the need for some employees to travel to work.

Objective: These programs help to reduce growth in single-occupant-vehicle commuting by providing incentives and alternative modes, timing, and locations for work.

Greenhouse gases affected: The principal greenhouse gas affected in the transportation sector is carbon dioxide. However, transportation actions also contribute to reductions of nitrous oxide and methane.

Type of policy or measure: Voluntary and negotiated agreements, tax incentives to employers and employees, information, education, and outreach.

Status of implementation: Launched in 2000, EPA's Commuter Choice Leadership Initiative intends to sign up 550 employers by end of 2002. The Taxpayer Relief Act of 1997 put Parking Cash Out and Transit Check into effective practice. A number of states, notably California, have implemented measures to encourage Parking Cash Out. NEPI is launching the Telecommuting Initiative effort in 2001 in five major metropolitan areas. EPA estimates greenhouse gas emission reductions of 3.5 teragrams of CO_2 in 2000, and projects reductions of more than 14 teragrams of CO_2 in 2010.

Implementing entities: Commuter Choice—EPA and DOT, in partnership with employers; Parking Cash Out and Transit Check—individual employers, through the revision in the Internal Revenue Service Code; Telecommuting Initiative—NEPI in collaboration with EPA, DOT, and DOE.

Costs of policy or measure: These programs impose modest, voluntarily borne costs on businesses, which are largely offset by other savings. Commuter option programs generate benefits through increased employee productivity, satisfaction, and lower taxes. Participants in the telecommuting initiative can sell emission credits on the open market or to states for use in state implementation plans. Educational and outreach programs pose no direct costs on businesses.

Non-GHG mitigation benefits of policy or measure: These programs will reduce energy use, traffic congestion, and criteria pollutant emissions, including nitrogen oxides and volatile organic compounds (which are ozone precursors considered to have indirect global warming potential).

Interaction with other policies or measures: These programs are synergistic with one another, with "smart growth" and transit programs, and with state implementation plans and other required measures under the Clean Air Act.

---

[6] Part of Action 20 in the 1997 *U.S. Climate Action Report.*
[7] These commuter options replace EPA's *Transportation Partners Program.*

Smart Growth and Brownfields Policies

Description: EPA began the Air-Brownfields Pilot Program in response to concerns that air regulations were preventing the redevelopment of brownfields, which are abandoned industrial properties that may be moderately contaminated. The program demonstrated that brownfield redevelopment and local land-use policies, such as infill and transit-oriented development, could help reduce vehicle miles traveled.

EPA issued *Improving Air Quality Through Land Use Activities* in 2001 on how to take credit in a state implementation plan (SIP) for local land-use policies that reduce emissions. Many cities have launched initiatives to encourage such development and plan to increase development beyond what was anticipated in their Clean Air Act SIP submissions.

Other brownfield initiatives include three types of grants: Assessment Demonstration Pilots, to assess brownfield sites and test cleanup and redevelopment models; Job Training Pilots, to train residents of affected communities to facilitate cleanup and work in the environmental field; and Cleanup Revolving Loan Fund Pilots, which capitalize loan funds for cleaning up brownfields.

The Smart Growth Network funds and facilitates a variety of smart growth-supportive activities and forums. The American Planning Association's Growing Smarter Program plans to target state government officials with a National Planning Statute Clearinghouse and Database, and the *Growing Smart™ Legislative Guidebook*, which will include model statutes for transportation demand management. Additionally, a consensus has emerged that shifting funding to transit, nonmotorized modes, and other alternatives more compatible with Smart Growth can increase demand for these alternatives, facilitate infill development, and decrease vehicle miles traveled and greenhouse gas emissions. Federal research and outreach have increased the inclusion of these induced demand/land-use issues into transportation models and planning processes.

Objective: These initiatives help to reduce the length and number of motorized trips.

Greenhouse gases affected: Primarily carbon dioxide, but also nitrous oxide and methane.

Type of policy or measure: Technical assistance, outreach, and voluntary acceptance of air- quality credits based on meeting guidance standards.

Status of implementation: The Air-Brownfields Pilot Program is complete, the land-use SIP guidance based on it has been pilot-tested in four cities, and credit issuance begins in 2001. Technical assistance underway includes over 350 Assessment Demonstration pilot programs, over 100 Loan Fund pilot programs, and nearly 50 Job Training and Development pilot programs. EPA estimates reductions of 2.7 teragrams of CO_2 in 2000, and projects 11 teragrams of CO_2 by 2010.

Implementing entity or entities: EPA, states, municipalities, and planning agencies.

Costs of policy or measure: Federal guidance for voluntary credits imposes no cost. Private infill and brownfields development remain voluntary market-based decisions, and so impose no private costs.

Non-GHG mitigation benefits of policy or measure: These initiatives reduce energy use, congestion, infrastructure costs, criteria air pollutants, and health threats from contaminated land; increase the tax base; and return contaminated land to productive use.

Interaction with other policies or measures: These initiatives interact with SIPs.

Ground Freight Transportation Initiative

Description: This initiative is a voluntary program aimed at reducing emissions from the freight sector through the implementation of advanced management practices and efficient technologies. It will focus on four areas: (1) assessing the most promising technology and management practices and identifying their savings potential; (2) inviting stakeholder participation (associations, independent truckers, fleet managers, state and local governments, manufacturers, etc.) to determine the feasibility of these opportunities and set program performance goals; (3) designing an emissions calculation tool that helps companies determine their environmental impact and identify cost-effective options for reaching the program's performance goals; and (4) developing, implementing, and publicizing a partnership initiative with these stakeholders.

Objective: This program facilitates reductions in the growth of emissions associated with ground freight (truck and rail) through the increased use of efficient management practices, such as speed management, intermodal use and load matching, and advanced technologies, such as idle control systems and aerodynamics.

Greenhouse gases affected: The principal greenhouse gas affected in the transportation sector is carbon dioxide. However, transportation actions also contribute to reductions of nitrous oxide.

Type of policy or measure: Voluntary and negotiated agreements, shipper policy changes, information, education, and outreach.

Status of implementation: The program was kicked off in December 2001; a full program launch will occur in the summer of 2002. EPA projects greenhouse gas emission reductions of 66 teragrams of CO_2 in 2010.

Implementing entities: EPA and possibly DOT. Other organizations, such as the American Trucking Association and the American Association of Railroads, will prove to be valuable allies in encouraging their members to join the initiative as member companies.

Costs of policy or measure: Similar programs impose modest, voluntarily borne costs on businesses, which are largely offset by other savings. Some options may have substantial financial investments, such as truck stop electrification. Three different stakeholder groups, including shippers, carriers, and manufacturers, will decide which strategies are most effective in their implementation and return on investments. Educational and outreach programs pose no direct costs on businesses.

Non-GHG mitigation benefits of policy or measure: The program will reduce energy use, traffic congestion, and criteria pollutant emissions, including nitrogen oxides and volatile organic compounds—ozone precursors considered to have indirect global warming potential.

Interaction with other policies or measures: This program is synergistic with state implementation plans and other required measures under the Clean Air Act.

Clean Automotive Technology

Description: EPA's Clean Automotive Technology (CAT) program is a research and partnership program with the automotive industry to develop advanced clean and fuel-efficient automotive technology.

Objectives: The program's objectives are to develop break-through engine and powertrain technologies to provide dramatic fuel economy improvement in cars and trucks—without sacrificing affordability, performance, or safety while meeting emissions standards.

Greenhouse gases affected: Primarily carbon dioxide, but also nitrous oxide and methane.

Type of policy or measure: Voluntary, research.

Status of implementation: EPA has demonstrated the CAT program's potential to meet its objectives. EPA is collaborating with its partners to transfer the unique EPA-patented highly efficient hybrid engine and powertrain components, originally developed for passenger car applications, to meet the more demanding size, performance, durability, and towing requirements of sport utility vehicles and urban delivery vehicle applications, while being practical and affordable with ultra-low emissions and ultra-high fuel efficiency. In 2001, the program signed a historic Cooperative Research and Development Agreement and License Agreement with the Ford Corporation to invest further develop hydraulic hybrid and high-efficiency engine technology with an aim toward putting a pilot fleet of vehicles on the road by the end of the decade.

Implementing entities: EPA and the National Vehicle and Fuel Emissions Laboratory working in collaboration with the Ford and Eaton Corporations.

Non-GHG mitigation benefits of policy or measure: This partnership increases energy efficiency, economic productivity, and competitiveness; reduces energy dependence; expands the nation's energy portfolio; has non-GHG environmental benefits; strengthens public–private cooperation and interaction; and creates jobs.

Interaction with other policies or measures: CAT could interact with state implementation plans and other Clean Air Act requirements.

DOT Emission-Reducing Initiatives

DOT provides funding for and oversees transportation projects and programs that are implemented by the states and metropolitan areas across the country. Funding is provided under numerous programs that have specific purposes broadly encompassed by DOT's five main goals in the areas of: safety, mobility, economic growth and trade, national security, and human and natural environment.

Flexibility exists under the law to use program funds for a variety of different project types that are consistent with the overall purposes of those funds. As such, highway funds may be used for transit, pedestrian improvements, bikeways, ride-sharing programs, and other transportation-demand-management projects, a well as system improvements on the road network. The approach is decentralized in that, based on their own needs assessments, state and local governments determine what projects should be implemented and use DOT funds in ways consistent with the purpose of the funding program.

From 1998 and through 2003, approximately $218 billion is available to the states and metropolitan areas under DOT's surface transportation programs. While none of these programs specifically targets greenhouse gas reduction, many of them reduce greenhouse gases as an ancillary benefit. Estimating the amount of greenhouse gases reduced is very difficult, since project selection is left to the individual states and metropolitan areas, and this benefit will vary among projects. Following is a sampling of some of the more significant DOT programs that are likely to have ancillary greenhouse gas-reduction benefits.

- *Transit Programs:* Under the current authorization, transit programs will receive $41 billion between fiscal years 1998 and 2003. Programs that allow funding for new starts of transit systems, fixed guideway modernization, bus system improvements and expansions, and high-speed rail development can have greenhouse gas reduction benefits. However, not all of the transit funding will have these benefits, since projects that help to operate or maintain the current system will probably not attract new riders.
- *Congestion Mitigation and Air Quality Improvement:* This program is targeted at reducing ozone, carbon monoxide, and particulate matter generated by transportation sources. As the most flexible program under the current law, it funds new transit services, bicycle and pedestrian improvements, alternative fuel projects, traffic flow improvements, and other emission-reducing projects. As such, several projects funded under the program will likely reduce greenhouse gases as well. This program provides about $1.35 billion a year to the states.
- *Transportation Enhancements:* Historically, about half of all Enhancement funding has been for bicycle and pedestrian improvements, which certainly have some greenhouse gas reduction benefits. The Transportation Equity Act for the 21st Century has authorized about $560 million a year is for Enhancement activities over a six-year period.
- *Transportation and Community System Preservation Pilot Program:* This unique pilot program helps develop more livable communities by addressing environmental, economic, and equity needs. States, local governments, and metropolitan planning organizations are eligible for discretionary grants to plan and implement strategies that improve the efficiency of the transportation system, reduce environmental impacts and the need for costly future public infrastructure investments, and ensure efficient access to jobs, services, and centers of trade. A total of $120 million is authorized for this program from fiscal years 1999 through 2003.
- *Corporate Average Fuel Economy (CAFE) Standards:* U.S. fuel economy standards for automobiles and light trucks were adopted primarily to save energy. Compliance is based on average performance, and additional credit toward compliance is available to alternatively fueled vehicles. New vehicles offered for sale are also required to display labels that give consumers a clear indication of fuel economy. DOT is currently examining other market-based approaches to increase the average fuel economy of new vehicles, and will review and provide recommendations on future CAFE standards.

Industry (Non-CO$_2$)

Natural Gas STAR[8]

Description: This a voluntary partnership between EPA and the U.S. natural gas industry is designed to overcome barriers to the adoption of cost-effective technologies and practices that reduce methane emissions.

Objective: The program's primary objective is to reduce methane emissions from U.S. natural gas systems.

Greenhouse gas affected: Methane.

Type of policy or measure: Voluntary/negotiated agreement.

Status of implementation: Launched in 1993 with the transmission and distribution sectors, Natural Gas STAR has since expanded twice—to the production sector in 1995 and the processing sector in 2000. The program includes 88 corporate partners representing 40 percent of U.S. natural gas production, 72 percent of transmission company pipeline miles, 49 percent of distribution company service connections, and 23 percent of processing throughput.

Natural Gas STAR has developed a range of tools designed to help corporate partners implement best management practices to reduce leakage. These include an implementation guide, streamlined electronic reporting, a series of "lessons learned" studies, focused workshops, and partner-to-partner information exchanges. Extensive partner support for and continued expansion of the program, combined with ongoing feedback from partners, demonstrate the effectiveness of these tools in promoting methane reduction activities.

EPA estimates that the program reduced 15 teragrams of CO$_2$ equivalent (38 Bcf methane) in 2000. Because of the expanded program's tremendous success, EPA projects the program will reduce 22 teragrams of CO$_2$ equivalent by 2010.

Implementing entities: EPA, in partnership with the U.S. natural gas industry.

Costs of policy or measure: Through Natural Gas STAR, partner companies implement only cost-effective methane reduction practices. Practices implemented since the program's launch have saved U.S. natural gas companies billions of dollars worth of gas that would otherwise have leaked to the atmosphere.

Non-GHG mitigation benefits of policy or measure: Many of the practices that partner companies undertake to reduce methane emissions also reduce emissions of air pollutants and improve safety.

Interaction with other policies or measures: None.

Contact: Paul Gunning, EPA, Climate Protection Partnerships Division, (202) 564-9736, gunning.paul@epa.gov.

---

[8] Action 32 in the 1997 *U.S.Climate Action Report;* continuing.

Coalbed Methane Outreach Program[9]

Description: This program reduces methane emissions associated with coal mining operations by (1) working with the coal industry and other stakeholders to identify and remove obstacles to increased investment in coalbed methane recovery projects, and (2) raising awareness of opportunities for profitable investments.

Objective: The program aims to cost-effectively reduce methane emissions from U.S. coal mining operations.

Greenhouse gas affected: Methane.

Type of policy or measure: Information, education, and outreach.

Status of implementation: EPA began working with the coal mining industry in1990 and officially launched the Coalbed Methane Outreach Program (CMOP) in 1994. In 1990, coal mines captured and utilized only 25 percent of the methane produced from their degasification systems. By 1999, the recovery fraction had grown to over 85 percent. To eliminate the remaining methane emitted from degasification systems, CMOP is working with industry to demonstrate the use of flare technology, which has never been employed at a U.S. mine.

With the program's tremendous success in reducing methane emissions from degasification systems, CMOP has expanded its focus to the methane emitted from coal mine ventilation systems. Ventilation air from coal mines typically contains methane at concentrations of just a few percent, yet accounts for 94 percent of the remaining methane emissions from underground coal mines—over 90 billion cubic feet of methane (about 36.6 teragrams of CO_2 equivalent) annually. CMOP is working with industry to demonstrate and deploy newly developed technologies that can reduce these emissions substantially over the next few years.

CMOP has developed a range of tools designed to overcome barriers to recovery and combustion of coal mine methane. These include numerous technical and economic analyses of technologies and potential projects; mine-specific project feasibility assessments; state-specific analyses of project potential; guides to state, local, and federal assistance programs; and market evaluations. CMOP has worked with operators of virtually every U.S. underground coal mine to apply these tools and nurture each project.

In 2000, EPA estimates that CMOP reduced methane emissions by more than 7 teragrams of CO_2 equivalent (19 Bcf methane). Because of unanticipated mine closures, EPA projections of reductions for the CMOP program have been reduced slightly since the 1997 submission, from 11 to 10 teragrams of CO_2 equivalent in 2010. However, CMOP's expected success in reducing ventilation air methane over the next few years may lead to an upward revision in the projected reductions for 2010 and beyond.

Implementing entities: EPA, in partnership with the U.S. coal industry.

Costs of policy or measure: Coal mines implement only cost-effective methane recovery and utilization projects. Projects implemented since the program's launch have earned U.S. coal companies million of dollars in energy sales.

Non-GHG mitigation benefits of policy or measure: CMOP improves both the efficiency of methane recovery from coal mines and mine safety.

Interaction with other policies or measures: None.

Contact: Karl Schultz, EPA, Climate Protection Partnerships Division, (202) 564-9468, schultz.karl@epa.gov.

---

[9] Action 35 in the 1997 *U.S.Climate Action Report*; continuing.

Significant New Alternatives Program[10]

Description: Section 612 of the Clean Air Act authorized EPA to develop a program for evaluating alternatives to ozone-depleting chemicals.

Objective: The Significant New Alternatives Program (SNAP) facilitates the smooth transition away from ozone-depleting chemicals in major industrial and consumer sectors, while minimizing risks to human health and the environment. Sectors that the program focuses on include air conditioning, refrigeration, aerosols, solvent cleaning, foams, fire suppression and explosion protection, adhesives, coatings and inks, and sterilants.

Greenhouse gases affected: Hydrofluorocarbons (HFCs) and perfluorocarbons (PFCs).

Type of policy or measure: While SNAP actions are regulatory, the program also serves as an information clearinghouse on alternative chemicals and technologies, and collaborates extensively with industry and other government partners on various research activities.

Status of implementation: Hundreds of alternatives determined to reduce overall risks to human health and the environment have been listed as acceptable substitutes for ozone-depleting chemicals. EPA has also used the authority under Section 612 to find unacceptable uses or narrow the scope of uses allowed for HFCs and PFCs with high global warming potentials for specific applications where better alternatives exist. EPA estimates that the program has reduced emissions by 50 teragrams of CO_2 equivalent in 2000 and projects reductions of 156 teragrams of CO2 equivalent in 2010.

Implementing entity: SNAP regulations are promulgated by EPA and enforced when needed at the national level.

Costs of policy or measure: Costs are considered to be neutral in aggregate. SNAP either expands lists of available alternatives to ozone-depleting chemicals that have been, or are being, phased out under the *Montreal Protocol*, or restricts the use of potential substitutes. In the first case, potential users are not required to use any one particular alternative listed as acceptable. Where SNAP finds the use of alternatives (e.g., PFCs) unacceptable, the decision is based on the fact that other viable (i.e., effective and affordable) alternatives are available that pose less risk to human health or the environment.

Non-GHG mitigation benefits of policy or measure: In addition to encouraging responsible use of greenhouse gases as substitutes for ozone-depleting chemicals, SNAP has increased worker and consumer safety by restricting the use of flammable or toxic chemicals, has encouraged the overall reduction in chemicals used in various applications (e.g., solvent cleaning), and, in some cases, has restricted the use of volatile organic chemicals that generate ground-level ozone.

Interaction with other policies or measures: SNAP compliments the phase-out of ozone-depleting chemicals mandated under the *Montreal Protocol* and Clean Air Act. The program has worked to maintain balance between the need to find safe and effective alternatives to ozone-depleting chemicals, while mitigating the potential effects of those alternatives on climate. HFCs, and in some cases, PFCs, have been listed as acceptable substitutes for specific end uses where safer or effective alternatives are not available. Depending on the end use, efficacy has been defined as effectiveness in suppressing or preventing fires and explosions, thermal insulation value, or heat transfer efficiency.

Contact: Jeff Cohen, EPA, Global Programs Division, (202) 564-0135, cohen.jeff@epa.gov.

---

[10] Action 40 in the 1997 *U.S. Climate Action Report.*

HFC-23 Partnership[11]

Description: This partnership works to cost-effectively reduce emissions of the potent greenhouse gas HFC-23, which is a by-product in the manufacture of HCFC-22.

Objective: Through this program, EPA encourages companies to develop and implement technically feasible, cost-effective processing practices or technologies to reduce HFC-23 emissions.

Greenhouse gas affected: HFC-23.

Type of policy or measure: Voluntary/negotiated agreement.

Status of implementation: This is an ongoing program with all the U.S. producers of HCFC-22. The program partners have effectively reduced emissions of HFC-23 through process optimization, reaching the total reductions that can likely be achieved through this technique. In addition, some companies have used thermal destruction to reduce or eliminate their emissions. The partnership has encouraged the industry to reduce the intensity of HFC-23 emissions (the amount of HFC-23 emitted per kilogram of HCFC-22 manufactured) by 35 percent. Thus, despite an estimated 35 percent increase in production since 1990, total emissions have declined by 15 percent. EPA estimates reductions of 17 teragrams of CO_2 equivalent in 2000 and projects reductions of 27 teragrams of CO_2 equivalent in 2010.

Implementing entity: EPA is the sole government entity implementing this program. The program is open to all producers of HCFC-22 operating in the United States.

Costs of policy or measure: Emission reductions achieved through process optimization are cost-effective.

Non-GHG mitigation benefits of policy or measure: None.

Interaction with other policies or measures: None.

Contact: Sally Rand, EPA, Global Programs Division, (202) 564-9739, rand.sally@epa.gov.

---

[11] Action 41 in the 1997 *U.S.Climate Action Report.*

Partnership with Aluminum Producers[12]

Description: This partnership program with the primary aluminum smelting industry is designed to reduce perfluorocarbons emitted as a by-product of the smelting process.

Objective: EPA is partnering with primary aluminum producers to reduce perfluoromethane and perfluoroethane where technically feasible and cost-effective. The overall goal of the partnership is to reduce emissions by 30–60 percent from 1990 levels by 2000. Future reduction goals are being set.

Greenhouse gas affected: Perfluoromethane and perfluoroethane.

Type of policy or measure: Voluntary/negotiated agreement.

Status of implementation: Since the partnership was formed in 1996, it has had great success in further characterizing the emissions from smelter operations and reducing overall emissions. As of 2000, a new agreement has been negotiated to continue to explore and implement emission reduction options through 2005. The overall goal for the program in 2000 has been met, with emissions reduced by about 50 percent relative to 1990 levels, on an emissions per unit of product basis. Absolute emissions have been reduced by an even greater percentage because some facilities have closed due to high energy costs in the Northwest. EPA estimates reductions of 7 teragrams of CO_2 equivalent in 2000 and projects reductions of 10 teragrams of CO_2 equivalent in 2010.

Implementing entity: EPA is the sole government entity implementing this program. The program is open to all U.S. primary aluminum producers.

Costs of policy or measure: Factors that cause these emissions are a sign of efficiency loss. Emission reductions result in process enhancements.

Non-GHG mitigation benefits of policy or measure: None.

Interaction with other policies or measures: None.

Contact: Sally Rand, EPA, Global Programs Division, (202) 564-9739, rand.sally@epa.gov.

---

[12] Action 42 in the 1997 *U.S. Climate Action Report*.

Environmental Stewardship Initiative[13]

Description: Environmental Stewardship Initiative was a new action proposed as part of the 1997 *U.S. Climate Action Report*, based on new opportunities to reduce emissions gases with high global warming potentials.

Objective: The objective initially was to limit emissions of hydrofluorocarbons, perfluororcarbons, and sulfur hexafluoride (which are potent greenhouse gases) in three industrial applications: semiconductor production, electric power systems, and magnesium production. Additional sectors are being assessed for the availability of cost-effective emission reduction opportunities and are being added to this initiative.

Greenhouse gases affected: Hydrofluorocarbons, perfluororcarbons, and sulfur hexafluoride.

Type of policy or measure: Voluntary/negotiated agreement.

Status of implementation: EPA launched the semiconductor partnership in 1996 and launched the electric power system and magnesium partnerships in 1999. Implementation of the magnesium and electric power system partnerships is ongoing, with no sunset date. The semiconductor partnership will be ongoing through 2010. EPA currently projects that the programs will reduce emissions by 93 teragrams of CO_2 equivalent in 2010. Because resource constraints delayed implementation of the electric power system and magnesium partnerships, EPA's estimate of the reduction in 2000, 3 teragrams of CO_2 equivalent, is less than expected.

Implementing entity: EPA is the sole government entity implementing this initiative. Partnerships are open to manufacturers operating in the United States and to electric power systems with equipment containing greater than 15 pounds of sulfur hexafluoride and all primary and die-casting magnesium operations.

Costs of policy or measure: Emission reductions are believed to be possible through inexpensive and cost-effective means.

Non-GHG mitigation benefits of policy or measure: None.

Interaction with other policies or measures: None.

Contact: Sally Rand, EPA, Global Programs Division, (202) 564-9739, rand.sally@epa.gov.

---

[13] New in the 1997 *U.S. Climate Action Report*.

Agriculture

Agriculture Outreach Programs: AgSTAR; Ruminant Livestock Efficiency Program[14]

Description: Specific practices aimed at directly reducing greenhouse gas emissions are developed, tested, and promoted through such outreach programs as AgSTAR and the Ruminant Livestock Efficiency Program (RLEP).

Objectives: Through outreach to the agricultural community, these programs aim to demonstrate the technical feasibility of the practices they promote.

Greenhouse gases affected: All greenhouse gases, but the focus has been on methane.

Type of policy or measure: Voluntary, information.

Status of implementation: These programs have been implemented. Their assessed impacts have changed since the 1997 submission. While their impact on greenhouse gas emissions has been small on a national scale, program stakeholders in the agricultural community have demonstrated that the practices promoted by the programs can be effective in reducing greenhouse gas emissions and increasing productivity.

Twelve digesters have been installed on AgSTAR charter farms, resulting in a 37,000 teragrams of CO_2 equivalent per year reduction of emissions. An additional 13 facilities are in various stages of planning, pending additional funding. Installations at charter farms have demonstrated the technical and economic feasibility of biogas production and utilization on livestock production facilities with a wide range of manure-handling systems. Workshops related to the program have been held around the country to further promote biogas production and utilization technology. In all, 31 systems are operating in the United States, resulting in a total annual reduction of approximately 110,000 teragrams of CO_2 equivalent.

The RLEP has funded the establishment of 50 demonstration farms throughout the Southeast. Production efficiency improvements have been recorded at these farms, and numerous field days have been held to transfer this knowledge to others. The RLEP has also supported the development of a cow/calf management course aimed at improving animal performance measures directly related to greenhouse gas emissions. In addition, with the support of state-level nongovernment organizations, such as the Virginia Forage and Grassland Council, the RLEP has helped to improve forage and pasture management by encouraging the effective use of rotational grazing practices.

EPA and the U.S. Department of Agriculture (USDA) will continue to evaluate these and other barriers and identify appropriate actions to address them.

Implementing entities: EPA and USDA.

Non-GHG mitigation benefits of policy or measure: Technologies used at certain confined animal feeding operations to reduce methane concentrations are achieving other environmental benefits, including odor control and nutrient management opportunities. In addition, many of the practices recommended by the RLEP for improving forage production remove carbon dioxide from the atmosphere by storing carbon in the soil as organic matter.

---

[14] Actions 38 and 39 in the 1997 *U.S. Climate Action Report*.

Nutrient Management Tools[15]

Description: The Nitrogen Leaching and Economic Assessment Package (NLEAP) was enhanced to include the ability to quantify nitrous oxide losses to the atmosphere. USDA began collaborating with partners on the development of two nutrient management tools that could be used to improve overall nitrogen fertilizer use efficiency at the farm level.

Objectives: This effort aims to build and make available to producers a database that documents nitrous oxide emissions from different types of nitrogen fertilizer management. These efforts are intended to improve the overall efficiency of nitrogen fertilizer use at the farm level and to reduce nitrous oxide emissions from the application of nitrogen fertilizer.

Greenhouse gas affected: Nitrous oxide.

Type of policy or measure: Research, information.

Status of implementation: The NLEAP model has been implemented. USDA is working with Purdue University to develop and implement the Manure Management Planner (MMP), a nutrient budgeting tool. MMP enables producers, and others who provide producers nutrient management assistance, to allocate nutrients based on a crop-specific nutrient budget that matches actual nutrient application rates with recommended application rates or crop removal rates. The combination of MMP and NLEAP will enable producers to both develop a detailed crop nutrient budget as well as assess its impact on nitrous oxide emissions. Proper use and crediting of the nitrogen contributed by legume crops, and the availability and use of both NLEAP and MMP, will assist in reducing nitrous oxide emissions. In the 1997 submission, projected reductions from this action were 18.3 teragrams of CO_2 equivalent. At this time, more analysis is needed to develop estimates and projections of emissions from this action.

Implementing entities: USDA, working with partners in 20 states.

---

[15] Part of Action 17 in the 1997 *U.S. Climate Action Report.*

USDA Commodity Credit Corporation Bioenergy Program

Description: USDA's Commodity Credit Corporation (CCC) Bioenergy Program pays U.S. commercial bioenergy producers to increase their bioenergy production from eligible commodities. Payments are based on the increase in bioenergy production compared to the previous year's production fiscal year to date. To receive payments, producers must provide CCC evidence of increased purchase of agricultural commodities and increased production of bioenergy. The program provides up to $150 million for fiscal years 2001 and 2002, which is paid out on a quarterly fiscal year-to-date basis. A payment limitation restricts the amount of funds any single producer may obtain annually under the program to 5 percent, or $7.5 million.

Objective: The program's goal is to expand industrial consumption of agricultural commodities by promoting their use in the production of bioenergy.

Greenhouse gas affected: Carbon dioxide.

Type of policy or measure: Economic.

Status of implementation: The program was implemented at $15 million in fiscal year 2001 and will receive $150 million in fiscal year 2002.

Implementing entities: The program is administered by USDA's Farm Service Agency and funded by CCC.

Non-GHG mitigation benefits of policy or measure: The program provides incentives for agriculture to be part of the nation's energy solutions by promoting the industrial consumption of agricultural commodities for bioenergy production; expands demand for corn and other grains used in ethanol production and creates new markets for oilseed crops; and increases net returns for ethanol and biodiesel processors, which will encourage expanded production capacity for these fuels and enhance rural development.

Conservation Reserve Program: Biomass Project

Description: USDA has implemented Section 769 of the Agriculture, Rural Development, Food and Drug Administration and Related Agencies Appropriations Act of 2000. This act authorizes Conservation Reserve Program (CRP) land for pilot biomass projects for the harvesting of biomass to be used for energy production. The program restricts all land subject to CRP contracts that participates in a biomass pilot project from being harvested for biomass more than once every other year. No more than 25 percent of the total acreage enrolled in any crop-reporting district may be harvested in any year. And participants in a project must agree to a 25 percent reduction in their normal CRP annual rental payment for each year in which the acreage is harvested.

Objective: The project's objective is to provide biomass for energy production.

Greenhouse gas affected: Carbon dioxide.

Type of policy or measure: Economic.

Status of implementation: The program has been implemented. The Secretary of Agriculture has approved four projects that will produce electricity using grasses in Iowa, hybrid poplar trees in Minnesota, willows in New York, and switchgrass in New York and Pennsylvania.

Implementing entity: USDA.

Non-GHG mitigation benefits of policy or measure: This project enhances rural development.

Forestry

Forest Stewardship[16]

Description: USDA's Forest Stewardship and Forest Stewardship Incentive Programs provide technical and financial assistance to nonindustrial, private forest owners. The Forest Stewardship Program helps such owners prepare integrated management plans, and the Stewardship Incentives Program cost-shares up to 75 percent of approved management practices, such as afforestation and reforestation. USDA's Forest Service manages both programs, in cooperation with state forestry agencies. A recent survey of landowners with Forest Stewardship Plans found that they were three times as likely to implement these plans if they received financial and technical assistance.

Objective: The programs' intent is to improve conservation of our lands through enhanced planning and management. An original goal of the Stewardship Incentive Program was to increase tree planting in the United States by over 94,000 hectares (232,180 acres) a year within five years and to maintain this expanded level of planting for another five years.

Greenhouse gas affected: Carbon dioxide.

Type of policy or measure: Voluntary, information.

Status of implementation: The programs have been implemented. During fiscal years 1991–99, 150,964 hectares (372,881 acres) of trees were planted.

Implementing entities: USDA Forest Service in cooperation with state forestry agencies.

Costs of policy or measure: The cost of the program during this same period was about $23.5 million. The program was not funded for fiscal years 1999 through 2001.

Non-GHG mitigation benefits of policy or measure: About 147 million hectares of U.S. forests are nonindustrial, private forestlands. Private forests provide many ecological and economic benefits. They currently provide about 60 percent of our nation's timber supply, with expectations of increases in the future. Improved planning and management on nonindustrial, private forestlands and marginal agricultural lands can help meet resource needs and provide important ancillary benefits that improve environmental quality—e.g., wildlife habitat, soil conservation, water quality protection and improvement, and recreation. Additionally, tree planting and forest management increase the uptake of carbon dioxide and the storage of carbon in living biomass, soils, litter, and long-life wood products.

---

[16] Action 44 in the 1997 *U.S.Climate Action Report.*

Waste Management
Climate and Waste Program[17]

Preserving Resources,
Preventing Waste

Description: This program encourages recycling and source reduction for the purpose of reducing greenhouse emissions. EPA is implementing a number of targeted efforts within this program to achieve its climate goals. WasteWise is EPA's flagship voluntary waste reduction program. EPA initiatives on extended product responsibility and biomass further reduction efforts through voluntary or negotiated agreements with product manufacturers and market development activities. The Pay-As-You-Throw initiative provides information to community-based programs on cost incentives for residential waste reduction.

Objective: The program aims to reduce greenhouse gas emissions through progressive waste management activities.

Greenhouse gases affected: The program takes a life-cycle perspective on greenhouse gas emissions from waste management practices, accounting for emissions and sinks from energy use, forest management, manufacturing, transportation, and waste management. The principal greenhouse gases affected are carbon dioxide and methane; nitrous oxide and perfluorocarbons are also affected.

Type of policy or measure: The program is a voluntary effort, using partnerships, information dissemination, technical assistance, and research to promote greenhouse gas reductions.

Status of implementation: WasteWise currently has over 1,200 partners, representing 53 civic and industrial sectors and ranging from Fortune 1000 companies to small local governments. Extended product responsibility is facilitating negotiations between industry and state leaders on product stewardship systems (e.g., carpets and electronics). The biomass effort includes a compost quality seal program, compost use for state highway projects, and market development for bio-based products. Over 5,000 communities are participating in the program's Pay-As-You-Throw educational initiative, which provides ongoing technical assistance to stakeholders ranging from industry to governments and international organizations. EPA estimates reductions of 8 teragrams of CO_2 equivalent in 2000 and projects reductions of 20 teragrams of CO_2 equivalent in 2010.

Implementing entities: EPA, working with government, industry, and nongovernment organizations, acts as the primary implementing agency.

Costs of policy or measure: Most of the waste-reduction measures result in cost savings or minimal costs when viewed from a full-cost accounting perspective.

Non-GHG mitigation benefits of policy or measure: Measures under this program yield collateral benefits, including energy savings, and reduced emissions from raw materials acquisition, virgin materials manufacturing, and waste disposal.

Interaction with other policies or measures: EPA's Climate and Waste Program has assisted organizations interested in quantifying and voluntarily reporting greenhouse gas emission reductions (e.g., through DOE's 1605b Program) from waste management activities. Also, EPA's activities under Initiative 16 complement its methane reduction programs (Actions 33 and 34), including the Landfill Methane Outreach Program.

---

[17] Part of Action 16 in the 1997 *U.S. Climate Action Report*; continuing.

Stringent Landfill Rule[18]

Description: Landfill gas, which is the largest contributor to U.S. anthropogenic methane emissions, also contains significant quantities of nonmethane organic compounds. Landfill New Source Performance Standards and Emissions Guidelines (Landfill Rule) require large landfills to capture and combust their landfill gas emissions. Due to climate concerns, this rule was made more stringent (i.e., by lowering the emissions level at which landfills must comply with the rule from 100 to 50 megagrams of non-methane organic compounds per year), resulting in greater landfill gas recovery and combustion.

The rule works hand-in-hand with EPA's Landfill Methane Outreach Program to promote cost-effective reductions in methane emissions at larger landfills. The Landfill Methane Outreach Program provides landfills with technical, economic, and outreach information to help them comply with the rule in a way that maximizes benefits to the environment while lowering costs.

Objective: The rule requires U.S. landfills to capture and combust their landfill gas emissions. This reduces their emissions of methane, as well as nonmethane organic compounds.

Greenhouse gas affected: Methane.

Type of policy or measure: Regulatory.

Status of implementation: The Landfill Rule was promulgated under the Clean Air Act in March 1996, and implementation began at the state level in 1998. Preliminary data on the impact of the rule indicate that increasing its stringency has significantly increased the number of landfills that must collect and combust their landfill gas. EPA estimates reductions in 2000 at 15 teragrams of CO_2 equivalent. The current projection for 2010 is 33 teragrams of CO_2 equivalent, although the preliminary data suggest that reductions from the more stringent rule may be even greater over the next decade.

Implementing entities: EPA promulgated the Landfill Rule, and individual states implement it.

Costs of policy or measure: The rule's objective is to reduce nonmethane organic compound emissions because of their contribution to local air pollution. Combustion of the of nonmethane organic compound-containing landfill gas also reduces the methane it contains, at no incremental cost.

Non-GHG mitigation benefits of policy or measure: Combusting landfill gas reduces emissions of nonmethane organic compounds as well as methane. It also can reduce odors and improve safety by stopping landfill gas migration.

Interaction with other policies or measures: The rule interacts with the Landfill Methane Outreach Program.

---

[18] Action 33 in the 1997 *U.S.Climate Action Report;* continuing.

Landfill Methane Outreach Program[19]

Description: Landfills are the largest source of U.S. anthropogenic methane emissions. Capture and use of landfill gas reduce methane emissions directly and carbon dioxide emissions indirectly by displacing the use of fossil fuels. The Landfill Methane Outreach Program (LMOP) works with landfill owners, state energy and environmental agencies, utilities and other energy suppliers, industry, and other stakeholders to lower the barriers to landfill gas-to-energy project development.

While LMOP works hand-in-hand with EPA's Landfill Rule to promote cost-effective reductions in methane emissions at larger landfills, it focuses its outreach efforts on smaller landfills not regulated by the rule, encouraging the capture and use of methane that would otherwise be emitted to the atmosphere. LMOP has developed a range of tools to help landfill operators overcome barriers to project development, including feasibility analyses, software for evaluation project economics, profiles of hundreds of candidate landfills across the country, a project development handbook, and energy end-user analyses.

Objective: The program aims to reduce methane emissions from U.S. landfills.

Greenhouse gases affected: Methane and carbon dioxide.

Type of policy or measure: Voluntary/negotiated agreements, information, education, and outreach.

Status of implementation: Launched in December 1994, LMOP has achieved significant reductions through 2000, reducing methane emissions from landfills by an estimated 11 teragrams of CO_2 equivalent in that year alone. The program includes over 240 allies and partners, and the number of landfill gas-to-energy projects has grown from less than 100 in the early 1990s to almost 320 projects by the end of 2000. EPA projects reductions of 22 teragrams of CO_2 equivalent in 2010.

Implementing entities: EPA, in partnership with landfills and the landfill gas-to-energy industry.

Costs of policy or measure: LMOP participants implement only cost-effective landfill gas-to-energy projects. Projects implemented since the program's launch have created millions of dollars of revenue for public and private landfill owners and others.

Non-GHG mitigation benefits of policy or measure: Combusting landfill gas reduces emissions of nonmethane organic compounds as well as methane. It also can reduce odors and improve safety by stopping landfill gas migration.

Interaction with other policies or measures: The program interacts with the Landfill Rule.

Contact: Paul Gunning, EPA, Climate Protection Partnerships Division, (202) 564-9736, gunning.paul@epa.gov.

---

[19] Action 34 in the 1997 *U.S.Climate Action Report;* continuing.

Cross-sectoral

Federal Energy Management Program

Description: The Federal Energy Management Program (FEMP) is a separate DOE sector. It reduces energy use in federal buildings, facilities, and operations by advancing energy efficiency and water conservation, promoting the use of renewable energy, and managing the utility choices of federal agencies. FEMP accomplishes its mission by leveraging both federal and private resources to provide technical and financial assistance to other federal agencies. FEMP helps agencies achieve their goals by providing alternative financing tools and guidance to use the tools, technical and design assistance for new construction and retrofit projects, training, technology transfer, procurement guidance, software tools, and reporting and evaluation of all agencies' programs.

Objective: The program aims to promote energy efficiency and renewable energy use in federal buildings, facilities, and operations.

Greenhouse gases affected: Carbon dioxide, carbon monoxide, nitrous oxide, methane, and volatile organic compounds.

Types of policy or measure: Economic, information, and education.

Status of implementation: Implemented.

Implementing entities: DOE and other federal agencies.

Non-GHG mitigation benefits of policy or measure: The program has non-greenhouse gas environmental benefits, improves energy efficiency, promotes interaction and information sharing across federal agencies, provides education and training to federal personnel, and supports technology development and deployment.

State and Local Climate Change Outreach Program

Description: This program provides a variety of technical and outreach/education services related to climate change, including guidance documents, impacts information, modeling tools, policy and technology case studies, electronic newsletters and communications, technical assistance, networking opportunities, and modest financial support for analysis and activities. The expected results are increased awareness about climate change, well-informed policy choices, and accelerated reductions in greenhouse gas emissions, as well as additional economic and clean air benefits achieved from lower emissions.

Objective: The program aims to enable state and local decision makers to incorporate climate change planning into their priority planning, so as to help them maintain and improve their economic and environmental assets.

Greenhouse gases affected: Carbon dioxide, carbon monoxide, nitrous oxide, methane, and volatile organic compounds.

Type of policy or measure: Information, education, and research (policy analysis).

Status of implementation: The program has been ongoing since the early 1990s and has recently expanded its focus to encourage comprehensive, multi-pollutant policy planning. The program's budget for fiscal year 2000 was $0.8 million; for fiscal year 2001, it was $1.23 million.

Implementing entities: EPA provides technical and financial support to state and local governments through this effort. The state and local governments, in turn, develop greenhouse gas inventories and action plans where they set reduction targets for themselves. They also conduct outreach and demonstration projects in their jurisdictions to increase awareness about climate change and facilitate replication of successful mitigation opportunities.

Costs of policy or measure: State and local governments have identified tremendous potential and actual opportunities from greenhouse gas emission reductions. For example, 12 of the state plans completed so far have forecast reductions of 2010 emissions by 13 percent (256 teragrams of CO_2 equivalent) cumulatively, with a cost savings exceeding $7.8 billion if the actions are implemented as recommended. Local governments are reporting actual savings of about 7 teragrams of CO_2 equivalent per year from their efforts, with cost savings of $70 million.

Non-GHG mitigation benefits of policy or measure: Local governments are reporting actual savings of 28,000 tons of air pollution and $70 million in energy and fuel costs each year. State plans have identified annual potential energy and fuel savings of almost $8 billion, plus the creation of more than 20,000 jobs from climate change mitigation policies. One state plan identified mitigation policies that would reduce cumulative acid rain precursors and ground-level ozone precursors by 24 and 30 percent, respectively, through 2020.

Interaction with other policies or measures: Rather than trying to be an expert at all levels, the program serves as a one-stop shop for state and local governments looking to reduce greenhouse gases. When governments express interest in particular activities and technologies that are covered under a national program, the program refers them to the appropriate program so they may acquire additional information and move forward under the guidance of national experts.

Contact: Julie Rosenberg, EPA, Global Programs Division, (202) 564-9154, rosenberg.julie@epa.gov.

Appendix C

Part 1: Selected Technology Transfer Activities

Part 2: Table 7.3—U.S. Direct Financial Contributions and Commercial Sales Related to Implementation of the UNFCCC

Renewable Energy Power Generation & Renewable Energy in Rural Areas

Purpose: Promote the appropriate and sustainable use of renewable-energy (RE) technologies in Mexico to (1) increase the quality and lower the costs of RE technologies and systems by expanding markets for, and providing feedback to, the U.S. and Mexican RE industry; (2) increase the use of clean energy sources to reduce greenhouse gas emissions and limit pollution; and (3) increase the economic, social, and health standards in rural, off-grid communities by utilizing energy for productive-use applications.

Recipient country: Mexico.

Sector: Mitigation: energy.

Total funding: $12 million.

Years in operation: 1994–present.

Description: The program implements RE projects that demonstrate the technology and its application on a larger scale by the Government of Mexico. The U.S. Agency for International Development (USAID) helped government counterpart agencies develop two national plans that promote RE through expansion/replication of USAID pilot efforts. For example in 1999, more than 100 RE systems were installed that will generate more than 14,000 megawatt-hours (MWh) of electricity over their lifetimes that would have otherwise been generated by fossil fuel plants. The majority of this power generation is taking place in the town of San Juanico, Baja California Sur, where a large USIJI-approved hybrid RE (wind and photovoltaic) project will avoid the generation of approximately 100,000 teragrams of CO_2 emissions over the project's estimated 30-year lifetime.

Factors that led to project's success: The model for the Mexico Renewable Energy Program is based on guidelines provided by the Photovoltaic System Assistance Center (15 years of domestic and international interactions) at Sandia National Laboratories and 8 years of experience gained by the program from working in Mexico. The fundamental principles of the model are (1) partnerships, (2) capacity building, (3) technical assistance, (4) pilot project implementation, (5) project replication, and (6) monitoring.

Technology transferred: Equipment to prevent the anthropogenic generation of GHGs: Photovoltaic and small wind electric systems tied to specific applications, such as water pumping, electrification, home lighting systems, and communications.

Impact on greenhouse gas emissions/sinks (optional): Not calculated.

EcoHomes Project/Sustainable Homes Initiative

Purpose: To promote energy-efficient housing design to increase savings in space heating and reduce carbon dioxide emissions. EcoHomes trains lenders, builders, and community groups to increase their awareness of low-cost, environmentally sound, energy-efficient housing that can be incorporated into South Africa's housing program.

Recipient country: South Africa.

Sector: Mitigation: energy.

Total funding: $750,000.

Years in operation: EcoHomes: 1997–2001; SHI: 1999–2002.

Description: USAID's approach to environmental programs in South Africa is aligned with the country's immediate development needs (i.e., housing) and the South African government's development policy, which seeks to implement environmentally benign development projects. USAID has supported these two sustainable housing initiatives that link renewable energy use with new housing developments as a solution for nonurban areas having no access to the power grid. These initiatives aim to promote the development of energy-efficient, environmentally sustainable, and affordable housing.

USAID co-sponsors the Sustainable Homes Initiative (SHI), a national effort to increase awareness and construction of environmentally sustainable, affordable housing. SHI provides professional assistance to housing designers and developers to increase energy efficiency and environmental sustainability; develop and market building materials and products; train builders, lenders, and policymakers; disseminate products, designs, and professional resources; and develop case studies documenting the cost and environmental benefits of energy-efficient housing.

Factors that led to project's success: Active participation of all stakeholders involved, ranging from the local and national governments to banks and builders; compatibility with the national government's development objectives.

Technology transferred: Equipment to prevent the anthropogenic generation of GHGs: Passive-solar building design, appropriate building materials for home insulation; landscape design with shade trees.

Impact on greenhouse gas emissions/sinks (optional):
1999: Kutlwanong Civic Association/Eco-Homes project saved an estimated 210 metric tons of CO_2 per year. The Gugulethu Community Development Corporation's construction of model homes led to a savings of 13 metric tons of CO_2.
2000: With the addition of 11,400 houses, 99.4 gigagrams of CO_2 will be avoided over the project's 25-year lifetime.

Coastal Resources Management Program

Purpose: To promote the essential elements of sustainable development—protecting the world's environment, fostering balanced economic growth, encouraging democratic participation in governance, and improving the health and well-being of people in developing countries—in the context of coastal resource management.

Recipient countries: Indonesia, Tanzania, Mexico.

Sector: Adaptation: coastal zone management, protection of coral reefs and other marine resources.

Total Funding: $31 million.

Years in operation: 1999–2003.

Description: This program promotes integrated coastal management. It includes such activities as development of watershed management plans; protection of marine areas; conservation of critical coastal habitats to protect from storm surge, sea level rise, and erosion; and development of best practices for coastal planning.

Factors that led to project's success: An integrated, participatory approach by stakeholders at the local and national levels to coastal management, which allows for effective response to development challenges, including those posed by climate vulnerability, variability, and sea level rise.

Technology transferred: Capacity building for integrated coastal management; geographic information systems (GIS) for mapping coastal resources.

Impact on greenhouse gas emissions/sinks (optional): Not applicable.

Philippine Climate Change Mitigation Program

Purpose: To mitigate greenhouse gas emissions through energy-sector initiatives without adversely affecting economic growth. The program focuses on four principal efforts to help the Government of the Philippines and the local private sector to (1) increase the use of clean fuels, including natural gas and renewable energy; (2) improve the policy environment for power-sector restructuring and privatization; (3) increase energy efficiency; and (4) strengthen the institutional capability of government agencies involved in the restructuring of the energy sector.

Recipient country: Philippines.

Sector: Mitigation: energy.

Total funding: $8.9 million.

Years in operation: 1998–2001.

Description: This joint program of USAID and the Government of the Philippines is a direct response to mitigate global climate change. It promotes more efficient generation, distribution, and consumption of electricity by expanding the use of clean fuels, building public and private-sector capacity for improved energy-sector development and management.

Factors that led to project's success: The development of policies that lead to the adoption of legislative and administrative actions that result in increased efficiency and/or cleaner energy production.

Technology transferred: Capacity building.

Impact on greenhouse gas emissions/sinks (optional): Avoidance of approximately 19.2 teragrams of CO_2 equivalent per year by 2002 through the use of cleaner fuels; and avoidance of at least 1.7 teragrams of CO_2 equivalent per year by 2002 through improvements in energy efficiency.

Caribbean Disaster Mitigation Program

Purpose: (1) To promote sustainable development by reducing vulnerability to natural hazards in existing and planned development; (2) to improve public awareness and development decision making by accurately mapping hazard-prone areas; (3) to improve hazard risk management by the insurance industry and help maintain adequate catastrophe protection for the region; (4) and to promote community-based disaster preparedness and prevention activities with support from the private sector.

Recipient countries: Caribbean region, Antigua & Barbuda, Barbados, Belize, Dominica, Dominican Republic, Grenada, Haiti, Jamaica, St. Kitts & Nevis, St. Vincent & the Grenadines, St. Lucia.

Sector: Adaptation: weather related disaster preparedness; vulnerability assessments.

Total funding: $5 million.

Years in operation: 1993–1999.

Description: Implemented for USAID's Office of Foreign Disaster Assistance by the Organization of American States' Unit of Sustainable Development and Environment, this program's activities target six major themes: (1) community-based preparedness, (2) hazard assessment and mapping, (3) hazard-resistant building practices, (4) vulnerability and risk audits for lifeline facilities, (5) promotion of hazard mitigation within the property insurance industry, and (6) incorporation of hazard mitigation into post-disaster recovery.

Factors that led to project's success: (1) Close coordination with development finance institutions; (2) training of Caribbean professionals, which raised awareness and provided potential long-term capacity for this type of work; (3) outreach to institutions that share a concern for disaster preparedness /loss reduction and have resources to contribute (e.g., financial services industry: banks and the property insurance industry); (4) USAID/OAS team approach to problem solving; (5) Technical Advisory Committee's ability to keep the project relevant to the needs of the region; and (6) implementation of the National Mitigation Policy and Planning Activity, which helped to facilitate the use of many mitigation tools, policies, and practices introduced by the project.

Impact on greenhouse gas emissions/sinks (optional): Not applicable.

Famine Early Warning System Network (FEWS NET)

Purpose: To help establish more effective, sustainable, and African-led food security and response networks that reduce vulnerability to food insecurity.

Recipient countries: Burkina Faso, Chad, Eritrea, Ethiopia, Kenya, Malawi, Mali, Mauritania, Mozambique, Niger, Rwanda, Somalia, Southern Sudan, Tanzania, Uganda, Zambia, Zimbabwe.

Sector: Adaptation: agriculture.

Total funding: $6.3 million.

Years in operation: FEWS: 1985–2000; FEWS NET: 2000–2005 (planned).

Description: FEWS NET assesses short- to long-term vulnerability to food insecurity with environmental information from satellites and agricultural and socioeconomic information from field representatives. The program conducts vulnerability assessments, contingency and response planning, and other activities aimed at strengthening the capacities of host country food security networks. Network members include host country and regional organizations that work on food security, response planning, environmental monitoring, and other relevant areas.

Factors that led to project's success: (1) The combined environmental monitoring expertise of the U.S. National Aeronautics and Space Administration (NASA), the National Oceanic and Atmospheric Administration (NOAA), and the U.S Geological Survey (USGS); (2) implementation by African field staff.

Technology transferred: Information networks: remote-sensing data acquisition, processing, and analysis; geographic information system (GIS) analytical skills. *Equipment to facilitate adaptation:* GIS hardware and software.

Impact on greenhouse gas emissions/sinks (optional): Not applicable.

Río Bravo Carbon Sequestration Pilot Project in Belize

Purpose: To reduce, avoid, and mitigate approximately 2.4 teragrams of carbon over the life of the project through the prevention of deforestation and sustainable forest management practices.

Recipient country: Belize.

Sector: Mitigation: forest conservation.

Total funding: $5.6 million (U.S.) for first 10 of 40 years.

Years in operation: January 1995–present. Project duration is 40 years.

Description: This project is one of the first fully funded forest-sector projects implemented under USIJI. It was developed by The Nature Conservancy in collaboration with Programme for Belize (PfB, a local NGO) and Winrock International. The project is underway at the Río Bravo Conservation and Management Area on 104,892 hectares (260,000 acres) of mixed lowland, moist subtropical broadleaf forest. PfB manages the project along with the entire private reserve. In addition to support from PfB, a number of energy companies provided $5.6 million to fund the first 10 years of the project, after which it is expected to be self-sustaining. These companies include Cinergy, Detroit Edison, PacifiCorp, Suncor, Utilitree Carbon Company, and Wisconsin Electric/Wisconsin Gas, and American Electric Power.

Factors that led to project's success: A well-designed forest conservation and management project can produce significant net carbon benefits that are scientifically valid and long lasting. The project also helps conserve biodiversity, improve local environmental quality, and meet a variety of sustainable development goals by enhancing local capacity to manage and secure the protected area. Management practices include (1) creation of undisturbed buffer areas and protection zones, (2) silvicultural treatments to boost biomass volume between cutting cycles, (3) reduced-impact harvesting techniques, (4) promotion of highly durable timber products, and (5) enhanced fire management and site security.

Technology transferred: Training: Jobs and training in forestry, forest management, and park security.

Impact on greenhouse gas emissions/sinks (optional): A total of 59,720 hectares (153,000 acres) of mixed lowland, moist subtropical broadleaf forest will be included under the project, leading to the protection of up to 240 tree species, 70 mammal species, and 390 bird species.

Coal Mine Methane Recovery Project in China

Purpose: To work with the Government of China and the Chinese coal industry (1) to identify opportunities to reduce methane emissions from coal mining and use these emissions as energy; (2) to develop the domestic capacity to implement coal mine methane technologies; and (3) to develop commercial partnerships between Chinese and foreign companies to realize profitable projects that reduce methane emissions.

Recipient country: China.

Sector: Mitigation: energy, industry.

Total funding: $150,000.

Years in operation: 1989–present.

Description: Chinese mines are the greatest global source of methane emissions from coal mining. This U.S. Environmental Protection Agency (EPA) project involves assessments and pilot projects for capturing abundant gas resources at Chinese mines, with concurrent mine safety, power production, and climate benefits. A Coalbed Methane Information Clearinghouse is housed at the China Coal Information Institute and has conducted considerable outreach to U.S. and other companies interested in this market. The clearinghouse has published journals in Chinese and English, has hosted several domestic and international seminars, and has developed with EPA an economic analysis model to identify profitable projects to reduce methane emissions. It currently is participating in studies that will ultimately lead to significant investment in commercial-scale projects. The clearinghouse and EPA signed an agreement in April 1999 at the Gore–Zhou Energy and Environment Forum, outlining a two-year market data development project, that, building on the Clearinghouse's experience, is providing information and analyses on specific coal mine methane project opportunities for Chinese and Western investors and developers.

Factors that led to project's success: (1) Interest generated by coal sector in methane recovery for safety, productivity, and energy value of coal mine methane recovery; (2) interest generated by the Government of China in developing coalbed methane resources for energy supply, energy security, and local/regional environmental benefits; (3) interest generated by international organizations and companies in the global environmental and energy benefits of coal mine methane; and (4) nurturing of partnerships with responsible Chinese organizations in developing the nation's coal mine methane resources.

Technology transferred: Equipment to reduce anthropogenic sources of GHGs: Coal mine methane gas production technologies (surface and in-mine); coal mine methane use technologies. Training/Capacity Building: Financial analysis and marketing to international companies.

Impact on greenhouse gas emissions/sinks (optional): Emission reductions have more than quadrupled since 1990 to approximately 500 million cubic meters of methane per year (more than 7 teragrams of CO_2 equivalent per year).

TABLE 7-3A: 1997

U.S. Direct Financial Contributions and Commercial Sales Related to Implementation of the UNFCCC (Millions of US Dollars)

Column groups: **MITIGATION** = Energy, Transport, Forestry (Forest Conservation, Biodiversity Conservation), Agriculture, Waste Management, Industry. **ADAPTATION** = Capacity Building (Water Supply, Disaster Preparedness, Droughts & Desertification), Coastal Zone Management (Coastal Resources, Coral Reefs), Other Vulnerability Studies. **OTHER GCC** = UNFCCC Participation, Cross-cutting Activities.

COUNTRY / REGION	Energy	Transport	Forest Conservation	Biodiversity Conservation	Agriculture	Waste Management	Industry	Water Supply	Disaster Preparedness	Droughts & Desertification	Coastal Resources	Coral Reefs	Other Vulnerability Studies	UNFCCC Participation	Cross-cutting Activities	TOTALS
WORLD	325.52	4.70	63.54	96.31	0.45	0.15	1.85	771.77	4.17	3.17	7.58	1.54	1.94	2.33	18.18	1,303.07
AFRICA	5.03	0.15	21.49	31.19	0.24		0.14	24.29	0.07	2.57	0.03	0.00		0.01	1.85	87.05
Africa Regional	0.06		2.26	10.88				0.42							0.15	13.76
Angola	0.04							1.13								1.17
Botswana	0.05							0.06								0.11
Burkina Faso								0.01								0.01
Cameroon	0.01							0.06								0.07
Cape Verde								0.03								0.03
Central African Republic								0.15								0.15
Chad								0.02		0.36						0.37
Congo (DROC)								0.15								0.15
Congo (ROC)	0.04							0.75								0.79
Cote d'Ivoire	0.12							2.27								2.39
Djibouti								0.06								0.06
East and Southern Africa Regional			0.08	0.11				0.00			0.03	0.00				0.22
Equatorial Guinea								1.37								1.37
Eritrea	0.02							0.30								0.30
Ethiopia			0.10	0.10	0.24			0.01		0.05						0.53
French Southern and Antarctic Lands								0.13								0.13
Gabon	0.00							0.05								0.05
Gambia	0.00							0.00								0.00
Ghana	0.03		2.74					0.42								3.20
Guinea	0.01														0.40	0.41
Kenya	0.01		0.41	0.77				0.37		1.86						3.42
Liberia								0.03								0.03
Madagascar			4.75	2.50					0.02							7.27
Malawi			6.49	0.50				0.50	0.03							7.53
Mali								0.07								0.07
Mauritania								0.01		0.03						0.03
Mauritius								0.00								0.00
Mozambique	0.00		0.48					0.17	0.02							0.67
Namibia								0.08								0.08
Niger								0.02								0.02
Nigeria	0.12							1.75								1.86
Sahel								0.22		0.28					0.19	0.69
Senegal	0.30		1.02	0.10				0.38								1.80
South Africa	3.36	0.15	0.02				0.14	10.04						0.01	0.06	13.77
Southern Africa Regional			1.06	8.22				2.96								12.24
Sudan	0.25							0.10								0.35
Swaziland								0.01								0.01
Tanzania	0.02							0.00								0.03
Uganda	0.18		1.17	4.42				0.00							1.05	6.82
Zambia								0.08								0.08
Zimbabwe	0.38		0.90	3.60				0.14								5.02
ASIA / NEAR EAST	128.83	2.03	10.13	38.26		0.15	0.97	465.81	0.53		0.30	0.31		0.40	2.20	649.91
Algeria	0.86							1.85								2.71
Asia / Near East Regional	0.65	0.59	0.67	1.60		0.15	0.97	0.42								5.05
Bahrain	0.41							0.45							0.01	0.86

TABLE 7-3A: 1997 U.S. Direct Financial Contributions and Commercial Sales Related to Implementation of the UNFCCC (Millions of US Dollars)

| COUNTRY / REGION | MITIGATION | | | | | | | ADAPTATION | | | | | | OTHER GCC | | TOTALS |
	Energy	Transport	Forest Conservation	Biodiversity Conservation	Agriculture	Waste Management	Industry	Water Supply	Disaster Preparedness	Droughts & Desertification	Coastal Resources	Coral Reefs	Other Vulnerability Studies	UNFCCC Participation	Cross-cutting Activities	
Bangladesh	1.17				0.45			0.43	0.16							2.20
Bhutan			0.38													0.38
Brunei								1.02								1.02
Burma (Myanmar)								0.00	0.03							0.03
Cambodia			1.20					0.03	0.05							1.28
China	19.35	0.04	0.35		0.12			42.23	0.05							62.13
Egypt	19.73							47.35						0.40		67.48
Federated States of Micronesia								0.00								0.00
Fiji	0.02							0.01								0.03
French Polynesia	0.01							0.14								0.15
Hong Kong	6.01							30.07								36.09
India	8.36	0.39	0.42		0.70			9.28								19.17
Indonesia	5.59		4.09	5.00				11.71	0.03		0.20	0.31			0.42	27.31
Iraq	0.01															0.01
Jordan	0.06							2.41								2.47
Kiribati	0.03															0.03
Kuwait	0.69							6.35								7.03
Laos			0.05						0.13							0.18
Lebanon	0.19							1.98								2.17
Macao	0.05							0.16								0.22
Malaysia	3.37							27.66	0.03							31.05
Maldive Islands	0.04							0.27								0.31
Marshall Islands								0.09								0.09
Mongolia	0.17							0.01	0.05							0.23
Morocco	1.09							4.13								5.22
Nepal	0.37		1.10		0.58			0.03							1.17	3.25
New Caledonia								0.08								0.08
Niue								0.86								0.86
Oman	0.88							4.21								5.09
Pakistan	4.34							5.12								9.46
Palau								0.05								0.05
Papua New Guinea	0.01		0.05					0.10								0.15
Philippines	4.95	0.50	1.69	2.50				22.14			0.03				0.60	32.40
Qatar	0.31							12.21								12.52
Reunion								0.05								0.05
Samoa								0.02								0.02
Saudi Arabia	13.61							53.46								67.07
Singapore	17.28							40.71								57.99
Sri Lanka	0.00		0.10					0.17			0.08					0.35
Syria	2.55							1.30								3.85
Taiwan	7.99							72.47								80.46
Thailand	5.54	0.50						37.32								43.36
Togo								0.04								0.04
Tonga								0.00								0.00
Tunisia								0.83								0.83
Turkey	2.07							15.01								17.08
Turkmenistan	0.01							0.29								0.30
Turks & Caic Islands	0.05							0.31								0.35

TABLE 7-3A: 1997

U.S. Direct Financial Contributions and Commercial Sales Related to Implementation of the UNFCCC (Millions of US Dollars)

COUNTRY / REGION	MITIGATION – Energy	MITIGATION – Transport	MITIGATION – Forestry: Forest Conservation	MITIGATION – Forestry: Biodiversity Conservation	MITIGATION – Agriculture	MITIGATION – Waste Management	MITIGATION – Industry	ADAPTATION – Capacity Building: Water Supply	ADAPTATION – Capacity Building: Disaster Preparedness	ADAPTATION – Capacity Building: Droughts & Desertification	ADAPTATION – Coastal Zone Management: Coastal Resources	ADAPTATION – Coastal Zone Management: Coral Reefs	ADAPTATION – Other Vulnerability Studies	OTHER GCC – UNFCCC Participation	OTHER GCC – Cross-cutting Activities	TOTALS
United Arab Emirates	0.91							9.91								10.82
Vietnam	0.13	0.04						0.93	0.03							1.13
West Bank/Gaza				27.32												27.32
Yemen								0.12								0.12
EUROPE / EURASIA	41.67	1.17	3.82	0.06			0.71	18.54	0.03	0.42				1.92	6.29	74.61
Albania															0.18	0.18
Armenia			3.64												2.74	6.39
Azerbaijan								0.70								0.70
Belarus								0.50								0.50
Bulgaria	1.23	0.07						0.03								1.33
Central Asia Regional	0.11		0.03					0.17								0.30
Croatia								1.94								1.94
Cyprus	0.02							0.46								0.48
Czech Rep	0.47							0.18								0.65
Estonia	0.17							0.19								0.36
Europe & Eurasia Regional	0.19		0.15	0.06				0.36						0.07	0.15	0.99
Georgia	6.29	0.72						0.06								7.07
Hungary	2.03							0.41								2.43
Kazakhstan	2.64	0.09						0.57								3.30
Kyrgyzstan	0.60							0.23								0.83
Latvia								0.48								0.48
Lithuania	0.10							1.05								1.15
Macedonia																0.00
Moldova	0.03															0.03
Poland		0.12						4.59	0.23						0.05	5.03
Romania	0.65							0.60								1.24
Russia	7.70	0.18					0.11	0.70						1.85		10.54
Slovakia																0.00
Slovenia	0.03							0.01								0.04
Tajikistan								1.09								1.09
Turkmenistan	2.58							0.25								2.82
Ukraine	14.48						0.60	3.86							3.17	22.11
Uzbekistan	2.14							0.45								2.59
Yugoslavia								0.01								0.01
LATIN AMERICA / CARIBBEAN	86.25	1.35	21.32	17.93	0.20		0.03	252.92	0.54		1.01	0.90		1.78		384.24
Anguilla	0.01							0.05								0.06
Antigua & Barbuda	0.00							0.06								0.06
Argentina	4.57	0.70						13.25								18.51
Aruba	0.06							0.09								0.15
Bahamas	0.39							4.50								4.88
Barbados	0.07							0.33								0.41
Belize	0.18		0.26	0.02				0.40								0.66
Bermuda									0.28							0.28
Bolivia	0.44	0.25	5.16	1.07				0.82								7.76
Brazil	17.14		1.52	1.53			0.03	27.48							0.14	48.08
British Virgin Islands	0.01							0.17								0.19
Cayman Islands	0.15							0.48								0.63
Central America Regional	0.40		9.12	8.88				0.08			0.82	0.19				19.50

TABLE 7-3A: 1997   U.S. Direct Financial Contributions and Commercial Sales Related to Implementation of the UNFCCC (Millions of US Dollars)

COUNTRY / REGION	MITIGATION		Forestry					ADAPTATION			Coastal Zone Management			OTHER GCC		TOTALS
	Energy	Transport	Forest Conservation	Biodiversity Conservation	Agriculture	Waste Management	Industry	Water Supply (Capacity Building)	Disaster Preparedness	Droughts & Desertification	Coastal Resources	Coral Reefs	Other Vulnerability Studies	UNFCCC Participation	Cross-cutting Activities	
Chile	2.39							16.65								19.04
Colombia	2.60		0.48		0.56			10.11	0.02							13.75
Costa Rica	0.67	0.29	0.05					2.01								3.03
Dominica Islands	0.00							0.01								0.01
Dominican Rep	0.59							2.49								3.08
Ecuador	1.08		0.99	1.75	0.05			4.90	0.03							8.80
El Salvador	0.30							1.98								2.28
French Guiana								0.04								0.04
Grenada Islands	0.00							0.10								0.10
Guadeloupe								0.10								0.10
Guatemala	0.80		1.18		1.03			2.92								5.93
Guyana	0.05							0.28								0.33
Haiti	0.03							0.22								0.25
Honduras	0.55		0.99	1.50				1.44	0.19							4.66
Jamaica	0.50		0.43	0.54				2.64			0.18	0.71				4.99
Latin America / Caribbean Regional	0.17	0.01	0.57												0.32	1.07
Martinique	0.00							0.02								0.02
Mexico	26.21	0.04						127.53							1.32	155.11
Montserrat Islands								0.09								0.09
Nicaragua	0.03							0.44								0.48
Panama	0.13		0.22	0.77				1.29								2.41
Paraguay	0.11			0.13				0.70	0.03							0.95
Peru	11.69		0.38	0.17	0.16			6.31								18.71
Saint Kitts-Nevis	0.01							0.01								0.01
Saint Lucia Islands	0.01							0.08								0.09
Saint Vincent & Grenadines	0.01							0.06								0.07
Suriname	0.02							0.35								0.37
Trinidad & Tobago	2.86							6.35								9.21
Uruguay	0.09	0.07						0.84								0.99
Venezuela	11.90							15.01								26.91
OTHER GLOBAL PROGRAMS	63.74		6.78	8.86				10.20	2.61	0.60	6.25	0.33	1.94		6.07	107.27

TABLE 7-3A: 1998

U.S. Direct Financial Contributions and Commercial Sales Related to Implementation of the UNFCCC (Millions of US Dollars)

| COUNTRY / REGION | MITIGATION | | | | | | | ADAPTATION | | | | | | OTHER GCC | | TOTALS |
	Energy	Transport	Forestry — Forest Conservation	Forestry — Biodiversity Conservation	Agriculture	Waste Management	Industry	Capacity Building — Water Supply	Capacity Building — Disaster Preparedness	Capacity Building — Droughts & Desertification	Coastal Zone Management — Coastal Resources	Coastal Zone Management — Coral Reefs	Other Vulnerability Studies	UNFCCC Participation	Cross-cutting Activities	
WORLD	390.75	8.60	8.11	75.21	0.09	38.96	4.07	729.74	17.97	6.89	5.29	10.19	1.97	18.05	103.44	1,419.20
AFRICA	46.83	0.08	0.39	17.67				25.97	4.61	5.96	0.77			2.79	42.27	147.32
Africa Regional	0.13	0.08	0.19	0.20				0.44		2.08				2.79	5.37	11.28
Angola	0.37							2.36							0.36	3.10
Cameroon	0.05							0.02								0.07
Chad								0.08	0.03							0.11
Congo (DROC)				0.00				0.03	0.19							0.23
Congo (ROC)	0.17							2.26								2.43
Cote d'Ivoire								0.12								0.12
Djibouti								0.06	0.03							0.09
East Africa Regional	0.10			0.33											5.11	5.53
Equatorial Guinea								1.03								1.03
Eritrea	0.38							0.13							0.22	0.73
Ethiopia	0.10							0.03		0.23					0.15	0.51
Gabon								0.27								0.27
Ghana	0.08			0.80				0.10							1.20	2.18
Guinea	0.02							0.59							2.91	3.52
Kenya	0.22			2.21				0.35	3.04	0.13	0.07				0.62	6.65
Liberia	0.09							0.04								0.13
Madagascar				4.50				0.02							2.50	7.02
Malawi								0.01							9.40	9.41
Mali								1.46							3.32	4.78
Mauritania								0.01								0.01
Mauritius								0.12								0.12
Mozambique								0.02							6.10	6.12
Namibia				0.04				0.01							0.32	0.36
Niger									0.03							0.03
Nigeria	32.43							5.21								37.64
Rwanda								0.01								0.01
Sahel										3.75					0.05	3.79
Senegal	0.06							0.04							1.69	1.79
Sierra Leone	0.15															0.15
Somalia															1.98	1.98
South Africa	12.08		0.20					9.02								21.30
Southern Africa Regional	0.13			4.67				1.92							0.38	7.10
Sudan									0.06							0.06
Swaziland									0.01							0.01
Tanzania				2.60				0.03	1.00		0.70				0.01	4.35
Uganda	0.01							0.06								0.07
Zambia								0.05								0.09
Zimbabwe	0.23			2.32				0.07							0.58	3.20
ASIA / NEAR EAST	146.61	2.56	0.48	17.50		38.96	4.07	394.42	3.90	0.93	1.63	2.38	0.03	15.26	24.77	653.48
Afghanistan								0.01								0.01
Algeria	0.04							1.96								2.00
Asia / Near East Regional	1.01	0.94		2.14		0.63	4.07						0.03		5.75	14.56
Bahrain								0.66								0.66
Bangladesh	2.33							0.29	1.56							4.17
Brunei	0.07							0.51								0.58

TABLE 7-3A: 1998

U.S. Direct Financial Contributions and Commercial Sales Related to Implementation of the UNFCCC (Millions of US Dollars)

Column groups — MITIGATION: Energy, Transport, Forestry (Forest Conservation, Biodiversity Conservation), Agriculture, Waste Management, Industry. ADAPTATION: Capacity Building (Water Supply, Disaster Preparedness, Droughts & Desertification), Coastal Zone Management (Coastal Resources, Coral Reefs), Other Vulnerability Studies. OTHER GCC: UNFCCC Participation, Cross-cutting Activities.

COUNTRY / REGION	Energy	Transport	Forest Conservation	Biodiversity Conservation	Agriculture	Waste Management	Industry	Water Supply	Disaster Preparedness	Droughts & Desertification	Coastal Resources	Coral Reefs	Other Vulnerability Studies	UNFCCC Participation	Cross-cutting Activities	TOTALS
Burma (Myanmar)	0.02							0.02								0.04
Cambodia								0.02								0.02
China	18.47	0.05						46.30	1.49			0.11			0.35	66.79
Cook Islands				0.03				0.00								0.00
Egypt	19.79			8.33				21.79			1.51	2.27		15.26	6.81	75.78
Federated States of Micronesia	0.01							0.27								0.28
Fiji								0.01								0.01
French Polynesia	0.01							0.20								0.21
Hong Kong	3.94							29.73								33.67
India	13.60	0.00						19.25	0.07							32.93
Indonesia	29.69		0.25	6.90				6.64		0.59	0.12				6.69	50.88
Jordan	0.05							4.14								4.18
Korea (ROK)								0.10	0.03							0.13
Kuwait	0.44							11.60								12.05
Laos								0.12	0.16							0.28
Lebanon	0.10	0.50						1.16							0.90	2.66
Macao								0.29								0.29
Malaysia	2.51	0.50						11.07	0.03							14.10
Maldive Islands	0.09							0.34								0.43
Marshall Islands	0.04							0.28								0.32
Mongolia	2.61							0.02								2.65
Morocco	1.67							4.15							1.00	6.83
Nauru								0.02								0.02
Nepal	0.48							0.64							1.94	3.07
New Caledonia								0.00								0.00
Oman	0.56							2.07								2.63
Pakistan	0.65							2.60	0.10							3.35
Palau								0.04								0.04
Papua New Guinea		0.22			0.10			0.04		0.13						0.50
Philippines	1.64		0.23					10.75							1.32	13.93
Qatar	1.43							1.24								2.67
Saudi Arabia	12.22							35.70								47.92
Seychelles								0.57								0.57
Singapore	5.83							34.57								40.40
Sri Lanka	0.13							0.35								0.48
Syria	1.96							1.51								3.47
Taiwan	12.33							94.91								107.24
Tajikistan									0.03							0.03
Thailand	2.89							20.14								23.04
Tokelau Islands								0.01								0.01
Tonga	0.01							0.00								0.01
Tunisia								1.87								1.87
Turkey	0.67	0.35						15.00								16.01
Turkmenistan	0.27							0.00								0.28
Turks & Caic Islands	0.02							0.40								0.42
United Arab Emirates	7.21							8.14								15.35
Uzbekistan		0.03														0.03
Vietnam	1.36							2.63	0.40	0.21						4.60

TABLE 7-3A: 1998 U.S. Direct Financial Contributions and Commercial Sales Related to Implementation of the UNFCCC (Millions of US Dollars)

COUNTRY / REGION	MITIGATION							ADAPTATION						OTHER GCC		TOTALS
	Energy	Transport	Forestry: Forest Conservation	Forestry: Biodiversity Conservation	Agriculture	Waste Management	Industry	Capacity Building: Water Supply	Capacity Building: Disaster Preparedness	Capacity Building: Droughts & Desertification	Coastal Zone Mgmt: Coastal Resources	Coastal Zone Mgmt: Coral Reefs	Other Vulnerability Studies	UNFCCC Participation	Cross-cutting Activities	
West Bank/Gaza							38.33	0.01		0.03						38.36
Yemen	0.45							0.26								0.71
EUROPE / EURASIA	60.74	4.36	0.15	0.11				16.27		0.08					9.12	90.83
Albania	4.55														4.75	9.30
Armenia	19.86															19.86
Azerbaijan	1.58							0.52								2.10
Belarus	0.01							0.11								0.12
Bosnia & Herzegovina		0.45														0.45
Bulgaria	0.20	0.30						0.13								0.63
Central Asia Regional	4.82							0.83							0.41	6.06
Croatia								0.13								0.13
Cyprus	0.96							0.30								1.26
Czech Rep	0.44															0.44
Estonia								0.05								0.05
Europe & Eurasia Regional	0.52														1.74	2.25
Europe Regional		0.55														0.55
Georgia	12.94															12.94
Greece	0.19															0.19
Hungary	1.43	0.07						0.53								2.04
Kazakhstan	0.31							0.92								1.22
Kyrgyzstan								0.38		0.03						0.42
Latvia	0.00							0.13								0.13
Lithuania	0.14							0.05							1.10	1.29
Macedonia		2.53						0.19							0.06	2.79
Moldova								0.55								0.55
Poland	5.41							4.11								9.53
Romania	1.56							0.31		0.03						1.89
Russia		0.45	0.15	0.11				5.45		0.03						6.19
Slovakia															1.04	1.04
Slovenia	0.01							0.47								0.48
Tajikistan	0.00							3.48								3.48
Ukraine								2.90								2.91
Uzbekistan	0.33							0.07								0.40
Yugoslavia	0.02							0.12								0.14
LATIN AMERICA / CARIBBEAN	115.05	1.59	6.28	27.25	0.09			288.73	6.68						18.97	464.65
Anguilla								0.00								0.00
Antigua & Barbuda	0.07							0.46	0.39							0.91
Argentina	11.24	0.34						17.64		0.41						29.63
Aruba	0.25							6.74								6.99
Bahamas	0.30							1.43								1.73
Barbados	0.05							0.31								0.36
Belize	0.02							0.44								0.46
Bermuda	0.06							0.63								0.69
Bolivia	0.51		5.13	5.94				1.69							0.04	13.31
Brazil	28.18	0.44	0.17	3.30				27.12							3.05	62.27
British Virgin Islands	0.06							0.12								0.18
Cayman Islands	0.00							0.47								0.47
Central America Regional	0.04			3.62											2.63	6.29

TABLE 7-3A: 1998   U.S. Direct Financial Contributions and Commercial Sales Related to Implementation of the UNFCCC (Millions of US Dollars)

Column groupings:
- **MITIGATION:** Energy; Transport; Forestry (Forest Conservation, Biodiversity Conservation); Agriculture; Waste Management; Industry
- **ADAPTATION:** Capacity Building (Water Supply, Disaster Preparedness, Droughts & Desertification); Coastal Zone Management (Coastal Resources, Coral Reefs); Other Vulnerability Studies
- **OTHER GCC:** UNFCCC Participation; Cross-cutting Activities

COUNTRY / REGION	Energy	Transport	Forest Conservation	Biodiversity Conservation	Agriculture	Waste Management	Industry	Water Supply	Disaster Preparedness	Droughts & Desertification	Coastal Resources	Coral Reefs	Other Vulnerability Studies	UNFCCC Participation	Cross-cutting Activities	TOTALS
Chile	1.89							14.46								16.35
Colombia	2.65							10.58								13.23
Costa Rica	1.66		0.15					1.89								3.69
Dominica Islands	0.01							0.00	0.03							0.04
Dominican Rep	0.98							4.54		1.93					0.20	7.66
Ecuador	0.81	0.30		0.12				5.51	0.08						2.42	9.24
El Salvador	0.12							1.31							2.27	3.69
French Guiana								0.08								0.08
Grenada Islands	0.02							3.11								3.12
Guadeloupe								0.29								0.29
Guatemala	0.53	0.19		1.90				3.60							2.04	8.26
Guyana	0.02							0.08								0.10
Haiti	0.31							0.34	0.34							0.99
Honduras	0.77			1.47				3.03							1.20	6.47
Jamaica	0.31			0.44				2.30							1.05	4.11
Latin America / Caribbean Regional	0.21		0.53	5.47											0.70	6.90
Martinique	0.00							0.09								0.09
Mexico	43.52	0.33	0.19	2.21	0.09			151.64	0.53						1.23	199.74
Nicaragua	0.15			1.09				0.40							0.73	2.37
Panama	7.25			0.40				1.32							0.68	9.64
Paraguay	0.11		0.01	0.53				0.62	0.03							1.30
Peru	3.46			0.77				6.81	2.95						0.73	14.72
Saint Kitts-Nevis	0.01							0.02								0.03
Saint Lucia Islands								0.12								0.12
Saint Vincent & Grenadines								0.06								0.06
Suriname	0.01							0.26								0.27
Trinidad & Tobago	0.89							2.58								3.48
Uruguay	0.04							0.33								0.36
Venezuela	8.53		0.10					16.31								24.94
OTHER GLOBAL PROGRAMS	21.53	0.01	0.81	12.68				4.35	2.71		2.89	7.81	1.94		8.31	62.92

TABLE 7-3A: 1999   U.S. Direct Financial Contributions and Commercial Sales Related to Implementation of the UNFCCC (Millions of US Dollars)

COUNTRY / REGION	MITIGATION: Energy	Transport	Forestry: Forest Conservation	Forestry: Biodiversity Conservation	Agriculture	Waste Management	Industry	ADAPTATION / Capacity Building: Water Supply	Disaster Preparedness	Droughts & Desertification	Coastal Zone Management: Coastal Resources	Coral Reefs	Other Vulnerability Studies	OTHER GCC: UNFCCC Participation	Cross-cutting Activities	TOTALS
WORLD	523.76	6.81	11.79	69.48	3.30	0.69	6.39	787.16	1,675.90	21.67	4.75	0.49	2.95	2.97	89.99	3,208.00
AFRICA	12.45	0.07	1.56	26.09	1.79			22.72	0.86	19.22	0.46				46.89	132.11
Africa Regional	0.51	0.07		4.09				1.84		2.01					9.31	17.83
Angola	0.23							1.61							0.50	2.34
Benin	0.00															0.00
Botswana	0.00							0.03								0.03
Cameroon			0.23					0.24								0.47
Cape Verde								0.27								0.27
Central African Republic								0.01								0.01
Chad									0.03							0.03
Congo (DROC)				4.00				0.03								4.03
Congo (ROC)								0.52								0.52
Cote d'Ivoire	0.28							0.01								0.29
Djibouti	0.15															0.15
East Africa Regional	0.03				0.08			0.42							5.65	5.73
Equatorial Guinea								0.11							0.16	0.46
Eritrea	0.04							0.11							0.27	0.27
Ethiopia	0.04				1.06			0.01		13.50					1.34	15.96
Gabon	0.01							0.00								0.01
Gambia									0.07							0.07
Ghana	0.21			0.90				0.32							3.00	4.42
Guinea								0.50							3.50	4.00
Kenya	0.19		0.63	0.66	0.19			0.19	0.71						0.19	2.77
Liberia								0.09								0.09
Madagascar				3.60				0.06								3.66
Malawi															7.90	7.90
Mali	0.01							2.59	0.01						3.34	5.94
Mauritania								0.00	0.03							0.04
Mauritius								0.18								0.18
Mozambique	0.01		0.06					0.01	0.03						4.84	4.93
Namibia	0.01		0.14	0.90	0.04			0.00							2.20	3.12
Nigeria	1.39							2.14								3.72
Sahel										3.71					0.03	3.74
Senegal	1.47							0.88							0.67	3.02
South Africa	7.74		0.21		0.21			7.17								15.33
Southern Africa Regional	0.11			6.64				3.36								10.11
Tanzania			0.13	1.98	0.06			0.02			0.46				0.07	2.70
Uganda				3.32				0.01							3.40	6.73
Zambia								0.06							0.80	0.86
Zimbabwe	0.05		0.15		0.15			0.04								0.40
ASIA / NEAR EAST	211.40	1.60	2.92	5.68	0.74	0.69	4.29	458.15	1.59		0.04		0.20		7.44	694.73
Afghanistan	0.01															0.01
Algeria				2.05				2.00								2.00
Asia / Near East Regional	0.88	0.91			0.61	0.61	3.95		0.35						3.05	11.80
Bahrain	0.08							1.67								1.75
Bangladesh	2.51			1.00			0.02	0.09								3.62
Brunei	0.06							0.63								0.69
Burma (Myanmar)								0.09								0.09

TABLE 7-3A: 1999

U.S. Direct Financial Contributions and Commercial Sales Related to Implementation of the UNFCCC (Millions of US Dollars)

COUNTRY / REGION	MITIGATION							ADAPTATION						OTHER GCC		TOTALS
	Energy	Transport	Forestry — Forest Conservation	Forestry — Biodiversity Conservation	Agriculture	Waste Management	Industry	Capacity Building — Water Supply	Capacity Building — Disaster Preparedness	Capacity Building — Droughts & Desertification	Coastal Zone Management — Coastal Resources	Coastal Zone Management — Coral Reefs	Other Vulnerability Studies	UNFCCC Participation	Cross-cutting Activities	
Cambodia			0.02						0.03							0.05
China	47.65	0.06	0.02					45.60	0.55						0.10	93.99
Egypt	20.07							33.53								53.60
Federated States of Micronesia	0.02							0.03								0.05
Fiji	0.04							0.10								0.14
French Polynesia	0.01							0.22								0.23
Hong Kong	4.98							22.81								27.80
India	76.41		0.69		0.02	0.08		14.87								92.06
Indonesia	7.38		1.05	2.63				7.37							1.31	19.77
Jordan								42.30			0.02					42.30
Korea (DROK)								0.01								0.01
Korea (ROK)	0.00						0.32		0.03							0.35
Kuwait	0.45							5.06								5.52
Lebanon	0.07	0.63						1.01								1.71
Macao								0.02								0.02
Malaysia	4.94							16.58								21.52
Maldive Islands	0.02							0.15								0.17
Marshall Islands								0.30								0.30
Mongolia								0.01								0.01
Morocco	1.60							2.18							2.70	6.48
Nauru								0.02								0.02
Nepal	1.35		0.48		0.63			0.05							0.25	2.76
New Caledonia								0.10								0.10
Oman	0.82							1.39								2.21
Pakistan	0.59							1.49		0.03						2.11
Palau	0.00							0.02								0.02
Papua New Guinea			0.40					0.10								0.50
Philippines	0.48							8.60		0.08						9.16
Qatar	0.25							1.00								1.25
Saudi Arabia	13.50							34.01								47.51
Seychelles								0.03								0.03
Singapore	4.28							27.98								32.26
Solomon Islands													0.20			0.20
South Asia Regional	0.00														0.02	0.02
Sri Lanka								0.89								0.89
Syria	1.70				0.10			0.94								2.74
Taiwan	14.24							63.07								77.31
Tajikistan										0.03						0.03
Thailand	4.05		0.26					15.03		0.03		0.02				19.37
Tunisia	0.01							0.40								0.41
Turkey	1.52							9.39								10.91
Turkmenistan								0.06								0.06
Turks & Caic Islands								0.35								0.35
United Arab Em	0.92							15.70								16.61
Vietnam	0.39							1.15		0.49						2.03
West Bank/Gaza								79.20								79.20
Yemen	0.13							0.47								0.60
EUROPE / EURASIA	73.45	2.88	0.28	0.36	0.09		2.11	16.30	0.05	1.42			0.01	2.21	9.31	108.46

TABLE 7-3A: 1999   U.S. Direct Financial Contributions and Commercial Sales Related to Implementation of the UNFCCC (Millions of US Dollars)

COUNTRY / REGION	MITIGATION		Forestry					ADAPTATION			Coastal Zone Mgmt			OTHER GCC		TOTALS
	Energy	Transport	Forest Conservation	Biodiversity Conservation	Agriculture	Waste Management	Industry	Capacity Building: Water Supply	Disaster Preparedness	Droughts & Desertification	Coastal Resources	Coral Reefs	Other Vulnerability Studies	UNFCCC Participation	Cross-cutting Activities	
Albania								0.07							0.34	0.40
Armenia	8.06															8.06
Azerbaijan	3.89						0.50	0.03								4.42
Belarus	0.02							0.40								0.42
Bosnia & Herzegovina							0.35	0.35								0.70
Bulgaria	0.36	0.65						0.04								1.04
Central Asia Regional	5.17							1.67							1.18	8.02
Croatia								0.11								0.11
Cyprus	0.41							0.52								0.93
Czech Rep	0.44	0.12						0.06								0.62
Estonia								0.03								0.03
Europe & Eurasia Regional	4.49			0.04												4.53
Europe Regional		1.26														1.26
Georgia	20.43															20.43
Hungary	0.02							0.74								0.78
Kazakhstan	0.58	0.14					0.90	0.12								1.72
Latvia								0.25								0.25
Lithuania	0.42							1.21								1.63
Macedonia	0.00	0.07						0.46							0.87	1.41
Moldova	2.30							0.53								2.83
Poland	7.42							3.48								10.91
Romania	1.96	0.19						0.23								2.38
Russia	4.07	0.45	0.28	0.22	0.09		0.15			1.42			0.01		1.31	7.99
Slovakia								0.05								0.05
Slovenia	0.04							0.13								0.17
Tajikistan								3.97								3.97
Turkmenistan							0.01									0.01
Ukraine	12.17			0.10			0.20	0.38	0.03					2.21	5.62	20.71
Uzbekistan	1.21							1.46								2.67
LATIN AMERICA / CARIBBEAN	203.87	2.27	5.84	30.90	0.68			286.21	1671.40						22.96	**2224.13**
Anguilla								0.21								0.21
Antigua & Barbuda	0.03							0.18								0.20
Argentina	12.51							17.27								29.78
Aruba	0.06							0.07								0.13
Bahamas	1.25							1.26								2.51
Barbados	0.04							1.60								1.64
Belize	0.02		0.40					1.31								1.74
Bermuda	0.24							1.87								2.10
Bolivia	0.20		1.57	2.20				1.83							2.80	8.59
Brazil	130.93		0.23	3.64	0.45			19.85							4.16	159.27
British Virgin Islands								0.13								0.13
Caribbean Regional										3.00					0.51	3.51
Cayman Islands								1.27								1.27
Central America Regional	0.22			3.19											1.32	4.73
Chile	1.50	0.36	0.02					9.89								11.76
Colombia	1.09							10.70		10.00						21.79
Costa Rica	0.91	0.19	0.13					2.38								3.61
Dominica Islands								0.00								0.00

TABLE 7-3A: 1999 — U.S. Direct Financial Contributions and Commercial Sales Related to Implementation of the UNFCCC (Millions of US Dollars)

COUNTRY / REGION	MITIGATION							ADAPTATION						OTHER GCC		TOTALS
			Forestry					Capacity Building			Coastal Zone Management					
	Energy	Transport	Forest Conservation	Biodiversity Conservation	Agriculture	Waste Management	Industry	Water Supply	Disaster Preparedness	Droughts & Desertification	Coastal Resources	Coral Reefs	Other Vulnerability Studies	UNFCCC Participation	Cross-cutting Activities	
Dominican Rep	0.45							4.06	29.50							34.01
Ecuador	0.68			1.51				2.58							2.85	7.63
El Salvador	0.30	0.20						2.28	20.50						2.25	25.53
Grenada Islands	0.01							0.16								0.16
Guadeloupe	0.11							0.06								0.17
Guatemala	0.47		0.12	5.99				3.18	25.00							34.76
Guyana	0.56		0.09					0.15								0.81
Haiti	0.14							0.16	9.80						2.90	13.01
Honduras	0.54		1.28	1.53	0.15			4.18	291.00						1.07	299.74
Jamaica	0.37			0.70				1.96							1.67	4.70
Latin America / Caribbean Regional	0.54	0.05	0.70	4.05				0.01	1188.50							1193.84
Martinique								0.02								0.02
Mexico	37.46		0.53	4.70	0.08			162.35								205.12
Montserrat Islands								0.00								0.00
Nicaragua	0.07		0.10	0.65				0.30	94.10						0.65	95.87
Panama	0.89	0.80	0.15	1.16				5.57							1.85	10.43
Paraguay	0.13			0.53				0.15								0.86
Peru	1.18		0.40	1.05				3.81							0.93	7.37
Saint Kitts-Nevis	0.12							0.02								0.14
Saint Lucia Islands	0.13							0.20								0.33
Saint Pierre & Miquelon								0.01								0.01
Saint Vincent & Grenadines	0.00							0.04								0.04
Suriname	0.35		0.12					0.25								0.71
Trinidad & Tobago	0.53							2.23								2.76
Uruguay	0.03							0.87								0.90
Venezuela	9.82	0.67						21.79								32.28
OTHER GLOBAL PROGRAMS	22.58	1.21		6.46				3.77	2.01	1.03	4.26	0.49	2.74	0.76	3.38	48.56

TABLE 7-3A: 2000 U.S. Direct Financial Contributions and Commercial Sales Related to Implementation of the UNFCCC (Millions of US Dollars)

COUNTRY / REGION	MITIGATION — Energy	Transport	Forestry — Forest Conservation	Forestry — Biodiversity Conservation	Agriculture	Waste Management	Industry	ADAPTATION — Capacity Building — Water Supply	Disaster Preparedness	Droughts & Desertification	Coastal Zone Management — Coastal Resources	Coral Reefs	Other Vulnerability Studies	OTHER GCC — UNFCCC Participation	Cross-cutting Activities	TOTALS
WORLD	624.43	5.19	13.29	101.66	27.88	1.04	6.74	903.03	20.15	20.06	20.70	1.47	3.37	2.07	86.80	1,837.78
AFRICA	28.63	0.12	1.93	20.92	18.21			21.57	14.50	12.62	0.63				39.26	158.39
Africa Regional	0.77			3.08	4.81			1.11		2.31					9.04	21.12
Angola	0.05							0.67		0.80					0.33	1.85
Botswana								0.01								0.01
Burkina Faso								0.20								0.20
Cameroon		0.12						0.10								0.23
Cape Verde	0.11															0.11
Central African Republic								0.01								0.01
Chad								0.04								0.04
Congo (DROC)				0.83				0.17								1.00
Congo (ROC)	0.98							0.33								1.31
Cote d'Ivoire	0.60							0.21								0.81
Djibouti								0.00								0.00
East Africa Regional				0.53											3.89	4.42
Equatorial Guinea	0.03							0.09								0.12
Eritrea								0.07							0.13	0.20
Ethiopia					1.83			0.05							0.42	2.30
Gabon								0.21								0.21
Ghana	0.18			0.66				0.30							1.00	2.15
Guinea	0.29							0.37							2.96	3.63
Kenya	0.36		0.37	1.69	0.04			0.13		5.70					0.72	9.01
Liberia	0.01															0.01
Madagascar				4.03				0.11							2.40	6.54
Malawi															6.39	6.39
Mali	0.27							2.18							1.97	4.42
Mauritania	0.02							0.03								0.05
Mauritius								0.02								0.02
Mozambique			1.05		7.52			0.05	13.28							21.90
Namibia			0.00	0.52	0.00										2.10	2.63
Niger	0.51															0.51
Nigeria	11.25							2.86								14.10
Rwanda					0.67			0.10								0.78
Sahel								0.46		3.81					1.03	5.30
Senegal	0.04				0.30			1.21							1.74	3.29
Sierra Leone								0.10								0.10
South Africa	12.96		0.14		0.85			8.07	1.20						0.25	23.45
Southern Africa Regional	0.15			2.66	0.36			1.33							1.33	5.81
Swaziland								0.81								0.81
Tanzania	0.00		0.05	2.46	0.05			0.01			0.63				0.06	3.27
Uganda	0.02		0.22	4.46	0.67			0.10							3.50	8.97
Zambia					0.91			0.03								0.94
Zimbabwe	0.03		0.10	0.10	0.20			0.04	0.03							0.39
ASIA / NEAR EAST	428.88	2.58	5.12	41.10	0.69	1.04	6.74	518.73	0.65	2.78	14.19	0.51			23.20	1046.21
Afghanistan								0.65		1.75						2.40
Algeria	4.90							6.38								11.28
Asia / Near East Regional	12.70	1.56		1.75		1.04	6.74								2.17	25.95
Bahrain	0.26							1.40								1.66

TABLE 7-3A: 2000

U.S. Direct Financial Contributions and Commercial Sales Related to Implementation of the UNFCCC (Millions of US Dollars)

COUNTRY / REGION	MITIGATION — Energy	Transport	Forestry: Forest Conservation	Forestry: Biodiversity Conservation	Agriculture	Waste Management	Industry	ADAPTATION — Capacity Building: Water Supply	Disaster Preparedness	Droughts & Desertification	Coastal Zone Management: Coastal Resources	Coral Reefs	Other Vulnerability Studies	OTHER GCC: UNFCCC Participation	Cross-cutting Activities	TOTALS
Bangladesh	4.28							0.12							0.26	4.66
Brunei								0.30								0.30
Burma (Myanmar)								0.01								0.01
Cambodia			0.02					0.00								0.02
China	29.76	0.08	0.26		0.08			61.22								91.40
Egypt	9.88			34.10				37.41			9.39				9.12	99.91
Federated States of Micronesia	0.08							0.52								0.61
Fiji								0.15			0.18					0.33
French Polynesia								0.06								0.06
Hong Kong	8.11							20.60								28.71
India	15.12	0.21	1.70		0.15			16.30		1.03						34.52
Indonesia	4.10		2.36	5.10	0.20			5.03			3.91	0.47			2.62	23.79
Jordan	0.04							52.93							5.20	58.18
Kiribati								0.00								0.00
Korea (ROK)	0.05							0.05								0.05
Kuwait	0.04							11.14								11.18
Lebanon	0.04							0.39							2.00	2.43
Macao	0.09							0.01								0.10
Malaysia	3.16							12.07								15.23
Maldive Islands								0.08								0.08
Marshall Islands	0.04							0.01								0.05
Mongolia								0.00								0.00
Morocco	1.97							1.01							1.74	4.73
Nauru								0.02								0.02
Nepal	1.72		0.50	0.15	0.05			0.02							0.09	2.52
New Caledonia								0.01								0.01
Oman	0.13							1.50								1.63
Pakistan	0.81							0.79								1.60
Palau								0.08								0.08
Papua New Guinea	0.01							0.01			0.11					0.12
Philippines	6.12		0.21		0.21			10.56			0.60					17.69
Qatar	0.30							2.19								2.49
Reunion								0.01								0.01
Samoa								0.04								0.04
Saudi Arabia	2.14							39.06								41.20
Seychelles								0.06								0.06
Singapore	9.15							36.00								45.15
Southeast Asia Regional												0.04				0.04
Sri Lanka	0.03							0.42								0.45
Syria	1.44							0.37								1.81
Taiwan	5.83	0.19						67.35								73.37
Thailand	3.56	0.32						16.83								20.71
Tokelau Islands								0.03								0.03
Tunisia	0.55							1.52								2.07
Turkey	300.33							8.06								308.41
Turkmenistan								0.07								0.07
Turks & Caic Islands	0.01							0.16								0.18
United Arab Emirates	0.50							13.49								13.99

TABLE 7-3A: 2000   U.S. Direct Financial Contributions and Commercial Sales Related to Implementation of the UNFCCC (Millions of US Dollars)

COUNTRY / REGION	Energy	Transport	Forest Conservation	Biodiversity Conservation	Agriculture	Waste Management	Industry	Water Supply	Disaster Preparedness	Droughts & Desertification	Coastal Resources	Coral Reefs	Other Vulnerability Studies	UNFCCC Participation	Cross-cutting Activities	TOTALS
Vanuatu								0.00								0.00
Vietnam	0.99	0.23	0.07					1.12								2.41
West Bank/Gaza								90.89								90.89
Yemen	0.65							0.88								1.53
EUROPE / EURASIA	51.89	0.93	0.86	0.10	0.04			18.00					0.03	1.08	6.18	79.10
Albania								0.02							0.11	0.12
Armenia	8.00															8.00
Azerbaijan	2.15							2.00								4.15
Belarus	0.01							0.03								0.04
Bosnia & Herzegovina								0.14								0.14
Bulgaria	1.84	0.43						0.03								2.30
Central Asia Regional	0.95														1.05	2.00
Croatia	0.03							1.00								1.04
Cyprus	0.00							2.70								2.71
Czech Rep	1.73	0.14						0.12								1.99
Estonia															0.12	0.12
Europe & Eurasia Regional	7.47			0.07	0.04										0.05	7.62
Georgia	4.04							0.01								4.05
Hungary	0.27	0.09						3.60								3.99
Kazakhstan	2.74	0.22						1.25								4.21
Kyrgyzstan	0.40															0.40
Latvia								0.39								0.39
Lithuania	0.15							0.28								0.43
Macedonia	0.25		0.20					0.31							0.40	1.16
Moldova	3.00															3.00
Poland	0.60	0.05						2.07								2.72
Romania	1.43							0.09								1.52
Russia	8.00		0.64	0.01				0.01					0.03		3.38	12.06
Slovakia								0.01								0.01
Slovenia	0.06							0.22								0.28
Tajikistan								2.00								2.00
Ukraine	8.75		0.02	0.02				0.63						1.08	1.08	11.58
Uzbekistan								1.07								1.07
Yugoslavia																0.00
LATIN AMERICA / CARIBBEAN	97.20	1.56	5.19	29.61	2.01			344.57	3.00	0.02					14.27	497.43
Anguilla	0.01							0.06								0.07
Antigua & Barbuda	0.05							0.20								0.25
Argentina	2.16	0.44						12.53								15.13
Aruba	0.12							0.28								0.40
Bahamas	0.22							2.91								3.14
Barbados	0.05							0.72								0.77
Belize	0.12		0.26					0.37								0.75
Bermuda	0.08							1.11								1.19
Bolivia	0.15		0.96	1.93				0.22							3.85	7.10
Brazil	13.25	0.28	2.48	3.23	0.09			25.53								44.85
British Virgin Islands	0.03							0.18								0.21
Caribbean Regional															0.54	0.54
Cayman Islands								0.11								0.11

TABLE 7-3A: 2000 U.S. Direct Financial Contributions and Commercial Sales Related to Implementation of the UNFCCC (Millions of US Dollars)

| COUNTRY / REGION | MITIGATION | | | | | | | ADAPTATION | | | | | | OTHER GCC | | TOTALS |
| | | | Forestry | | | | | Capacity Building | | | Coastal Zone Management | | | | | |
	Energy	Transport	Forest Conservation	Biodiversity Conservation	Agriculture	Waste Management	Industry	Water Supply	Disaster Preparedness	Droughts & Desertification	Coastal Resources	Coral Reefs	Other Vulnerability Studies	UNFCCC Participation	Cross-cutting Activities	
Central America Regional	0.20			2.90											1.20	4.30
Chile	1.75	0.21	0.25					12.09								14.30
Colombia	2.45							5.17								7.62
Costa Rica	0.96	0.02	0.20		0.15			3.21								4.54
Dominica Islands								0.00								0.00
Dominican Rep	0.26							2.87								3.13
Ecuador	0.71			1.16				1.80								3.67
El Salvador	0.05				0.82			1.90								2.76
French Guiana								0.01								0.01
Grenada Islands								0.53								0.53
Guadeloupe	0.01							0.04								0.05
Guatemala	0.33			4.74				1.92								7.00
Guyana								0.11								0.11
Haiti	0.39							0.34							4.14	4.87
Honduras	0.25		0.08	1.20	0.85			0.87								3.24
Jamaica	1.45			0.87				1.99							1.74	6.06
Latin America / Caribbean Regional	0.60		0.60	4.50				0.10							0.05	5.85
Martinique	0.00							0.01								0.01
Mexico	40.45	0.47	0.30	5.47	0.11			223.30		0.02					0.30	270.41
Montserrat Islands								0.01								0.01
Nicaragua	0.06			1.20				0.66							0.30	2.22
Panama	0.36	0.15		1.00				2.36							1.00	4.87
Paraguay	0.00			0.53				2.58								3.10
Peru	0.33		0.06	0.90				8.71							1.15	11.14
Saint Kitts-Nevis	0.00							0.08								0.08
Saint Lucia Islands	0.12							0.16								0.28
Saint Pierre & Miquelon								0.01								0.01
Saint Vincent & Grenadines	0.05							0.02								0.07
Suriname	0.05							0.37								0.42
Trinidad & Tobago	1.18							2.13								3.31
Uruguay	0.09							0.34								0.44
Venezuela	28.85							26.67	3.00							58.52
OTHER GLOBAL PROGRAMS	17.83	0.19		9.92	6.95			2.01	0.15	4.63	5.89	0.96	3.34	0.99	3.89	56.65

TABLE 7-3B: 1997 - 2000   U.S. Indirect Financial Contributions Related to Implementation of the UNFCCC (Millions of US Dollars)

| COUNTRY / REGION | MITIGATION | | | | | | | ADAPTATION | | | | | | | OTHER GCC | | TOTALS |
| | Energy | Transport | Forestry | | Agriculture | Waste Management | Industry | Capacity Building | | Droughts & Desertification | Coastal Zone Management | | Other Vulnerability Studies | | UNFCCC Participation | Cross-cutting Activities | |
			Forest Conservation	Biodiversity Conservation				Water Supply	Disaster Preparedness		Coastal Resources	Coral Reefs					
WORLD (1997 - 2000)	954.33																954.33
1997																	
ASIA / NEAR EAST	127.30																127.30
China	58.30																58.30
Israel	55.00																55.00
Turkey	14.00																14.00
LATIN AMERICA / CARIBBEAN	182.10																182.10
Argentina	182.10																182.10
1998																	
ASIA / NEAR EAST	240.40																240.40
Bahrain	21.00																21.00
Thailand	185.00																185.00
Turkey	34.40																34.40
EUROPE / EURASIA	58.60																58.60
Croatia	58.60																58.60
LATIN AMERICA / CARIBBEAN	260.00																260.00
Argentina	60.00																60.00
Brazil	200.00																200.00
1999																	
ASIA / NEAR EAST	57.80																57.80
India	32.00																32.00
Turkey	25.80																25.80
EUROPE / EURASIA	3.13																3.13
Bulgaria	3.13																3.13
2000																	
LATIN AMERICA / CARIBBEAN	25.00																25.00
Dominican Republic	25.00																25.00

Appendix D

Climate Change Science: An Analysis of Some Key Questions[1]

This National Research Council study originated from a May 11, 2001, White House request to help inform the Administration's review of U.S. climate change policy. In particular, the written request asked for the National Academies' "assistance in identifying the areas in the science of climate change where there are the greatest certainties and uncertainties," and "views on whether there are any substantive differences between the IPCC [Intergovernmental Panel on Climate Change] Reports and the IPCC summaries." In addition, based on discussions with the Administration, a number of specific questions were incorporated into the statement of task for the study.

SUMMARY

Greenhouse gases are accumulating in Earth's atmosphere as a result of human activities, causing surface air temperatures and subsurface ocean temperatures to rise. Temperatures are, in fact, rising. The changes observed over the last several decades are likely mostly due to human activities, but we cannot rule out that some significant part of these changes is also a reflection of natural variability. Human-induced warming and associated sea level rises are expected to continue through the 21st century. Secondary effects are suggested by computer model simulations and basic physical reasoning. These include increases in rainfall rates and increased susceptibility of semi-arid regions to drought. The impacts of these changes will be critically dependent on the magnitude of the warming and the rate with which it occurs.

The mid-range model estimate of human-induced global warming by the Intergovernmental Panel on Climate Change (IPCC) is based on the premise that the growth rate of climate forcing[2] agents such as carbon dioxide will accelerate. The predicted warming of 3°C (5.4°F) by the end of the 21st century is consistent with the assumptions about how clouds and atmospheric relative humidity will react to global warming. This estimate is also consistent with inferences about the sensitivity[3] of climate drawn from comparing the sizes of past temperature swings between ice ages and intervening warmer periods with the corresponding changes in the climate forcing. This predicted temperature increase is sensitive to assumptions concerning future concentrations of greenhouse gases and aerosols. Hence, national policy decisions made now and in the longer-term future will influence the extent of any damage suffered by vulnerable human populations and ecosystems later in this century. Because there is considerable uncertainty in current understanding of how the climate system varies naturally and reacts to emissions of greenhouse gases and aerosols, current estimates of the magnitude of future warming should be regarded as tentative and subject to future adjustments (either upward or downward).

Reducing the wide range of uncertainty inherent in current model predictions of global climate change will require major advances in understanding and modeling of both (1) the factors that determine atmospheric concentrations of greenhouse gases and aerosols, and (2) the so-called "feedbacks" that determine the sensitivity of the climate system to a prescribed increase in greenhouse gases. There also is a pressing need for a global observing system designed for monitoring climate.

The committee generally agrees with the assessment of human-caused climate change presented in the IPCC Working Group I (WGI) scientific report, but seeks here to articulate more clearly the level of confidence that can be ascribed to those assessments and the caveats that need to be attached to them. This articulation may be helpful to policy makers as they consider a variety of options for mitigation and/or adaptation. In the sections that follow, the committee provides brief responses to some of the key questions related to climate change science. More detailed responses to these questions are located in the main body of the text.

What is the range of natural variability in climate?

The range of natural climate variability is known to be quite large (in excess of several degrees Celsius) on local and regional spatial scales over periods as short as a decade. Precipitation also can vary widely. For example, there is evidence to suggest that droughts as severe as the "dust bowl" of the 1930s were much more common in the central United States during the 10th to 14th centuries than they have been in the more recent record. Mean temperature variations at local sites have exceeded 10°C (18°F) in association with the repeated glacial advances and retreats that occurred over the course of the past million years. It is more difficult to estimate the natural variability of global mean temperature because of the sparse spatial coverage of existing data and difficulties in inferring temperatures from various proxy data. Nonetheless, evidence suggests that global warming rates as large as 2°C (3.6°F) per millennium may have occurred during retreat of the glaciers following the most recent ice age.

---

[1] The text in this appendix is from the foreword and summary of NRC 2001a, found at http://books.nap.edu/html/climatechange.
[2] A climate forcing is defined as an imposed perturbation of the Earth's energy balance. Climate forcing is typically measured in watts per square meter (W/m^2).
[3] The sensitivity of the climate system to a prescribed forcing is commonly expressed in terms of the global mean temperature change that would be expected after a time sufficiently long for both the atmosphere and ocean to come to equilibrium with the change in climate forcing.

Are concentrations of greenhouse gases and other emissions that contribute to climate change increasing at an accelerating rate, and are different greenhouse gases and other emissions increasing at different rates? Is human activity the cause of increased concentrations of greenhouse gases and other emissions that contribute to climate change?

The emissions of some greenhouse gases are increasing, but others are decreasing. In some cases the decreases are a result of policy decisions, while in other cases the reasons for the decreases are not well understood.

Of the greenhouse gases that are directly influenced by human activity, the most important are carbon dioxide, methane, ozone, nitrous oxide, and chlorofluorocarbons (CFCs). Aerosols released by human activities are also capable of influencing climate. Table D-1 lists the estimated climate forcing due to the presence of each of these "climate-forcing agents" in the atmosphere.

Concentrations of carbon dioxide (CO_2) extracted from ice cores drilled in Greenland and Antarctica have typically ranged from near 190 parts per million by volume (ppmv) during the ice ages to near 280 ppmv during the warmer "interglacial" periods, like the present one that began around 10,000 years ago. Concentrations did not rise much above 280 ppmv until the Industrial Revolution. By 1958, when systematic atmospheric measurements began, they had reached 315 ppmv. They are currently ~370 ppmv and rising at a rate of 1.5 ppmv per year (slightly higher than the rate during the early years of the 43-year record). Human activities are responsible for the increase. The primary source, fossil fuel burning, has released roughly twice as much CO_2 as would be required to account for the observed increase. Tropical deforestation also has contributed to CO_2 releases during the past few decades. The oceans and land biosphere have taken up the excess CO_2.

Like CO_2, methane (CH_4) is more abundant in Earth's atmosphere now than at any time during the 400,000-year ice core record, which dates back over a number of glacial/interglacial cycles. Concentrations increased rather smoothly by about 1 percent per year from 1978 until about 1990. The rate of increase slowed and became more erratic during the 1990s. About two-thirds of the current CH_4 emissions are released by human activities, such as rice growing, the raising of cattle, coal mining, use of landfills, and natural gas handling—all of which have increased over the past 50 years.

A small fraction of the ozone (O_3) produced by natural processes in the stratosphere mixes into the lower atmosphere. This "tropospheric ozone" has been supplemented during the 20th century by additional O_3, created locally by the action of sunlight upon air polluted by exhausts from motor vehicles, emissions from fossil fuel burning power plants, and biomass burning.

Nitrous oxide (N_2O) is formed by many microbial reactions in soils and waters, including those acting on the increasing amounts of nitrogen-containing fertilizers. Some synthetic chemical processes that release N_2O have also been identified. Its concentration has increased approximately 13 percent in the past 200 years.

Atmospheric concentrations of chlorofluorocarbons rose steadily following their first synthesis in 1928 and peaked in the early 1990s. Many other industrially useful fluorinated compounds—e.g., carbon tetrafluoride (CF_4), and sulfur hexafluoride (SF_6)—have very long atmospheric lifetimes, which is of concern, even though their atmospheric concentrations have not yet produced large radiative forcings. Hydrofluorocarbons (HFCs), which are replacing CFCs, have a greenhouse effect, but it is much less pronounced because of their shorter atmospheric lifetimes. The sensitivity and generality of modern analytical systems make it quite unlikely that any currently significant greenhouse gases remain to be discovered.

What other emissions are contributing factors to climate change (e.g., aerosols, carbon monoxide, black carbon soot), and what is their relative contribution to climate change?

Besides greenhouse gases, human activity also contributes to the atmospheric burden of aerosols, which include both sulfate particles and black carbon (soot). Both are unevenly distributed, owing to their short lifetimes in the atmosphere. Sulfate particles scatter solar radiation back to space, thereby offsetting the greenhouse effect to some degree. Recent "clean coal technologies" and use of low-sulfur fuels have resulted in decreasing sulfate concentrations, especially in North America, reducing this offset. Black carbon aerosols are end-products of the incomplete combustion of fossil fuels and biomass burning (forest fires and land clearing). They impact radiation budgets both directly and indirectly; they are believed to contribute to global warming, although their relative importance is difficult to quantify at this point.

How long does it take to reduce the buildup of greenhouse gases and other emissions that contribute to climate change? Do different greenhouse gases and other emissions have different drawdown periods?

A removal time of 100 years means that much, but not all, of the climate-forcing agent would be gone in 100 years. Typically, the amount remaining at the end of 100 years is 37 percent; after 200 years, 14 percent; after 300 years, 5 percent; and after 400 years, 2 percent (see Table D-1).

TABLE D-1 Removal Times and Climate-Forcing Values for Specified Atmospheric Gases and Aerosols

A removal time of 100 years means that much, but not all, of the climate-forcing agent would be gone in 100 years. Typically, the amount remaining at the end of 100 years is 37 percent; after 200 years, 14 percent; after 300 years, 5 percent; and after 400 years, 2 percent.

Climate-Forcing Agents	Approximate Removal Times	Climate Forcing Up to the Year 2000 (Watts/m²)
Greenhouse Gases		
Carbon Dioxide	>100 years	1.3–1.5
Methane	10 years	0.5–0.7
Tropospheric Ozone	10–100 days	0.25–0.75
Nitrous Oxide	100 years	0.1–0.2
Perfluorocarbon Compounds (including SF_6)	>1,000 years	0.01
Fine Aerosols		
Sulfate	10 days	-0.3 to -1.0
Black Carbon	10 days	0.1–0.8

Is climate change occurring? If so, how?

Weather station records and ship-based observations indicate that global mean surface air temperature warmed between about 0.4° and 0.8°C (0.7° and 1.5°F) during the 20th century. Although the magnitude of warming varies locally, the warming trend is spatially widespread and is consistent with an array of other evidence detailed in this report. The ocean, which represents the largest reservoir of heat in the climate system, has warmed by about 0.05°C (0.09°F) averaged over the layer extending from the surface down to 10,000 feet, since the 1950s.

The observed warming has not proceeded at a uniform rate. Virtually all the 20th-century warming in global surface air temperature occurred between the early 1900s and the 1940s and during the past few decades. The troposphere warmed much more during the 1970s than during the two subsequent decades, whereas Earth's surface warmed more during the past two decades than during the 1970s. The causes of these irregularities and the disparities in the timing are not completely understood. One striking change of the past 35 years is the cooling of the stratosphere at altitudes of ~13 miles, which has tended to be concentrated in the wintertime polar cap region.

Are greenhouse gases causing climate change?

The IPCC's conclusion that most of the observed warming of the last 50 years is likely to have been due to the increase in greenhouse gas concentrations accurately reflects the current thinking of the scientific community on this issue. The stated degree of confidence in the IPCC assessment is higher today than it was 10—or even 5—years ago. However, uncertainty remains because of (1) the level of natural variability inherent in the climate system on time scales of decades to centuries, (2) the questionable ability of models to accurately simulate natural variability on those long time scales, and (3) the degree of confidence that can be placed on reconstructions of global mean temperature over the past millennium based on proxy evidence. Despite the uncertainties, there is general agreement that the observed warming is real and has been particularly strong within the past 20 years. Whether it is consistent with the change that would be expected in response to human activities is dependent upon what assumptions one makes about the time history of atmospheric concentrations of the various forcing agents, particularly aerosols.

By how much will temperatures change over the next 100 years and where?

Climate change simulations for the period of 1990 to 2100 based on the IPCC emissions scenarios yield a globally averaged surface temperature increase by the end of the century of 1.4–5.8°C (2.5–10.4°F) relative to 1990. The wide range of uncertainty in these estimates reflects both the different assumptions about future concentrations of greenhouse gases and aerosols in the various scenarios considered by the IPCC and the differing climate sensitivities of the various climate models used in the simulations. The range of climate sensitivities implied by these predictions is generally consistent with previously reported values.

The predicted warming is larger over higher latitudes than over lower latitudes, especially during winter and spring, and larger over land than over sea. Rainfall rates and the frequency of heavy precipitation events are predicted to increase, particularly over the higher latitudes. Higher evaporation rates would accelerate the drying of soils following rain events, resulting in lower relative humidities and higher daytime temperatures, especially during the warm season. The likelihood that this effect could prove important is greatest in semi-arid regions, such as the U.S. Great Plains. These predictions in the IPCC report are consistent with current understanding of the processes that control local climate.

In addition to the IPCC scenarios for future increases in greenhouse gas concentrations, the committee considered a scenario based on an energy policy designed to keep climate change moderate in the next 50 years. This scenario takes into account not only the growth of carbon emissions, but also the changing concentrations of other greenhouse gases and aerosols.

Sufficient time has elapsed now to enable comparisons between observed trends in the concentrations of CO_2 and other greenhouse gases with the trends predicted in previous IPCC reports. The increase of global fossil fuel CO_2 emissions in the past decade has averaged 0.6 percent per year, which is somewhat below the range of IPCC scenarios, and the same is true for atmospheric methane concentrations. It is not known whether these slowdowns in growth rate will persist.

How much of the expected climate change is the consequence of climate feedback processes (e.g., water vapor, clouds, snow packs)?

The contribution of feedbacks to climate change depends upon "climate sensitivity," as described in the report. If a central estimate of climate sensitivity is used, about 40 percent of the predicted warming is due to the direct effects of greenhouse gases and aerosols; the other 60 percent is caused by feedbacks.

Water vapor feedback (the additional greenhouse effect accruing from increasing concentrations of atmospheric water vapor as the atmosphere warms) is the most important feedback in the models. Unless the relative humidity in the tropical middle and upper troposphere drops, this effect is expected to raise the temperature response to increases in human-induced greenhouse gas concentrations by a factor of 1.6. The ice–albedo feedback (the reduction in the fraction of incoming solar radiation reflected back to space as snow and ice cover recede) also is believed to be important. Together, these two feedbacks amplify the simulated climate response to the greenhouse gas forcing by a factor of 2.5. In addition, changes in cloud cover, in the relative amounts of high versus low clouds, and in the mean and vertical distributions of relative humidity could either enhance or reduce the amplitude of the warming.

Much of the difference in predictions of global warming by various climate models is attributable to the fact that each model represents these processes in its own particular way. These uncertainties will remain until a more fundamental understanding of the processes that control atmospheric relative humidity and clouds is achieved.

What will be the consequences (e.g., extreme weather, health effects) of increases of various magnitudes?

In the near term, agriculture and forestry are likely to benefit from CO_2 fertilization and an increased water efficiency of some plants at higher atmospheric CO_2 concentrations. The optimal climate for crops may change, requiring significant regional adaptations. Some models project an increased tendency toward drought over semi-arid regions, such as the U.S. Great Plains. Hydrologic impacts could be significant over the western United States, where much of the water supply is dependent on the amount of snow pack and the timing of the spring runoff. Increased rainfall rates could impact pollution runoff and flood control. With higher sea level, coastal regions could be subject to increased wind and flood damage, even if tropical storms do not change in intensity. A significant warming also could have far-reaching implications for ecosystems. The costs and risks involved are difficult to quantify at this point and are, in any case, beyond the scope of this brief report.

Health outcomes in response to climate change are the subject of intense debate. Climate is one of a number of factors influencing the incidence of infectious disease. Cold-related stress would decline in a warmer climate, while heat stress and smog-induced respiratory illnesses in major urban areas would increase, if no adaptation occurred. Over much of the United States, adverse health outcomes would likely be mitigated by a strong public health system, relatively high levels of public awareness, and a high standard of living.

Global warming could well have serious adverse societal and ecological impacts by the end of this century, especially if globally-averaged temperature increases approach the upper end of the IPCC projections. Even in the more conservative scenarios, the models project temperatures and sea levels that continue to increase well beyond the end of this century, suggesting that assessments that examine only the next 100 years may underestimate the magnitude of the eventual impacts.

Has science determined whether there is a "safe" level of concentration of greenhouse gases?

The question of whether there exists a "safe" level of concentration of greenhouse gases cannot be answered directly because it would require a value judgment of what constitutes an acceptable risk to human welfare and ecosystems in various parts of the world, as well as a more quantitative assessment of the risks and costs associated with the various impacts of global warming. In general, however, risk increases with increases in both the rate and the magnitude of climate change.

What are the substantive differences between the IPCC reports and the summaries?

The Committee finds that the full IPCC Working Group I (WGI) report is an admirable summary of research activities in climate science, and the full report is adequately summarized in the *Technical Summary*. The full WGI report and its *Technical Summary* are not specifically directed at policy. The *Summary for Policymakers* reflects less emphasis on communicating the basis for uncertainty and stronger emphasis on areas of major concern associated with human-induced climate change. This change in emphasis appears to be the result of a summary process in which scientists work with policymakers on the document. Written responses from U.S. coordinating and lead scientific authors to the committee indicate, however, that (a) no changes were made without the consent of the convening lead authors (this group represents a fraction of the lead and contributing authors) and (b) most changes that did occur lacked significant impact.

It is critical that the IPCC process remain truly representative of the scientific community. The committee's concerns focus primarily on whether the process is likely to become less representative in the future because of the growing voluntary time commitment required to participate as a lead or coordinating author and the potential that the scientific process will be viewed as being too heavily influenced by governments which have specific postures with regard to treaties, emission controls, and other policy instruments. The United States should promote actions that improve the IPCC process, while also ensuring that its strengths are maintained.

What are the specific areas of science that need to be studied further, in order of priority, to advance our understanding of climate change?

Making progress in reducing the large uncertainties in projections of future climate will require addressing a number of fundamental scientific questions relating to the buildup of greenhouse gases in the atmosphere and the behavior of the climate system. Issues that need to be addressed include (1) the future use of fossil fuels; (2) the future emissions of methane; (3) the fraction of the future fossil-fuel carbon that will remain in the atmosphere and provide radiative forcing versus exchange with the oceans or net exchange with the land biosphere; (4) the feedbacks in the climate system that determine both the magnitude of the change and the rate of energy uptake by the oceans, which together determine the magnitude and time history of the temperature increases for a given radiative forcing; (5) details of the regional and local climate change consequent to an overall level of global climate change; (6) the nature and causes of the natural variability of climate and its interactions with forced changes; and (7) the direct and indirect effects of the changing distributions of aerosols. Maintaining a vigorous, ongoing program of basic research, funded and managed independently of the climate assessment activity, will be crucial for narrowing these uncertainties.

In addition, the research enterprise dealing with environmental change and the interactions of human society with the environment must be enhanced. This includes support of (1) interdisciplinary research that couples physical, chemical, biological, and human systems; (2) an improved capability of integrating scientific knowledge, including its uncertainty, into effective decision-support systems; and (3) an ability to conduct research at the regional or sectoral level that promotes analysis of the response of human and natural systems to multiple stresses.

An effective strategy for advancing the understanding of climate change also will require (1) a global observing system in support of long-term climate monitoring and prediction; (2) concentration on large-scale modeling through increased, dedicated supercomputing and human resources; and (3) efforts to ensure that climate research is supported and managed to ensure innovation, effectiveness, and efficiency.

Appendix E
Bibliography

Academy for Educational Development, "Let Kids Lead." <www.letkidslead.org>

American Planning Association, *Growing Smart^sm Legislative Guidebook.*

Bove et al. 1998—Bove, M.C., J.B. Elsner, C.W. Landsea, X. Niu, and J.J. O'Brien, "Effect of El Niño on U.S. Landfalling Hurricanes, Revisited," *Bulletin of the American Meteorological Society,* vol. 79, pp. 2477–82.

Business Roundtable 2001—*Unleashing Innovation: The Right Approach to Climate Change, Turning the Promise of Technology into Reality* (Washington, DC: Business Roundtable). <http://www.brtable.org/pdf/524.pdf>

Changnon et al. 1996—Changnon, S.S., K.E. Kunkel, and B.C. Reinke, "Impacts and Response to the 1995 Heat Wave: A Call to Action," *Bulletin of the American Meteorological Society,* vol. 77, pp. 1497–1506.

EOP 1993—Executive Office of the President, *Climate Change Action Plan* (Washington, DC).

EOP 2001a—Executive Office of the President, *Action on Climate Change Review Initiatives,* Washington, DC.

EOP 2001b—Executive Office of the President, *Climate Change Review—Initial Report,* Washington, DC. <http://www.whitehouse.gov/news/releases/2001/06/climatechange.pdf>

Florida State University, Center for Ocean–Atmospheric Prediction Studies. <http://www.coaps.fsu.edu>

4-H Youth Curriculum, "Going Places, Making Choices." <www.fourhcouncil.edu>

Groisman et al. 2001—Groisman, P. Ya, R.W. Knight, and T.R. Karl, "Heavy Precipitation and High Streamflow in the Contiguous United States: Trends in the Twentieth Century," *Bulletin of the American Meteorological Society,* vol. 82, pp. 219–46.

Hartmann, Dennis, *Reports to the Nation on Our Changing Planet: Our Changing Climate* (Boulder, CO: University Center for Atmospheric Research and National Oceanic and Atmospheric Administration, Fall 1997). <http://www.ogp.noaa.gov/library/rtn4.pdf>

IGFA 2000—International Group of Funding Agencies for Global Change Research, *National Updates* (Oslo, Norway: IGFA Secretariat). <http://www.igfagcr.org/PDF/2000Nat'l.pdf>

IPCC 1991—Intergovernmental Panel on Climate Change, *Climate Change: The IPCC Response Strategies* (Washington, DC: Island Press).

IPCC 1996a—Intergovernmental Panel on Climate Change, *Climate Change 1995: Impacts, Adaptation and Mitigation of Climate Change: Scientific-Technical Analyses,* R.T. Watson et al., eds. (Cambridge, U.K.: Cambridge University Press).

IPCC 1996b—Intergovernmental Panel on Climate Change, *Climate Change 1995: The Science of Climate Change,* J.T. Houghton, L.G. Meira Filho, B.A. Callandar, N. Harris, A. Kattenberg, and K. Maskell, eds. (Cambridge, U.K.: Cambridge University Press). <http://www.ipcc.ch/pub/sarsum1.htm>

IPCC 1998—Intergovernmental Panel on Climate Change, *The Regional Impacts of Climate Change: An Assessment of Vulnerability,* R.T. Watson et al., eds. (Cambridge, U.K.: Cambridge University Press).

IPCC 1999—Intergovernmental Panel on Climate Change, *Aviation and the Global Atmosphere: A Special Report of IPCC Working Groups I and III in Collaboration with the Scientific Assessment Panel to the Montreal Protocol on Substances that Deplete the Ozone Layer,* J.E. Penner, D.H. Lister, D.J. Griggs, D.J. Dokken, and M. McFarland, eds. (Cambridge, U.K.: Cambridge University Press). <http://www.ipcc.ch/pub/av(E).pdf>

IPCC 2000a—Intergovernmental Panel on Climate Change, *Good Practice Guidance and Uncertainty Management in National Greenhouse Gas Inventories*, J. Penman, D. Kruger, I. Galbally, T. Hiraishi, B. Nyenzi, S. Emmanul, L. Buendia, R. Hoppaus, T. Martinsen, J. Meijer, K. Miwa, and K. Tanabe, eds. (Japan: Institute for Global Environmental Strategies). <http://www.ipcc-nggip.iges.or.jp/public/gp/gpgaum.htm>

IPCC 2000b—Intergovernmental Panel on Climate Change, "Trends in Technology Transfer: Financial Resource Flows," in *Methodological and Technological Issues in Technology Transfer* (Cambridge, U.K.: Cambridge University Press).

IPCC 2001a—Intergovernmental Panel on Climate Change, *Climate Change 2001: Impacts, Adaptation, and Vulnerability. Contribution of Working Group II to the Third Assessment Report of the Intergovernmental Panel on Climate Change*, J.J. McCarthy, O.F. Canziani, N.A. Leary, D.J. Dokken, and K.S. White, eds. (Cambridge, U.K., and New York, NY: Cambridge University Press).

IPCC 2001b—Intergovernmental Panel on Climate Change, *Climate Change 2001: Mitigation. Contribution of Working Group III to the Third Assessment Report of the Intergovernmental Panel on Climate Change*, B. Metz, O. Davidson, R. Swart, and J. Pan, eds. (Cambridge, U.K., and New York, NY: Cambridge University Press).

IPCC 2001c—Intergovernmental Panel on Climate Change, *Climate Change 2001: Synthesis Report. Contribution to the Third Assessment Report of the Intergovernmental Panel on Climate Change*, R.T. Watson, D.L. Albritton, T. Barker, I.A. Bashmakov, O. Canziani, R. Christ, U. Cubasch, O. Davidson, H. Gitay, D. Griggs, J. Houghton, J. House, Z. Kundzewicz, M. Lal, N. Leary, C. Magadza, J.J. McCarthy, J.F.B. Mitchell, J.R. Moreira, M. Munasinghe, I. Noble, R. Pachauri, B. Pittock, M. Prather, R.G. Richels, J.B. Robinson, J. Sathaye, S. Schneider, R. Scholes, T. Stocker, N. Sundararaman, R. Swart, T. Taniguchi, and D. Zhou, eds. (Cambridge, U.K., and New York, NY: Cambridge University Press).

IPCC 2001d—Intergovernmental Panel on Climate Change, *Climate Change 2001: The Scientific Basis. Contribution of Working Group I to the Third Assessment Report of the Intergovernmental Panel on Climate Change*, J.T. Houghton, Y. Ding, D.J. Griggs, M. Noguer, P.J. van der Linden, and D. Xiasou, eds. (Cambridge, U.K., and New York, NY: Cambridge University Press).

IPCC/UNEP/OECD/IEA 1997—Intergovernmental Panel on Climate Change, United Nations Environment Programme, Organization for Economic Cooperation and Development, International Energy Agency, *Revised 1996 IPCC Guidelines for National Greenhouse Gas Inventories* (Paris, France: IPCC/UNEP/OECD/IEA). <http://www.ipcc-nggip.iges.or.jp/public/gl/invs1.htm>

Karl et al. 1995—Karl, T.R., V.E. Derr, D.R. Easterling, C.K. Folland, D.J. Hofmann, S. Levitus, N. Nicholls, D.E. Parker, and G.W. Withee, "Critical Issues for Long-Term Climate Monitoring," *Climatic Change*, vol. 31.

Keeling, C.D., and T.P. Whorf, "Atmospheric CO_2 Records from Sites in the SIO Air Sampling Network," in *Trends: A Compendium of Data on Global Change* (Oak Ridge, TN: Carbon Dioxide Information Analysis Center, Oak Ridge National Laboratory, 2000).

McCabe, G.J., and D.M. Wolock, "General-Circulation-Model Simulations of Future Snowpack in the Western United States, *Journal of the American Water Resources Association*, vol. 35 (1999), pp. 1473–84.

Mills et al. 2001—Mills, Evan, Eugene Lecomte, and Andrew Peara, *U.S. Insurance Industry Perspectives on Global Climate Change* (Berkeley, CA: U.S. Department of Energy). <http://eetd.lbl.gov/insurance>

NAAG 2002—National Agriculture Assessment Group, *Agriculture: The Potential Consequences of Climate Variability and Change*, J. Reilly et al., eds. (Cambridge, U.K.: Cambridge University Press and U.S. Department of Agriculture, for the U.S. Global Change Research Program). <http://www.usgcrp.gov>

NASA 2001—National Aeronautics and Space Administration, *Earth Observing System Global Change Media Directory 2001* (Greenbelt, MD: Earth Observing System Project Science Office, Goddard Space Flight Center). <http://earthobservatory.nasa.gov/Newsroom/>

NAST 2000—National Assessment Synthesis Team, *Climate Change Impacts on the United States: The Potential Consequences of Climate Variability and Change: Overview* (Cambridge, U.K.: Cambridge University Press and U.S. Global Change Research Program). <http://www.usgcrp.gov>

NAST 2001—National Assessment Synthesis Team, *Climate Change Impacts on the United States: The Potential Consequences of Climate Variability and Change: Foundation* (Cambridge, U.K.: Cambridge University Press and U.S. Global Change Research Program). <http://www.usgcrp.gov>

NCAG 2000—National Coastal Assessment Group, *Coastal: The Potential Consequences of Climate Variability and Change* (Washington, DC: U.S. Department of Commerce, National Oceanic and Atmospheric Administration, for the U.S. Global Change Research Program). <http://www.usgcrp.gov>

NEPD Group 2001—National Energy Policy Development Group, *National Energy Policy* (Washington, DC: U.S. Government Printing Office). <http://www.whitehouse.gov/energy>

NFAG 2001—National Forest Assessment Group, *Forests: The Potential Consequences of Climate Variability and Change* (Washington, DC: U.S. Department of Agriculture, for the U.S. Global Change Research Program). <http://www.usgcrp.gov>

NHAG 2000—National Health Assessment Group (J.A. Patz, M.A. McGeehin, S.M. Bernard, K.L. Ebi, P.R. Epstein, A. Grambsch, D.J. Gubler, P. Reiter, I. Romeiu, J.B. Rose, et al.), "The Health Impacts of Climate Variability and Change for the United States: Executive Summary of the Report of the Health Sector of the U.S. National Assessment," *Environmental Health Perspectives*, vol. 108, pp. 367–76. <http://www.usgcrp.gov>

NHAG 2001—National Health Assessment Group, *Health: The Potential Consequences of Climate Variability and Change* (Washington, DC: Johns Hopkins University, School of Public Health, and U.S. Environmental Protection Agency, for the U.S. Global Change Research Program). <http://www.usgcrp.gov>

NJ 2000—State of New Jersey, Department of Environmental Protection, *New Jersey Sustainability Greenhouse Gas Reduction Plan* (December 1999, 1st reprint May 2000). <http://www.state.nj.us/dep/dsr/gcc/gcc-download.htm>

NRC 1999—National Research Council, *Adequacy of Climate Observing Systems* (Washington, DC: National Academy Press).

NRC 2001a—National Research Council, Committee on the Science of Climate Change, *Climate Change Science: An Analysis of Some Key Questions* (Washington, DC: National Academy Press). <http://books.nap.edu/html/climatechange>

NRC 2001b—National Research Council, Committee on Climate, Ecosystems, Infectious Disease, and Human Health, *Under the Weather: Climate, Ecosystems, and Infectious Disease* (Washington, DC: National Academy Press).

NSC 2000—National Safety Council, *Reporting on Climate Change: Understanding the Science* (Washington, DC: National Safety Council, Environmental Health Center). <http://www.nsc.org/ehc/guidebks/climtoc.htm>

NSTC 2000—National Science and Technology Council, Committee on Environmental and Natural Resources, Subcommittee on Global Change Research, *Our Changing Planet: The FY 2001 U.S. Global Change Research Program* (U.S. Government Printing Office: Washington, DC).

NWAG 2000—National Water Assessment Group, *Water: The Potential Consequences of Climate Variability and Change* (Washington, DC: U.S. Geological Survey, Department of the Interior, and Pacific Institute, for the U.S. Global Change Research Program). <http://www.usgcrp.gov>

OECD 2000—Organization of Economic Cooperation and Development, *Environmental Goods and Services: An Assessment of the Environmental, Economic and Development Benefits of Further Global Trade Liberalisation* (Paris, France: OECD, Trade Directorate & Environment Directorate).

OMB 2001—Office of Management and Budget, *Report to Congress on Federal Climate Change Expenditures* (Washington, DC).

OOSDP 1995—Ocean Observing System Development Panel, *Scientific Design for the Common Module of the Global Ocean Observing System and the Global Climate Observing System: An Ocean Observing System for Climate.*
< http://www-ocean.tamu.edu/OOSDP/FinalRept/t_of_c.html>

Powell et al. 1993—Powell, D.S., J.L. Faulkner, D.R. Darr, Z. Zhu, and D.W. MacCleery, *Forest Resources of the United States–1992,* Gen. Tech. Rep. RM-234. (Fort Collins, CO: Rocky Mountain Forest and Range Experiment Station, Forest Service, U.S. Department of Agriculture).

Timmermann et al. 1999—Timmermann, A., J. Oberhuber, A. Bacher, M. Esch, M. Latif, and E. Roeckner, "Increased El Niño Frequency in a Climate Model Forced by Future Greenhouse Warming," *Nature,* vol. 398, pp. 694–97.

UNFCCC—United Nations Framework Convention on Climate Change, *UNFCCC Guidelines on Reporting and Review.*

US–AEP/USAID 2000—United States–Asia Environmental Partnership/U.S. Agency for International Development, *U.S. Environment Industry Export Competitiveness in Asia.*

USAID 2000a—U.S. Agency for International Development, *Annual Report 2000: EcoLinks* (Washington, DC: U.S. Government Printing Office).

USAID 2000b—U.S. Agency for International Development, *Market Opportunities for Climate Change Technologies and Services in Developing Countries* (Washington, DC: U.S. Government Printing Office).

USAID 2000c—U.S. Agency for International Development, *Partnership Grants 2000: EcoLinks* (Washington, DC: U.S. Government Printing Office).

USAID 2001a—U.S. Agency for International Development, *EcoLinks* (fact sheet) (Washington, DC: U.S. Government Printing Office).

USAID 2001b—U.S. Agency for International Development, "Towards a Water Secure Future: USAID's Obligations in Water Resources Management for FY2000," Part II (Draft, May 18, 2001).

U.S. Congress 1993—U.S. Congress, Office of Technology Assessment, *Preparing for an Uncertain Climate,* vols. I and II (Washington, DC: U.S. Government Printing Office), OTA-O-567 and 568.

U.S. CSP 1997—U.S. Country Studies Program, *Global Climate Change Mitigation Assessment Results for Fourteen Transition and Developing Countries.*

U.S. CSP 1998—U.S. Country Studies Program, *Climate Change Assessments by Developing and Transition Countries.*

USDA 2000—U.S. Department of Agriculture, *Submission to the United Nations Framework Convention on Climate Change on Methodological Issues Related to Carbon Sinks.*

USDA 2001—U.S. Department of Agriculture, *Food and Agricultural Policy: Taking Stock for the New Century* (Washington, DC: U.S. Government Printing Office), SN-001-000-04696-9. <http://www.usda.gov/farmpolicy/farmpolicy.htm>

USDA/ERS 2000—U.S. Department of Agriculture, Economic Research Service, *Agricultural Resources and Environmental Indicators: 2000* (Washington, DC: USDA). <http://www.ers.usda.gov/emphases/harmony/issues/arei2000/>

USDA/ERS 2001a—U.S. Department of Agriculture, Economic Research Service, "Major Uses of Land in the United States," Marlow Vesterby and Kenneth S. Krupa, eds., ERS Statistical Bulletin No. 973 (Washington, DC: USDA/ERS). <http://www.ers.usda.gov/data/majorlanduses/>

USDA/ERS 2001b—U.S. Department of Agriculture, Economic Research Service, Major Land Uses data files, October 2001. <http://www.ers.usda.gov/data/majorlanduses/>

USDA/NRCS 2000—U.S. Department of Agriculture, Natural Resources Conservation Service, *Summary Report: 1997 National Resources Inventory* (Ames, IA: Iowa State University Statistical Laboratory). <http://www.nhq.nrcs.usda.gov/NRI/1997>

USDA/NRCS 2001—U.S. Department of Agriculture, Natural Resources Conservation Service, *Food and Agricultural Policy: Taking Stock for the New Century*. <http://www.usda.gov>

U.S. DOC/BEA 2000—U.S. Department of Commerce, Bureau of Economic Analysis, *National Income and Product Accounts* (Washington, DC: DOC/BEA). <http://www.bea.doc.gov/bea/dn/gdplev.htm>

U.S. DOC/Census 2000—U.S. Department of Commerce, Bureau of the Census, *Statistical Abstract of the United States: 2000*, 120th edition (Washington, DC: U.S. Government Printing Office). <http://www.census.gov/statab/www/>

U.S. DOC/Census 2001—U.S. Department of Commerce, Bureau of the Census, "States Ranked by Numeric Population Change: 1990 to 2000," data released on April 2, 2001. <http://www.census.gov/population/cen2000/phc-t2/tab02.txt>

U.S. DOC/NOAA 1998a—U.S. Department of Commerce, National Oceanic and Atmospheric Administration, National Climatic Data Center, *Historical Climatology Series 5-1* (Asheville, NC: NOAA). <http://www.ncdc.noaa.gov/ol/documentlibrary/hcs/hcs.html#overview5-1>

U.S. DOC/NOAA 1998b—U.S. Department of Commerce, National Oceanic and Atmospheric Administration, National Climatic Data Center, *Historical Climatology Series 5-2* (Asheville, NC: NOAA). <http://www.ncdc.noaa.gov/ol/documentlibrary/hcs/hcs.html#overview5-2>

U.S. DOC/NOAA 1999a—U.S. Department of Commerce, National Oceanic and Atmospheric Administration, National Climatic Data Center, *Historical Climatology Series 5-1* (Asheville, NC: NOAA). <http://www.ncdc.noaa.gov/ol/documentlibrary/hcs/hcs.html#overview5-1>

U.S. DOC/NOAA 1999b—U.S. Department of Commerce, National Oceanic and Atmospheric Administration, National Climatic Data Center, *Historical Climatology Series 5-2* (Asheville, NC: NOAA). <http://www.ncdc.noaa.gov/ol/documentlibrary/hcs/hcs.html#overview5-2>

U.S. DOC/NOAA 2001a—U.S. Department of Commerce, National Oceanic and Atmospheric Administration, National Climatic Data Center, *Historical Climatology Series 5-1* (Asheville, NC: NOAA). <http://www.ncdc.noaa.gov/ol/documentlibrary/hcs/hcs.html#overview5-1>

U.S. DOC/NOAA 2001b—U.S. Department of Commerce, National Oceanic and Atmospheric Administration, National Climatic Data Center, *Historical Climatology Series 5-2* (Asheville, NC: NOAA). <http://www.ncdc.noaa.gov/ol/documentlibrary/hcs/hcs.html#overview5-2>

U.S. DOC/NOAA 2001c—U.S. Department of Commerce, National Oceanic and Atmospheric Administration, National Climatic Data Center, *The U.S. Detailed National Report on Systematic Observations for Climate* (Silver Spring, MD: NOAA). <http://www.eis.noaa.gov/gcos>

U.S. DOE/EIA 1999—U.S. Department of Energy, Energy Information Administration, *A Look at Residential Energy Consumption: 1997* (Washington, DC: U.S. DOE), DOE/EIA-0632(97). <http://www.eia.doe.gov/emeu/recs>

U.S. DOE/EIA 2000a—U.S. Department of Energy, Energy Information Administration, *Annual Energy Review 1999* (Washington, DC: U.S. DOE), DOE/EIA-0384(99). <http://www.eia.doe.gov/emeu/aer/contents.html>

U.S. DOE/EIA 2000b—U.S. Department of Energy, Energy Information Administration, *Electric Power Annual 1999*, vols. II and III (Washington, DC: U.S. DOE), DOE/EIA-0348(99)/2. <http://www.eia.doe.gov/cneaf/electricity/epav2/epav2.pdf>

U.S. DOE/EIA 2000c—U.S. Department of Energy, Energy Information Administration, *Emissions of Greenhouse Gases in the United States, 1999* (Washington, DC: U.S. DOE), DOE/EIA-0573(99).

U.S. DOE/EIA 2000d—U.S. Department of Energy, Energy Information Administration, *Short-Term Energy Outlook* (Washington, DC: U.S. DOE), DOE/EIA-0202(00). <http://www.eia.doe.gov/emeu/steo/pub/contents.html>.

U.S. DOE/EIA 2001a—U.S. Department of Energy, Energy Information Administration, *Annual Energy Outlook, 2002* (Washington, DC: U.S. DOE), DOE/EIA-0384(2000). <http://www.eia.doe.gov/oiaf/aeo>

U.S. DOE/EIA 2001b—U.S. Department of Energy, Energy Information Administration, *Annual Energy Review, 2000* (Washington, DC: U.S. DOE), DOE/EIA-0384(2000). <http://www.eia.doe.gov/emeu/aer/contents.html>

U.S. DOE/EIA 2001c—U.S. Department of Energy, Energy Information Administration, *Emissions of Greenhouse Gases in the United States, 2000* (Washington, DC: U.S. DOE), DOE/EIA-0573(2000).

U.S. DOE/OPIA 2001—U.S. Department of Energy, Office of Policy and International Affairs, preliminary data.

U.S. DOL/BLS—U.S. Department of Labor, Bureau of Labor Statistics, "Current Population Survey: Household Data (2000)–Annual Averages," Table 17. <http://www.bls.gov/cps/cps_over.htm>

U.S. DOS 1994—U.S. Department of State, Office of Global Change, *U.S. Climate Action Report: Submission of the United States of America Under the United Nations Framework on Climate Change* (Washington, D.C.: U.S. DOS).

U.S. DOS 1997—U.S. Department of State, Office of Global Change, *Climate Action Report: 1997 Submission of the United States of America Under the United Nations Framework on Climate Change* (Washington, D.C.: U.S. DOS).

U.S. DOS 2000—U.S. Department of State, *United States Submission on Land Use, Land Use Change and Forestry*, U.S. submission to the UN Framework Convention on Climate Change.
<http://www.state.gov/www/global_issues/climate/climate_2000_submission.html>

U.S. DOT/BTS 2000a—U.S. Department of Transportation, Bureau of Transportation Statistics, *Air Carrier Traffic Statistics Monthly*, Dec. 2000/1999, Dec. 1999/1998, Dec. 1998/1997 (Washington, D.C.: U.S. DOT).

U.S. DOT/BTS 2000b—U.S. Department of Transportation, Bureau of Transportation Statistics, *National Transportation Statistics: 2000* (Washington, D.C.: U.S. DOT), BTS01-01. <http://www.bts.gov/btsprod/nts/>

U.S. DOT/FAA 1998—U.S. Department of Transportation, Federal Aviation Administration, *FAA Statistical Handbook of Aviation 1996* (Washington, DC: U.S. DOT), BTS99-03. <http://www.api.faa.gov/handbook96/toc96.htm>

U.S. DOT/FHWA 1999—U.S. Department of Transportation, Federal Highway Administration, *Draft 1998 Highway Statistics* (Washington, DC: DOT/FHWA), report FHWA-PL-96-023-annual.

U.S. DOT and U.S. EPA—U.S. Department of Transportation and U.S. Environmental Protection Agency, "It All Adds Up To Cleaner Air." <www.epa.gov/otaq/traq/traqpedo/italladd>

U.S. EPA 1989—U.S. Environmental Protection Agency, *The Potential Effects of Global Climate Change on the United States*, J.B. Smith and D.A. Tirpak, eds. (Washington, DC: U.S. EPA), 230-05-89-050.

U.S. EPA 1999—U.S. Environmental Protection Agency, *U.S. Methane Emissions 1990–2020: Inventories, Projections, and Opportunities for Reductions* (Washington, DC: U.S. EPA). <www.epa.gov/ghginfo>

U.S. EPA 2000—U.S. Environmental Protection Agency, Office of Air Quality Planning and Standards, *National Air Pollutant Emissions Trends Report, 1900–1999* (Research Triangle Park, NC: U.S. EPA). <http://www.epa.gov/air/data/net.html>

U.S. EPA 2001a—U.S. Environmental Protection Agency, *Draft Addendum to U.S. Methane Emissions 1990–2020: Inventories, Projections, and Opportunities for Reductions* (Washington, DC: U.S. EPA). <www.epa.gov/ghginfo>

U.S. EPA 2001b—U.S. Environmental Protection Agency, *Draft U.S. Nitrous Oxide Emissions 1990–2020: Inventories, Projections, and Opportunities for Reductions* (Washington, DC: U.S. EPA). <www.epa.gov/ghginfo>

U.S. EPA 2001c—U.S. Environmental Protection Agency, *Improving Air Quality Through Land Use Activities.* <http://www.epa.gov/ncepi/Catalog/EPA420R01001.html>

U.S. EPA 2001d—U.S. Environmental Protection Agency, *Inventory of U.S. Greenhouse Gas Emissions and Sinks: 1990–1999* (Washington, DC: U.S. EPA), 236-R-01-001. <http://www.epa.gov/globalwarming/emissions/national>

U.S. EPA 2001e—U.S. Environmental Protection Agency, *U.S. High GWP Emissions 1990–2010: Inventories, Projections, and Opportunities for Reductions* (Washington, DC: U.S. EPA). <www.epa.gov/ghginfo>

U.S. EPA, NASA, and NOAA 1999—U.S. Environmental Protection Agency, National Aeronautics and Space Administration, and National Oceanic and Atmospheric Administration, *Climate Change Presentation Kit,* <http://www.epa.gov/ncepihom/Catalog/EPA236C99001.html>

U.S. EPA and NPS 2001—U.S. Environmental Protection Agency and National Park Service, *Climate Change, Wildlife, and Wildlands: A Toolkit for Teachers and Interpreters.* <http://www.epa.gov/globalwarming/publications/outreach/orwkit.html>

USGCRP 1998–2000—U.S. Global Change Research Program, *Acclimations* (on-line newsletter of the National Assessment of the Potential Consequences of Climate Variability and Change). <http://www.usgcrp.gov/usgcrp/nacc>

U.S. IJI 2000—U.S. Initiative on Joint Implementation, *Activities Implemented Jointly: Fifth Report to the Secretariat of the United Nations Framework Convention on Climate Change.*

WMO 1995—World Meteorological Organization, *GCOS Plan for Space-based Observations,* GCOS-14, WMO Technical Document No. 681 (Geneva, Switzerland: WMO). <http://www.wmo.ch/web/gcos/gcoshome.html>

WMO 1997—World Meteorological Organization, GCOS/GTOS *Plan for Terrestrial Climate-related Observations,* version 2.0, GCOS-32, WMO Technical Document No. 796 (Geneva, Switzerland: WMO). <http://www.wmo.ch/web/gcos/gcoshome.html>

World Bank 2000—*World Development Indicators 2000* (Washington, DC: World Bank). <http://www.worldbank.org/data/wdi/home.html>